Fit-for-Purpose Land Administration-Providing Secure Land Rights at Scale. Volume 1: Conceptual Innovations

Fit-for-Purpose Land Administration-Providing Secure Land Rights at Scale. Volume 1: Conceptual Innovations

Editors

Stig Enemark
Robin McLaren
Christiaan Lemmen

MDPI • Basel • Beijing • Wuhan • Barcelona • Belgrade • Manchester • Tokyo • Cluj • Tianjin

Editors

Stig Enemark
Aalborg University
Denmark

Robin McLaren
University of Glasgow
United Kingdom

Christiaan Lemmen
University of Twente
The Netherlands

Editorial Office
MDPI
St. Alban-Anlage 66
4052 Basel, Switzerland

This is a reprint of articles from the Special Issue published online in the open access journal *Land* (ISSN 2073-445X) (available at: https://www.mdpi.com/journal/land/special_issues/FFPLA).

For citation purposes, cite each article independently as indicated on the article page online and as indicated below:

LastName, A.A.; LastName, B.B.; LastName, C.C. Article Title. *Journal Name* **Year**, *Volume Number*, Page Range.

Volume 1
ISBN 978-3-0365-2083-4 (Hbk)
ISBN 978-3-0365-2084-1 (PDF)

Volume 1-2
ISBN 978-3-0365-2087-2 (Hbk)
ISBN 978-3-0365-2088-9 (PDF)

Cover image courtesy of Stig Enemark

Contents

About the Editors

Stig Enemark is Professor Emeritus of Land Management at Aalborg University, Denmark, where he was Head of School of Surveying and Planning for 15 years. He holds an MSc in Surveying, Planning and Land Management from the Royal Agricultural Academy in Copenhagen and, before joining Aalborg University in 1982, he worked for ten years as a licensed surveyor in private practice. He is Honorary President of the International Federation of Surveyors, FIG, where he served as President 2007-2010. He is a well-known international expert in the areas of land governance, land administration and spatial planning, and related issues of education and capacity development. He has consulted widely in these areas especially in Eastern Europe, South East Asia and Sub-Sahara Africa. He has published a number of books and more than 400 scientific and professional papers.

Robin McLaren is an Honorary Fellow at the School of GeoSciences, University of Edinburgh where his research interests are focused on how crowdsourcing can be used to support land administration. He holds degrees from the University of New Brunswick and the University of Glasgow where he was awarded an honorary doctorate for contributions to geomatics and land administration. He is also the owner and director of Know Edge Ltd a UK based, independent management consulting company formed in 1986 specialising in the application of geospatial information and is a prominent consultant in land administration. He has been at the forefront of the GIS/LIS revolution and is recognised as an expert in Spatial Data Infrastructures and Land Policy. He works extensively with United Nations agencies, World Bank and EU on land policy/land administration / NSDI programmes in Africa, Far East, Asia and Europe.

Christiaan Lemmen is a full professor at ITC, the Faculty of Geo-information Science and Earth Observation, University of Twente, the Netherlands. He is a geodetic advisor with Kadaster International, the international branch of the Netherlands Cadastre, Land Registry and Mapping Agency. Christiaan is co-editor of the the Land Administration Domain Model (LADM), an international standard published as ISO–19152. He is a co-chair of the Land Administration Domain Working Group of the Open GeoSpatial Consortium. Christiaan is director of the OICRF, the Office International du Cadastre et du Regime Foncier (the international Office for Cadastre and Land Records), a documentation center for land administration, and a permanent institution of the International Federation of Surveyors (FIG). He holds a Ph.D. in Land Administration from Delft University of Technology, The Netherlands.

land

Editorial

Fit-for-Purpose Land Administration—Providing Secure Land Rights at Scale

Stig Enemark [1,*], Robin McLaren [2] and Christiaan Lemmen [3,4]

1 Department of Planning, Aalborg University, 9000 Aalborg, Denmark
2 Know Edge Ltd., Edinburgh EH14 1AY, UK; Robin.McLaren@KnowEdge.com
3 School for Land Administration Studies, ITC Faculty, University of Twente, 7500 AE Enschede, The Netherlands; chrit.lemmen@kadaster.nl
4 Kadaster International, Cadastre, Land Registry and Mapping Agency, 7311 KZ Apeldoorn, The Netherlands
* Correspondence: enemark@plan.aau.dk

This Special Issue provides an insight, collated from 26 articles, focusing on various aspects of the Fit-for-Purpose Land Administration (FFPLA) concept and its application. It presents some influential and innovative trends and recommendations for designing, implementing, maintaining and further developing FFP solutions for providing secure land rights at scale. The first group of 14 articles is published in Volume One and discusses various conceptual innovations related to spatial, legal and institutional aspects of FFPLA and its wider applications within land use management. The second group of 12 articles is published in Volume Two and focuses on case studies from various countries throughout the world, providing evidence and lessons learned from the FFPLA implementation process. However, in order to facilitate a more global understanding of the issues and their interrelationship, this editorial embraces both volumes. It should be noted, though, that the online version of this Special Issue presents the articles in a different order, namely in the chronology of their publication.

Citation: Enemark, S.; McLaren, R.; Lemmen, C. Fit-for-Purpose Land Administration—Providing Secure Land Rights at Scale. *Land* **2021**, *10*, 972. https://doi.org/10.3390/land10090972

Received: 6 September 2021
Accepted: 9 September 2021
Published: 15 September 2021

1. Evolution of the FFPLA Approach

The term "land administration" is rooted in cadastral and land registration systems originally developed for providing information about land value, land ownership and types of land use [1]. Historically, these systems were designed for slightly different purposes in various cultures, judicial systems and regions throughout the world. The key difference is whether the transaction alone is recorded (deeds systems) or the title itself is recorded and secured (title systems). The cultural and judicial aspects relate to whether a country is based on Roman law (deeds systems) or German or Anglo common law (title systems). This difference is also apparent in relation to the legacy of colonization.

A couple of decades ago, land administration emerged as a more generic term referred to as "the processes of determining, recording, and disseminating information about ownership, value, and use of land when implementing land management policies" [2,3]. This focus on information is still present, but, in recent years, the type and quality of information needed have changed and pushed the design of land administration systems (LASs) towards "an enabling infrastructure for implementing land policies and land management strategies in support of sustainable development" [4]. The operational components of this land governance infrastructure are the four functions of land tenure, land value, land use and land development. These four functions ensure the proper management of rights, restrictions, and responsibilities in relation to property, land use and natural resources, including the marine environment. However, the basis or the backbone of such solutions is the land tenure component related to the individual land parcel and establishing the relationship between people and land.

In most developed countries, security of tenure is taken for granted. Over centuries, these countries developed mature land institutions and laws that protect the people to land relationship and provide the services needed for supporting an efficient land market and effective land use management. However, an educated estimate indicates that for 70 percent of the world's population, this is not the case [5]. In most developing countries, people cannot register and safeguard their land rights, nor it is too costly. The majority of these people are the poor and the most vulnerable in society. Therefore, over recent years, LASs have developed to also capture and include more informal and social types of tenure. This is encouraged and supported through the development of concepts such as the continuum of land rights [6], the social tenure domain model [7], and aspects of responsible governance of tenure [8,9].

The key driver behind this evolution has been the overall global agenda focusing on poverty eradication, food security, gender equity, human rights, etc., as adopted by the Millennium Development Goals (MDGs) in 2000, and followed by the Sustainable Development Goals (SDGs) in 2015. This agenda has put a strong focus on security of land rights and provided targets and indicators for monitoring the progress of achieving the goals. Another key driver is technology development that has enabled easy access to new, innovative mapping and surveying techniques, such as satellite and drone imagery, mobile phones and handheld GPS, as well as techniques for the storage and management of huge datasets [10].

Over time, these evolutionary endeavors have been conceptualized in the FFPLA approach in 2014 designed to meet the challenges of providing secure land rights at scale [11,12]. This FFP approach indicates that is appropriate and of necessary standard for its main purpose . . . namely providing secure land rights at scale within a given jurisdiction. The concept, as illustrated in Figure 1, includes three interrelated frameworks that work together to deliver the FFP approach: the spatial, legal and institutional frameworks.

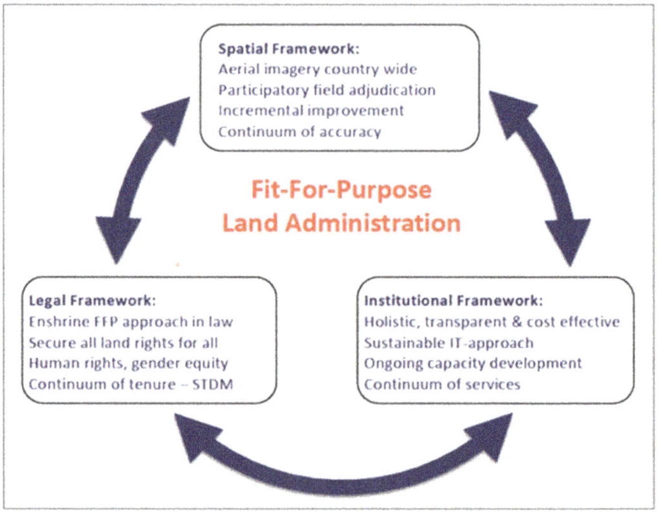

Figure 1. The FFPLA concept [12] (p. 17).

The spatial framework geospatially defines the way in which land is occupied and used. The scale and accuracy of this representation are not determined by rigid, high accuracy regulations, but instead by the users' requirements for effectively identifying the land parcels as a basis for securing the various kinds of legal and legitimate rights and tenure forms recognized through the legal framework. The institutional framework and partnerships are designed to manage these rights, the use of land and natural resources, and to deliver inclusive and accessible services. The approach is flexible, affordable, participatory, and the outcome is upgradeable over time [12].

2. The FFPLA Special Issue

In recent years, the FFPLA concept has been introduced in many countries throughout the world for providing secure land rights at scale, within a short timeframe and at affordable costs. Figure 2 shows the range of countries where FFPLA assessment and implementation are addressed in this Special Issue.

Figure 2. FFPLA Country assessments and implementations addressed in this Special Issue.

The first aim of this Special Issue is to present some recent innovations in the design and use of the FFPLA concept that are shaping new, more efficient approaches to FFPLA. This first group of 14 articles is published in Volume One. The second aim is to synthesize the experiences and lessons learned from country implementation in different cultures and jurisdictions throughout the world. This second group of 12 articles is published in Volume Two.

The conceptual innovations (Volume One) include issues such as:

(i) Assessing procedures of maintenance of conventional as well as unconventional systems;
(ii) Assessing adjudication and quality assurance for legal and geospatial data collected in the participatory processes of field work;
(iii) Applying innovative geospatial tools to FFPLA;
(iv) Using decentralization as a strategy for scaling FFPLA;
(v) Assessing the role of FFPLA for providing security of tenure in violent conflict settings;
(vi) Applying the FFP approach to wider land management functions and to urban resilience in times of climate change and the COVID-19 pandemic; and:
(vii) Exploring the role and opportunities of the private financial sector and public private partnerships within FFPLA.

These innovations are making the implementation of the FFPLA approach more efficient and widening the use in land management applications. Overall, these conceptual innovations are making the approach more attractive for countries to implement and allow the social and economic benefits to be realized more quickly for a sustainable future.

The experiences and lessons learned from country implementation (Volume Two) include cases such as:

(i) Assessing the development impacts of the processes used in China and Vietnam for providing secure land rights at scale;
(ii) Analyzing the strategy and implementation processes for applying an FFPLA approach in Indonesia, Nepal, Uganda and Mozambique;
(iii) Evaluating demonstrative cases of piloting a FFPLA approach and applying FFPLA tools for land recordation in Ghana, Kenya, Uganda, Zambia and Namibia;
(iv) Investigating the impact of applying the FFPLA approach to South Africa;

(v) Using a FFP approach for upscaling of land administration in Benin;
(vi) Applying the FFPLA approach in response to post disasters in the Caribbean; and:
(vii) Assessing FFPLA applications in Colombia and Ecuador.

This wide range of country cases clearly demonstrates that the FFPLA approach is applicable within different contexts by reflecting the specific cultural, legal and institutional settings. The pilot cases validate that the FFPLA methodology for recording land rights in the field is flexible and is working effectively.

3. Concluding Remarks

The main motivation for this special edition was to share experiences and research into the FFPLA approach to help accelerate its implementation at scale and quickly resolve the global insecurity of tenure crisis. The articles indeed illustrate the significant progress that has been achieved over the past decade. They provide some very encouraging lessons learned, as well as exciting, innovative technologies to inspire land professionals to achieve the challenging objectives of the SDGs.

These new highlighted opportunities for going to scale include a clearer understanding of how to decentralize roles and responsibilities and manage organizational change. It also includes a better comprehension of how to obtain political support, gain consensus and formulate national FFPLA strategies, and new insights into implementing robust and the sustainable maintenance of land rights. The articles provide examples of obtaining alternative sources of financing for FFPLA through new types of PPPs, and a pioneering use of private social enterprises for embracing FFP land financing to support the regularization and upgrading of informal settlements. A range of technical innovations is presented, including greater efficiencies derived from the use of machine learning to extract information from drone imagery. Finally, and very importantly, the articles provide a rich set of experiences from FFPLA national scale implementations, as well as pilot projects from developing countries in three continents.

The articles also indicate that the impact of the FFP approach is unfolding beyond its initial focus on security of tenure. UN agencies are widely adopting FFPLA as a tool to mitigate underlying land issues in violent conflict settings, and it is being embraced in wider, urban land management functions to support housing resilience, property valuations, and mitigate the impact of climate change and pandemics. The articles confirm that the FFPLA approach is growing in acceptance across the land professional community, is gaining considerable momentum and is a game changer in achieving key elements of the global agenda, the SDGs, and the globally accepted policies and guidelines around responsible governance of tenure. The FFPLA is already triggering a change in society towards greater social equity, leaving no one behind.

4. Overview of Contributions

This Special Issue includes 26 articles divided into two groups: conceptual innovations (Volume One) and country implementations (Volume Two). The full list of articles is presented in Tables 1 and 2, each of which is followed by a short synthesis of the individual articles.

Bennett et al. [13] use an impressive range of contemporary sources to review FFPLA approaches from the perspective of system maintenance. The "fit-for-purpose" era is producing a wide range of new social and technological innovations; however, large-scale and sustainable implementations still struggle with system maintenance. They present a consolidated model summarizing the story of the maintenance concept in land administration—in terms of key terminology, typologies, approaches, aspects and options. Then, they provide an overview of maintenance problems and related solutions. Finally, they identify that new solutions, as yet unpublished, and newly identified challenges, are emerging.

Table 1. Conceptual Innovations.

	Title	Country Focus	Application
Bennett et al.	Land Administration Maintenance: A review of the Persistent Problem and Emerging Fit-for-Purpose Solutions	Global	Methodologies of maintenance
Lengoiboni et al.	Initial Insights on Land Adjudication in a Fit-for-Purpose Land Administration	Global	Methodologies of adjudication
Augustinus and Tempra	Fit-for-Purpose Land Administration in Violent Conflict Settings	Sudan, Iraq, DRC, Honduras, Peru, Somalia,	Addressing land rights in conflict settings
Ho et al.	Decentralization as a Strategy to Scale Fit-for-Purpose Land Administration: An Indian Perspective on Institutional Challenges	India	Decentralization as a FFPLA tool
Mitchell et al.	The Benefits of Fit-for-Purpose Land Administration for Urban Community Resilience in a Time of Climate Change and COVID-19 Pandemic	Solomon Islands	FFPLA in support of Improving urban resilience
Kelm et al.	Applying the FFP Approach to Wider Land Management Functions	Global	The wider use of the FFPLA approach
Childress et al.	Fit-for-Purpose, Private-Sector Led Land Regularization and Financing of Informal Settlements in Brazil	Brazil	Applying a private sector led approach
Moran et al.	Exploring PPPs in Support of Fit-for-Purpose Land Administration: A Case Study from Côte d'Ivoire	Ivory Coast	Applying a PPP in support of FFPLA
Reydon et al.	The Amazon Forest Preservation by Clarifying Property Rights and Potential Conflicts: How Experiments Using Fit-for-Purpose Can Help	Brazil	Applying a FFP approach in support of forest preservation
Rocha et al.	Quality Assurance for Spatial Data Collected in Fit-for-Purpose Land Administration Approaches in Colombia	Colombia	Assessing the FFPLA data quality
Hall and Whittal	Do Design Science Research and Design Thinking Processes Improve the 'Fit' of the Fit-for-Purpose Approach to Securing Land Tenure for All in South Africa?	South Africa	Exploring the use of design science research and design thinking within FFPLA
Koeva et al.	Geospatial Tool and Geocloud Platform Innovations: A Fit-for-Purpose Land Administration Assessment	Rwanda, Kenya, Ethiopia, and Zanzibar	Assessing the use of geospatial tools in Africa
Chipofya et al.	SmartSkeMa: Scalable Documentation for Community and Customary Land Tenure	Global	Spatial documentation of community land tenure
Biraro et al.	Good Practices in Updating Land Information Systems that Used Unconventional Approaches in Systematic Land Registration	Global	Updating practices in unconventional land registration

Lengoiboni et al. [14] explore how primary ('ownership') and secondary, overlapping ('non-ownership') interests in land are being adjudicated and recorded in a FFPLA context. They prepared questionnaires and developed criteria that organizations invited to answer the questionnaires must meet. They define the components of land adjudication. Then, the processes used by the organizations to achieve these components are described and insights are gained on how land tenure and land rights are framed and how this influences the outcomes of what is recorded. Results show that the legal perspective of land rights intersects with the perspective of communities regarding legitimate rights.

Augustinus and Tempra [15] tackle the challenging subject of UN peace keeping in violent conflict settings and examine how FFPLA interventions have become an integral part of the dispute and conflict resolution. They discuss seven cases across multiple countries and conflict sites where UN-Habitat either supported, or was directly involved in, UN peace keeping. They identify that land governance is of importance because of the way in which land-related power dynamics play out across the conflict cycle; the UN uses its power and capacity to strengthen land governance. They recommend that the FFPLA approaches provide practical options to support peace-building operations by the UN and other stakeholders.

Ho et al. [16] analyze the socio-political and institutional consequences of using decentralization as a scaling strategy for implementing FFPLA through three case studies in India. They review how decentralization is coordinated and governed across multiple levels. Their cases demonstrate a reduced role for the state, and a need for increased collaboration across a diverse set of stakeholders, including a greater number of non-state actors at multiple levels. Although decentralization can work to effectively kickstart the implementation of FFPLA at scale, there is significant work required to ensure that implementation is "fit-for-people" to introduce a trustworthy system that redistributes power and distributes critical social justice.

Mitchell et al. [17] investigate the interlinkages between land tenure, climate vulnerability, and pandemics. Through research in Honiara, Solomon Islands, they consider how improving tenure security at scale through FFPLA can enhance climate resilience to mitigate vulnerability to both climate and pandemic impacts. They contend that this can be achieved at both the city and settlement levels by including tenure in vulnerability and risk assessments (VRA) and the development of resilience action plans. Their proposed FFPLA process, informed by participatory enumeration of the complexities of urban land tenure, can support scaling up efforts to improve tenure security, and deliver more effective and equitable climate resilience actions for vulnerable urban communities.

Kelm et al. [18] analyze how the FFP approach, which has predominantly been applied to the land tenure aspects, can be expanded into a wider set of land management functions. They test their hypothesis through three World Bank urban case studies focusing on land valuation, housing resilience and waste management. Machine learning techniques extract information from drone and street-level imagery to produce minimum viable product models. Their analysis has revealed that there is a common set of geospatial datasets that can be captured once and shared across many other land management functions in an urban environment. This will allow single land intervention projects to be holistically integrated into a wider program of land management functions.

Childress et al. [19] use the analysis of an innovative private social enterprise in Brazil, called Terra Nova, to demonstrate that the concept of FFP land regularization can be widened to include FFP land financing with relevance for wider efforts in informal settlement regularization and upgrading. Their analysis of parcel-level repayment and price data provides some evidence of the sustainability of the business model and the increase of property values of the regularized parcels (pre-COVID-19). Since 2001, Terra Nova has regularized over 20,000 parcels, primarily in São Paolo and Curitiba. They contend that the approach is widely replicable.

Morán et al. [20] introduce a new, innovative form of public private partnership (PPP) being piloted in the coffee growing areas of Côte d'Ivoire, that includes a partnership between the Government and a consortium of cocoa industry leaders and Meridia, the Dutch private land documentation firm. The private sector companies provide funding and service delivery, while the Government enables a political environment for interventions, provides in-kind contributions, collaborates in the execution of projects, and operates the land information system/registry to which the consortium's service connects. This PPP approach could potentially provide countries with alternatives to donor funding/loans.

Reydon et al. [21] blame deforestation in Brazil on the absence of cadastral mapping, land registration, and an effective regulation of property rights. This already involves some 200 million hectares, mostly on public land or undesignated land. This land is easy to grab, deforest and to be used for speculative purposes. The availability of well-defined land rights can reduce the process of deforestation. Participatory determination of land rights based on FFPLA methodologies promotes forest preservation. They hope that their methodology for determining the land rights of small landholders and of traditional population landholders will become mainstream. This will require some legal and institutional adjustments in order to improve the sustainability of the Amazon rainforest.

Rocha et al. [22] present a FFPLA quality assurance model for the evaluation of the quality of the geospatial data collected in the Municipality of Apartadó in Colombia. The FFPLA approach allowed the right holders to walk their parcel boundaries using a smartphone application connected to a GPS receiver to collect their boundary data points. The project evaluated how well the FFPLA dataset conformed to its product specification and was able to determine whether FFPLA data were of sufficient quality, specifically in the case of positional accuracy and logical consistency. The model supported the creation of a product quality life cycle and a quality model in the Colombian context.

Hull and Whittal [23] conduct a reflective retrospective of the processes of land administrative reform in South Africa to determine how land administration systems should be reshaped and new land tenure reforms to be developed. They adopt a thematic framework that innovatively combines FFPLA, design science research, and design thinking processes to help to unlock sensitive and empathetic innovations in land administration systems reform initiatives that will deliver restorative justice. These approaches should be embedded into FFPLA at the start. They admit that this is new and untested and they encourage case studies explicitly implementing these additional approaches in land administration reform.

Koeva et al. [24] assess a series of innovative tools recently developed within the framework of a European Commission Horizon 2020 project. The tools they review are designed to effectively apply FFPLA approaches and are based on requirements from FFPLA projects in Rwanda, Kenya, Ethiopia and Zanzibar. The study was conducted under the appealing title "its4land". The tools include software that implements the smart sketch mapping concept, a workflow for data acquisition based on unmanned aerial vehicles, and a boundary delineator tool based on semi-automatic feature extraction from UAV images. The 'Publish and Share' platform enables the integration of all the outputs of tool sharing and publishing of land information through geocloud web services.

Chipofya et al. [25], with a reference to the 'landmarkmap database', point out that fifty percent of habitable land on this planet is held by indigenous communities. There are no proper tools to document these rights quickly and effectively. Existing software and facilities for documentation of these rights still assume parcel-oriented thinking with statutory rights. The Smart Sketch Map (SmartSkeMa) allows people to document their land rights using concepts from their everyday experiences. SmartSkeMa supports both the legibility of customary land tenure to government authorities and the preservation of the customs within which the tenure relations operate.

Biraro et al. [26] study the maintenance of data in land information systems for which the data were obtained using unconventional approaches. The paper proves that there are good, recommendable practices in the selected countries, including infrastructure for updating; simplified systems; reasonable registration fees; decentralized services; accessible and secured digital databases; awareness raising about registration; availability of a legal framework; incentives to motivate people to report transactions on time; and trained staff and political support. The authors conclude that efforts are still needed to shorten updating procedures, introduce data-sharing platforms, ensure financial and technical sustainability, and reduce the number of involved institutions.

Table 2. Country Implementations.

	Title	Country Focus	Application
Byamugisha	Experiences and Development Impacts of Securing Land Rights at Scale in Developing Countries: Case Studies of China and Vietnam	China, Vietnam	Securing land rights at scale in China and Vietnam
Martono et al.	The Legal Element of Fixing the Boundary for Indonesian Complete Cadastre	Indonesia	Applying FFPLA in Indonesia
Panday et al.	Securing Land Rights for All through Fit-for-Purpose Land Administration Approach: The Case of Nepal	Nepal	Applying FFPLA in Nepal
Musinguzi et al.	Fit for Purpose Land Administration: Country Implementation Strategy for Addressing Uganda's Land Tenure Security Problems	Uganda	Applying FFPLA in Uganda
Chigbu et al.	Fit-for-Purpose Land Administration from Theory to Practice: Three Demonstrative Case Studies of Local Land Administration Initiatives in Africa	Ghana, Kenya, Namibia	Applying FFPLA approaches in Africa
Antonio et al.	Transforming Land Administration Practices through the Application of Fit-for-Purpose Technologies: Country Case Studies in Africa	Uganda, Kenya, Zambia	Applying the STDM in Africa
Mekking et al.	Fit-for-Purpose Upscaling Land Administration—A Case Study from Benin	Benin	Applying FFPLA in Benin
Balas et al.	The Fit for Purpose Land Administration Approach-Connecting People, Processes and Technology in Mozambique	Mozambique	Applying FFPLA in Mozambique
Williams-Wynn	Applying the Fit-for-Purpose Land Administration Concept to South Africa	South Africa	Assessment of applying FFPLA in South Africa
Griffith-Charles	Application of FFPLA to Achieve Economically Beneficial Outcomes Post Disaster in the Caribbean	Caribbean Islands	Applying FFPLA in the Caribbean
Becerra et al.	Fit-for-Purpose Applications in Colombia: Defining Land Boundary Conflicts between Indigenous Sikuani and Neighbouring Settler Farmers	Colombia	Applying FFPLA in Colombia
Todorovski et al.	Assessment of Land Administration in Ecuador Based on the Fit-for-Purpose approach	Ecuador	Assessment of applying FFPLA in Ecuador

Byamugisha [27] details, for the first time, the journeys that China and Vietnam embraced to register all land rights within their countries. This formidable task was triggered when the countries decollectivized agricultural production and allocated rural land to farming households in the 1980s and 1990s; about 1.5 billion rural arable land parcels in China and about 70 million in Vietnam. This was in addition to registering the urban land rights. In both countries, the registration of rural land was done in two rounds and the FFPLA approach was adopted. He distills an excellent set of lessons learned and challenges, and these should inform other countries embarking on similar security of tenure journeys to eradicate extreme poverty.

Martono et al. [28] aim to establish the distinction between physical and legal elements in determining cadastral boundaries in Indonesia. Interviews were conducted for this purpose, and six "cadastral elements" have been investigated and assessed: the parties that locate the boundary, the agreement between the adjoining landowners, the use of boundary markers, the role of the determination officer, the survey method, and the accuracy of the base map. Agreements could be obtained using aerial imagery instead of a field survey. Fixed boundary with exact coordinates based on prescribed survey methods, and with the accuracy of base map, as required by regulations, is not important to people in rural areas.

Panday et al. [29] analyze two pilot studies designed for testing the implementation of a FFPLA National Strategy in Nepal for providing security of tenure for 10 million land parcels currently outside the formal land registration system. They present the methodological workflow of this action-oriented research using one pilot study in an urban setting, including about 1500 spatial units of informal settlements, while the other is rural with 3400 arable land parcels. They explain how the results validate the FFPLA national implementation strategy approach designed for the specific and complex Nepal country context, and they argue that this methodology may be applicable for other low-income countries, where a large amount of the land is informally occupied.

Musinguzi et al. [30] explore, in great depth, the process of developing a national strategy for providing secure land rights at scale in Uganda, covering 23 million land parcels. They describe the current tenure types in Uganda and examine three representative pilot projects, in order to identify how lessons learned from these case studies informed a FFPLA implementation strategy in terms of building the spatial, legal and institutional frameworks. They highlight how pilot projects can provide opportunities for explaining benefits to obtain the necessary political, community and stakeholder support. In conclusion, they argue that a country implementation strategy, if developed as a result of a national dialogue and consensus among all stakeholders, is a promising way of advancing the FFPLA concept.

Chigbu et al. [31] provide evidence that the FFPLA approach represents an unprecedented opportunity to provide tenure security in Africa. They use three country case studies based on hands-on, local land administration projects to demonstrate how the features of the FFPLA guidelines were adopted. Support is provided for the understanding that high-precision measurements are not necessary for legal certainty. They conclude that local people, including youth and women, can be used for data collection and cadastral mapping purposes that are both inexpensive and can be used as necessary documentation for the promotion of tenure security.

Antonio et al. [32] investigate whether the STDM (social tenure domain model) tool facilitates the improvement of land tenure security. The STDM provides a flexible way to meet local needs in capturing human–country relationships. The tool is internationally recognized as practical, fast, and affordable. Their study shows that the STDM can be effectively applied to establish the spatial framework for land administration and to facilitate the implementation of land tenure security of poor communities. Focusing on one pillar of FFPLA can influence positive changes in other frameworks of the FFPLA concept through the use and application of technology, such as the STDM tool.

Mekking et al. [33] present a case study from Benin, with a focus on upscaling the FFPLA approach. At present, only 60,000 of the estimated 5 million plots are registered. For a parcel of 500 m^2, the cost of a title amounts to 540 USD, which is unaffordable for the vast majority of the population. The Benin Government wants legal security for all, and the FFPLA approach offers the opportunity to achieve this. The core of their approach is the introduction of a tenure system based on presumed ownership in parallel to the existing title system. Right holders then have the option to move from "presumed" ownership to state-guaranteed ownership.

Balas et al. [34] provide evidence from Mozambique that the FFPLA approach works. The former "Terra Segura" programme lasted four years and provided only 220,000 parcels and boundaries of 400 communities out of a set target of 5 million plots. The average costs were 50 USD per parcel and 10,000 USD per community. This is too expensive. A fundamentally different FFPLA approach was needed. The FFPLA-MOZ approach was developed and resulted in a better performance and a cost per parcel of 15 USD and 2000 USD for a community boundary definition. From the end of 2017 to March 2020, almost 1.4 million parcels and 826 community delimitations have been processed.

Williams-Wynn [35], the Surveyor General, investigates whether South Africa can adopt FFPLA to provide security of tenure for the five million land occupants that exist outside the formal land tenure system. As Surveyor General, he uses South Africa as a case study to demonstrate how adjustments to institutional, legal and spatial frameworks will develop a fully inclusive, sufficiently accurate land administration system that fits the purpose for which it is envisioned. He is optimistic that the adoption of the FFPLA approach with political support, trust built through community participation, and endorsement of the approach by land professionals will provide security of tenure that is beneficial to all.

Griffith-Charles [36] reviews the experiences with adjudication and titling being undertaken by countries of the Caribbean, with specific examples from Trinidad and Tobago, Barbados, and Jamaica, and others. Her assessment identifies that many countries had spent a lot of time, money, and effort, but were still without a predicted time of completion. The unhurried progress in some countries can be accelerated through the adoption of the FFPLA principles. She reasons that an essential aspect of achieving economically beneficial results is for a country to first identify and publicize a clear vision and objective of what is to be achieved, which requires land-related solutions to be efficient and inclusive.

Becerra et al. [37] introduce a FFPLA approach to support conflict resolution related to overlapping land claims. Indigenous people in Coumaribo, Colombia encounter land-related conflicts with newly arrived and established farmers. The methodology involves both parties independently surveying their land claims. This results in representations of the claims in georeferenced polygons, making any overlaps visible. In a public inspection, the results of the field measurements are displayed, with the presence of the cadastral authority. Discussing the results with all stakeholders helps to clarify the conflicts, to reduce the conflict to specific, relatively small, geographical areas, and to define concrete steps towards solutions.

Todorovski et al. [38] present an assessment of the existing land administration in Ecuador based on the spatial, legal and institutional frameworks—and related principles—of the FFPLA approach. This assessment is used to make recommendations for the improvement of the existing land administration to make the Government's plans for the implementation of a country-wide land administration system more feasible. They identify principles in a developed score table with a low and medium alignment that need to be addressed and adapted to a FFPLA approach; specifically with interventions in the current requirements for the precise measurement of fixed boundaries and a large number of text attributes collected in rural areas.

Author Contributions: S.E.: visualization, conceptualization and writing—original draft preparation. R.M.: conceptualization, writing—review and editing. C.L.: conceptualization, writing—review and editing. All authors have read and agreed to the published version of the manuscript.

Funding: The funding of the publication costs for this editorial, as well as the 26 articles included in this Special Issue, has kindly been provided by the School of Land Administration Studies, Faculty of ITC from the University of Twente, in combination with Kadaster International, The Netherlands.

Institutional Review Board Statement: Not applicable.

Informed Consent Statement: Not applicable.

Data Availability Statement: Not applicable.

Acknowledgments: The authors, as guest editors, would like to acknowledge the contributions of the authors of the individual articles included in this Special Issue, as well as all the external reviewers involved in assessing the scientific level of the manuscripts. We would also like to acknowledge the support of colleagues of academic and government originations that supported this work, in particular the School of Land Administration, Faculty of ITC from the University of Twente, in combination with Kadaster International, The Netherlands.

Conflicts of Interest: The authors declare no conflict of interest.

References

1. International Federation of Surveyors (FIG). *The FIG Statement on the Cadastre*; FIG Publication No. 11; FIG: Copenhagen, Denmark, 2015.
2. United Nations Economic Commission for Europe (UNECE). *Land Administration Guidelines: With Special Reference to Countries in Transition*; UNECE: Geneva, Switzerland; New York, NY, USA, 1996.
3. Dale, P.; McLaughlin, J. *Land Administration*; Oxford University Press: Oxford, UK, 1999.

4. Williamson, I.; Enemark, S.; Wallace, J.; Rajabifard, A. *Land Administration for Sustainable Development*; ESRI Academic Press: Redlands, CA, USA, 2010.
5. McLaren, R. *How Big Is Global Insecurity of Tenure?* Geomares Publishing: Lemmer, The Netherlands, 2015.
6. Global Land Tool Network (GLTN). *Secure Land Rights for All*; UN-Habitat: Nairobi, Kenya, 2008.
7. FIG/GLTN. *The Social Tenure Domain Model*; FIG Publication No. 52; FIG: Copenhagen, Denmark, 2010.
8. Food and Agriculture Organization of the United Nations (FAO). *Voluntary Guidelines on the Responsible Governance of Tenure of Land, Fisheries and Forests in the Context of Food Security*; FAO: Rome, Italy, 2012.
9. Zevenbergen, J.; De Vries, W.; Bennett, R. (Eds.) *Advances in Responsible Land Administration*; CRC Press: Boca Raton, FL, USA, 2016.
10. World Bank. *New Technology and Emerging Trends: The State of Play for Land Administration*; World Bank: Washington, DC, USA, 2017.
11. Enemark, S.; Bell, K.C.; Lemmen, C.; McLaren, R. *Fit-for-Purpose Land Administration*; FIG Publication No 60. Joint Publication; International Federation of Surveyors (FIG), World Bank: Copenhagen, Denmark, 2014.
12. Enemark, S.; McLaren, R.; Lemmen, C. *Fit-for-Purpose Land Administration—Guiding Principles for Country Implementation*; GLTN; UN-Habitat: Nairobi, Kenya, 2016.
13. Bennett, R.M.; Unger, E.-M.; Lemmen, C.; Dijkstra, P. Land Administration Maintenance: A Review of the Persistent Problem and Emerging Fit-for-Purpose Solutions. *Land* **2021**, *10*, 509. [CrossRef]
14. Lengoiboni, M.; Richter, C.; Aspersen, P.; Zevenbergen, J. Initial Insights on Land Adjudication in a Fit-for-Purpose Land Administration. *Land* **2021**, *10*, 414. [CrossRef]
15. Augustinus, C.; Tempra, O. Fit-for-Purpose Land Administration in Violent Conflict Settings. *Land* **2021**, *10*, 139. [CrossRef]
16. Ho, S.; Choudhury, P.R.; Haran, N.; Leshinsky, R. Decentralization as a Strategy to Scale Fit-for-Purpose Land Administration: An Indian Perspective on Institutional Challenges. *Land* **2021**, *10*, 199. [CrossRef]
17. Mitchell, D.; Barth, B.; Ho, S.; Sait, M.S.; McEvoy, D. The Benefits of Fit-for-Purpose Land Administration for Urban Community Resilience in a Time of Climate Change and COVID-19 Pandemic. *Land* **2021**, *10*, 563. [CrossRef]
18. Kelm, K.; Antos, S.; McLaren, R. Applying the FFP Approach to Wider Land Management Functions. *Land* **2021**, *10*, 723. [CrossRef]
19. Childress, M.; Carter, S.; Barki, E. Fit-for-Purpose, Private-Sector Led Land Regularization and Financing of Informal Settlements in Brazil. *Land* **2021**, *10*, 797. [CrossRef]
20. Morán, A.G.; Ulvund, S.; Unger, E.-M.; Bennett, R.M. Exploring PPPs in Support of Fit-for-Purpose Land Administration: A Case Study from Côte d'Ivoire. *Land* **2021**, *10*, 892. [CrossRef]
21. Reydon, B.; Molendijk, M.; Porras, N.; Siqueira, G. The Amazon Forest Preservation by Clarifying Property Rights and Potential Conflicts: How Experiments Using Fit-for-Purpose Can Help. *Land* **2021**, *10*, 225. [CrossRef]
22. Rocha, L.A.; Montoya, J.; Ortiz, A. Quality Assurance for Spatial Data Collected in Fit-for-Purpose Land Administration Approaches in Colombia. *Land* **2021**, *10*, 496. [CrossRef]
23. Hull, S.; Whittal, J. Do Design Science Research and Design Thinking Processes Improve the 'Fit' of the Fit-for-Purpose Approach to Securing Land Tenure for All in South Africa? *Land* **2021**, *10*, 484. [CrossRef]
24. Koeva, M.; Humayun, M.I.; Timm, C.; Stöcker, C.; Crommelinck, S.; Chipofya, M.; Bennett, R.; Zevenbergen, J. Geospatial Tool and Geocloud Platform Innovations: A Fit-for-Purpose Land Administration Assessment. *Land* **2021**, *10*, 557. [CrossRef]
25. Chipofya, M.; Jan, S.; Schwering, A. SmartSkeMa: Scalable Documentation for Community and Customary Land Tenure. *Land* **2021**, *10*, 662. [CrossRef]
26. Biraro, M.; Zevenbergen, J.; Alemie, B.K. Good Practices in Updating Land Information Systems that Used Unconventional Approaches in Systematic Land Registration. *Land* **2021**, *10*, 437. [CrossRef]
27. Byamugisha, F.F.K. Experiences and Development Impacts of Securing Land Rights at Scale in Developing Countries: Case Studies of China and Vietnam. *Land* **2021**, *10*, 176. [CrossRef]
28. Martono, B.; Aditya, T.; Subaryono and, S.; Nugroho, P. The Legal Element of Fixing the Boundary for Indonesian Complete Cadastre. *Land* **2021**, *10*, 49. [CrossRef]
29. Panday, U.S.; Chhatkuli, R.R.; Joshi, J.R.; Deuja, J.; Antonio, D.; Enemark, S. Securing Land Rights for All through Fit-for-Purpose Land Administration Approach: The Case of Nepal. *Land* **2021**, *10*, 744. [CrossRef]
30. Musinguzi, M.; Enemark, S.; Mwesigye, S.P. Fit for Purpose Land Administration: Country Implementation Strategy for Addressing Uganda's Land Tenure Security Problems. *Land* **2021**, *10*, 629. [CrossRef]
31. Chigbu, U.E.; Bendzko, T.; Mabakeng, M.R.; Kuusaana, E.D.; Tutu, D.O. Fit-for-Purpose Land Administration from Theory to Practice: Three Demonstrative Case Studies of Local Land Administration Initiatives in Africa. *Land* **2021**, *10*, 476. [CrossRef]
32. Antonio, D.; Njogu, S.; Nyamweru, H.; Gitau, J. Transforming Land Administration Practices through the Application of Fit-for-Purpose Technologies: Country Case Studies in Africa. *Land* **2021**, *10*, 538. [CrossRef]
33. Mekking, S.; Kougblenou, D.V.; Kossou, F.G. Fit-for-Purpose Upscaling Land Administration—A Case Study from Benin. *Land* **2021**, *10*, 440. [CrossRef]
34. Balas, M.; Carrilho, J.; Lemmen, C. The Fit for Purpose Land Administration Approach-Connecting People, Processes and Technology in Mozambique. *Land* **2021**, *10*, 818. [CrossRef]
35. Williams-Wynn, C. Applying the Fit-for-Purpose Land Administration Concept to South Africa. *Land* **2021**, *10*, 602. [CrossRef]

36. Griffith-Charles, C. Application of FFPLA to Achieve Economically Beneficial Outcomes Post Disaster in the Caribbean. *Land* **2021**, *10*, 475. [CrossRef]
37. Becerra, L.; Molendijk, M.; Porras, N.; Spijkers, P.; Reydon, B.; Morales, J. Fit-for-Purpose Applications in Colombia: Defining Land Boundary Conflicts between Indigenous Sikuani and Neighbouring Settler Farmers. *Land* **2021**, *10*, 382. [CrossRef]
38. Todorovski, D.; Salazar, R.; Jacome, G. Assessment of Land Administration in Ecuador Based on the Fit-for-Purpose Approach. *Land* **2021**, *10*, 862. [CrossRef]

land

MDPI

Review

Land Administration Maintenance: A Review of the Persistent Problem and Emerging Fit-for-Purpose Solutions

Rohan Mark Bennett [1,2,*], Eva-Maria Unger [1], Christiaan Lemmen [1,3] and Paula Dijkstra [1]

1 Kadaster, The Netherlands Cadastre, Land Registry and Mapping Agency, 7311 KZ Apeldoorn, The Netherlands; Eva-Maria.Unger@kadaster.nl (E.-M.U.); Chrit.Lemmen@kadaster.nl (C.L.); Paula.Dijkstra@kadaster.nl (P.D.)
2 Swinburne Business School, Swinburne University of Technology, Hawthorn, VIC 3122, Australia
3 ITC Faculty, University of Twente, 7500 AE Enschede, The Netherlands; c.h.j.lemmen@utwente.nl
* Correspondence: rohanbennett@swin.edu.au

Abstract: A contemporary review of land administration, from the perspective of systems maintenance, is provided. A special emphasis is placed on emerging fit-for-purpose land administration solutions. The research synthesis uses reputable sources from the contemporary era. Results show the challenges of maintaining land administration systems and the data held are long recognized. The 1970s–1980s gave the issue impetus as data and processes moved from paper-based and manual to digital and automated. The 1990s recognized concerns on maintenance, albeit as a secondary issue: system establishment was the primary concern. The 2000s placed more emphasis on more holistic sociotechnical systems but, again maintenance was supplementary. The fit-for-purpose era deliveres a vast range of new social and technological innovations; however, scaled and sustainable implementations still struggle with system maintenance. From the findings, a consolidated model for analyzing maintenance problems and solutions at jurisdictional level is developed. Maintenance of a land administration system can be understood by identifying the level of change, method for change, components to change, and options for what to change to. The United Nations-endorsed Framework for Effective Land Administration is then used to identify specific maintenance challenges and available solutions. It is suggested that due to the scope and size of what can be considered maintenance issues, there exists no single solution—instead the country should identify its persistant maintenance problems, and the most appropriate solution set from the suite of available options. Emerging solutions and challenges include ensuring interlinkage to maintenance of spatial planning, land valuation, and marine administration; exploiting survey data 'back capture' initiatives; supporting grassroots IT; and giving serious attention to cybersecurity concerns.

Keywords: maintenance; update; upgrade; upkeep; renewal; cadastre; land registration; data quality

Citation: Bennett, R.M.; Unger, E.-M.; Lemmen, C.; Dijkstra, P. Land Administration Maintenance: A Review of the Persistent Problem and Emerging Fit-for-Purpose Solutions. *Land* **2021**, *10*, 509. https://doi.org/10.3390/land10050509

Academic Editor: Volker Beckmann

Received: 1 April 2021
Accepted: 5 May 2021
Published: 11 May 2021

1. Introduction

While much attention is afforded the establishment of land administration systems, less attention is given to ensuring they can be sustained and maintained [1]. That is, after an initial recording occurs of the parcel boundaries, associated rights, and the parties involved, processes for recording changes to those entities are ill-defined, poorly thought through, or inadequately resourced.

This creates many problems, not the least being that the significant upfront investment put into system establishment is wasted, and none of the intended longer-term benefits flow to the community [2].

The bias against maintenance has long been recognized in both the theory and practice of land administration [3], and the reasons for its occurrence are at least acknowledged anecdotally. Considerable effort has even been afforded to both conceptual and applied academic work on the issue [4,5].

The arrival of the fit-for-purpose land administration (hereafter FFPLA) philosophy supports the new wave of technical solutions [6]; ones that could also assist in dealing with the persistent problems of land administration maintenance. The elements of FFPLA are flexibility, inclusivity, participation, affordability, reliability, attainability, and perhaps most important to this work, upgradability. These can be achieved via establishment of appropriate legal, governance, and spatial frameworks. Despite the significant developments around FFPLA, evidence suggests that maintenance concerns are continuing to receive less attention in practice—even in more high-profile FFPLA projects [7]. Indeed, overwhelmingly, demonstrators, pilots, and implementations continue to focus on establishment procedures [8].

Therefore, emerging FFPLA approaches deserve appraisal, if not critique, in the context of land administration maintenance. Given the quantity and novelty of new approaches, there appears room for a comprehensive contemporary review of land administration maintenance solutions more generally.

To this end this paper undertakes a full review of available land administration literature, with a specific focus on maintenance and FFPLA developments. The aim is to reveal the state of play with regards to theoretical, methodological, and technological developments. The work acts as a reference and stepping off point for theoreticians and practitioners dealing with the various challenges associated with land administration maintenance, updates, upgrades, and other related terminology. Further, it invites these same actors to continue to explore more novel and emerging solutions to dealing with the maintenance issues.

Following this introductory section, an overview of the materials and methods underpinning the paper is provided. This is premised on a comprehensive review and synthesis of land administration literature. In terms of results, first the conventional challenges and solutions relating to land data maintenance are provided chronologically. Subsequently, the more recent fit-for-purpose approaches are introduced and evaluated. This leads to a synthesis of the entire domain of approaches, the relative merits and drawbacks, and persistent challenges and opportunities still requiring redress. The conclusion section provides an overarching summary and suggested areas for further work, both in practice and research.

2. Materials and Methods

The research activities underpinning this work can fundamentally be classified as research synthesis [9]. Research synthesis involves undertaking a form of systematic analysis of previous research outputs, with set constraints with regards to the scope of the body of literature, in terms of inclusion criteria, and reassembling the findings to create a novel contribution. Research synthesis is widely used across many domains, including the study area of land administration, with examples, inspiration, and justification for the approach in this particular work being found in [10–15].

The review involved a search bounded between January 1950 and December 2020. It was felt this rather expansive time period would surely include all identified challenges and opportunities relating to land administration maintenance in the contemporary era. In practical terms the majority of focus is post-1990: this is generally nominated as the starting point of increased global discourse on the widespread mainstreaming of cadastral and land registry data digitization and process digitalization. The digital computing approach was a paradigm shift for land administration—offering newer and more rapid methods of data capture, processing and storage [16], but also challenges in terms of ensuring data maintenance.

This is not to suggest that data maintenance issues did not exist prior to 1990. On the contrary, by the early 1990s, many colonial-era systems in 'developing' contexts were in states of decay or disrepair on account of open conflicts or administrative vacuums created in the post-colonial era [17,18]. The issue of maintenance and update was therefore pertinent; however, the amount of literature on the topic is markedly less, only growing in the 1990s, thanks largely to the discourse spawned by the International Federation of Surveyors (FIG), and an increased focus on 'pracademic' publishing, more generally.

In terms of searchable repositories, initial use was made of Google Scholar followed by Scopus and Science Direct searches. Additionally, the OICRF (International Office of Cadastre and Land Records; https://www.oicrf.org/search) website, the dedicated land administration repository maintained by the Dutch Cadastre, Land Registry and Mapping Agency (Kadaster) was also searched. Standard Google searches also supplemented the academic searches, incorporating gray literature considered relevant into the study. Primary search terms and combinations included 'maintenance', 'update', 'upkeep', 'upgrade', and 'renewal' linked to 'land administration', 'cadastre', 'land registry', and 'fit for purpose'. While the above repositories and search terms resulted in returns of thousands of results in total, a process of snowballing [19] was used to ascertain when the list of relevant papers and reports was comprehensive. This resulted in approximately one hundred works being included in the final synthesis compilation. The analysis of the literature was both undertaken and reported chronologically. In this regard a complete and contemporary overview of the discourse on land administration maintenance is presented. The results of this work are presented in Section 3.

A synthesis in terms of the state of play and related challenges and opportunities identified was subsequently undertaken (presented in Section 4). The results of this part of the review were compiled descriptively, additionally via the development of a conceptual model and synthesis table. More details on the construction of these are provided in the relevant section, but it is worth noting that internationally recognized frameworks were used to guide the analysis and presentation (i.e., the Framework for Effective Land Administration or FELA, as endorsed via the United Nations Committee of Experts on Global Geospatial Information Management or UNGGIM).

3. Results

The land administration maintenance problem is not new, neither in academic literature, nor practice. The review work undertaken here confirms this notion. In terms of presentation of results, for simplicity, the results from the review of pre fit-for-purpose literature (i.e., <2010s) are presented in chronological sequence, generally organized by decade. The review results from the fit-for-purpose era are presented subsequently.

Effort is made to outline key terminology as it emerged in the literature, although it is necessary to define several key constructs upfront. In this work, 'land administration' is taken to be those activities used to determine and/or record the land tenures, land values, land use, and land development relationships between people and land. This aligns with the prevailing discourse [16,20,21]. It is taken to incorporate the concepts of 'cadastre' and 'land registration'. A cadastre is typically understood as the 'where' component of land administration, identifying the location of the land being administered. Meanwhile, 'land registration' tends to consider the 'who' and 'how' components of land administration [3]. However, there is much overlap with regards to these definitions, mostly stemming from practical historical developments in various jurisdictions— and these are not considered important to the discussion here. Finally, 'fit-for-purpose' land administration is an approach to land administration system development premised on the pragmatism and evolutionary improvement [6]. Other key terminology includes 'maintenance', 'update', 'upkeep', 'upgrade', and 'renewal': definitions on these are provided within the subsequent analysis.

3.1. *The 1950s to 1980s*

While the scope of the review was primarily beyond the year 1990, it is important to note that scholarship countenanced the issue prior to this arbitrarily set date. The identifiable mentions tend to be brief, or implicit, stemming from works examining colonial-era administration regimes, although not always. For example, Singh [22] examines the context of India (circa 1834–1900), outlining the imperative for maintenance of land records for the Patwari in the northwestern provinces for revenue generation. Meanwhile, Hanstad [23], albeit writing in the 1990s, quoting the much earlier seminal work of [24] (p. 683), provides a more obvious example:

"A careful study of the long and abundant history of registration of rights to land shows that the difficulties that have so frequently been encountered ... due to defects in, or handicaps to, the **daily working of the service**, not to any extraneous disturbing conditions, and certainly not to fraud" (emphasis added).

Other pre-1990s works that touch upon the issue of maintenance are those written in the context of emerging digital database technologies, enabling the multipurpose cadastre concept, and coordinated cadastres. For example, Williamson [25] reveals the central importance of digital data and databases to support cadastral maintenance in common law jurisdictions. Others [26–31] provide positive examples from other jurisdictions. Flemming [28] provides the most obvious example of recognition that continuous maintenance is a core aspect of any cadastral system, providing a four-phase maintenance model with regards to different scales of map and survey, ranging from control surveys, to base mapping, to property ownership, to other types of land information. Braasch [29] speaks of phases of cadastral development, from graphical, to numerical, to coordinated, to computational—a process of improvement that remains relevant in the contemporary era. Through these two later examples, one begins to see the importance of distinguishing between updating the system and/or the qualities inherent in the system (i.e., system development/upgrade), versus updates to the data held within the system (i.e., data updates). However, the majority of works from pre-1990s, while making clear mention of the need for ongoing maintenance, do not do so in a way suggesting the need to differentiate between processes of establishment and ongoing upkeep.

3.2. 1990 to 1999

In the 1990s, the opportunity for entry of countries into the European Union from the former Soviet Union, following the end of the Cold War, was a boon for global discourse on the combined areas of cadastre and land registration, ultimately resulting in the broad adoption of the term 'land administration'. The International Federation of Surveyors (FIG) was particularly active in these developments, helping to record and formalize much of the discourse in conference proceedings and meeting reports. The narrower technology focus on digital cadastral databases (DCDBs) from the 1970s and 1980s gave way to a broader discourse on systems, including sociotechnical processes, institutional arrangements, and capacity development. Further aiding the discourse was the mainstreaming of the internet, enabling much easier access to scholarship than previously, and the globalization (or standardization) of academic scholarship of the 1990s.

This period ultimately resulted in creation of many foundational documents and references of the discipline [20,32–35]. The issue of maintenance is included, although again often as a tag-on to statements on establishment. Dale and McLaughlin [20] did go further to provide a standalone chapter on the processes of managing land administration, although they tend to limit the focus to technical reform, system reengineering, and project management (i.e., upgrades as opposed to day-to-day management or updates).

The 1990s marked an era when land administration systems were being challenged to undergo modernization, or renewal [36], due to the opportunities of digitalization, developments in geospatial technologies (e.g., geographic information systems or GIS, and global navigation satellite systems or GNSS), and demands for a more customer-oriented focus. This created opportunities for improving maintenance processes [37]. On this, Lemmen and Van Osterom [38] provided an overview of the specific case for the Netherlands, while [39] provided other examples from common law counties. On the impact of GNSS, others [40,41] outlined that the geodetic basis for modern cadastres were becoming more dynamic (due to always-on GNSS stations, later known as CORS (continuously operating reference stations)), and that by association, cadastral boundaries also needed to be understood as dynamic, even in the physical sense, even if legal agreements were not changing. These sorts of developments resulted in the need to distinguish the day-to-day maintenance of land administration data (i.e., land transactions), from the larger-scale maintenance or improvement of land administration infrastructure (i.e., GNSS control networks, DCDB development and quality improvement, land regulation reform, etc.). In general, 'update' was used to refer to the former and 'upgrade' to the latter, although the distinction was not used universally [5,42].

3.3. 2000–2009

Up until this point, the majority of the discourse on 'updating' or 'upgrading' had focused on developed contexts. These contexts had long-established land administration systems with the capacity and resources to keep the data in those systems up-to-date. It was therefore natural for most attention to be afforded to the more 'modern' business of system upgrade, transformation or renewal. In the early 2000s, there were many works in this regard (e.g., [43–47]), with attention focusing on large-scale cadastral database management [48], remodeling property transactions [49], e-land administration, the role of cadastres in SDIs, and early work on 3D cadastres.

Meanwhile, less attention had been afforded to so-called developing contexts with regards to maintenance. In these contexts, inadequate land rights recognition was understood, among other factors, to entrench social inequality, increase poverty levels, disadvantage women, enable various forms of land grabbing from vulnerable groups, degrade land and natural resources, and ultimately impede economic growth [6]. This is not to say these contexts were not the focus of significant land administration efforts previously (or literature for that matter). Indeed, these contexts had been the focus for significant loans for land administration system development from the World Bank, and other international donors, for decades. By the late 1990s and early 2000s, however, it was increasingly apparent that these interventions and projects were failing to deliver on the expected social and economic benefits, and were not often being sustained: simple transfers of approaches and technology from the West were not appropriate [50]. The reasons for these failures are many, varied, and are documented elsewhere—but the failure to put in place sustainable maintenance processes, after the completion of the donor project, and the importance of doing so, was certainly recognized [23].

During this period in the 2000s, therefore, more urgency can be observed in calls to better understand [1] and find solutions to the data maintenance issue in developing contexts (albeit usually alongside calls for establishment or renewal of the systems in the first place [51]). This included seeking better ways to conceptualize, describe, and understand maintenance processes [4,5,52–54], identifying generalized best practices [55], and incorporating criteria and indicators into land administration systems, specifically relating to maintenance [56,57]. In these works, longstanding land administration terminology was reassembled into frameworks for better understanding the temporal aspects relating to land administration. This included grouping and redefining spatial updates (i.e., land subdivision, readjustment, or consolidation), updates to rights (i.e., land transfer, inheritance), and the length of time that a land right, restriction, or responsibility may apply (i.e., one-off, ad-hoc, repeat, indefinite, 'sunset clause')—and differentiating these day-to-day activities from full system redevelopment. Rearticulation of the difference between systematic and sporadic [20] adjudication and demarcation was also provided: systematic referred to covering a whole jurisdiction in a methodical fashion, while sporadic referred to capturing the data as transactions occurred. These approaches and terminologies sought to bring more clarity to the maintenance issue, providing a basis upon which to build more sustainable land data updating mechanisms, although embedding the ideas from theory into practice, at scale, would be more challenging.

3.4. Fit-for-Purpose Land Administration Era (2010s Onwards)

The 2010s continued the trends from the previous decade, crystalizing technology innovations into international standards and tangible tools. Terminology relating to 'upgrade' and 'renewal' focused increasingly, if not almost exclusively, on tools to enable transition from 2D to 3D cadastres [58–63].

Another key tool development that could support both upgrade and updating issues was ISO 19152, the Land Administration Domain Model (LADM) endorsed in 2012, a generic data model standard for land administration. The model was developed, in part, with the problems of data updating and system upgrading as core drivers [64]. The model found application and profile development (or comparison against existing systems) in many contexts [65–68]. Accompanying LADM was the Social Tenure Domain Model (STDM), intended as a more generalized version of LADM, sympathetic to the needs of contexts with customary, communal, and other non-Western land tenure types [69]. The Land Administration Domain Model and STDM represented truly tangible tools, at least at a technical level, for dealing with the maintenance issue: a generic data model might not only reduce system development costs, it could persist beyond the life of the any technology architecture, could be extended and be adapted, and be transferable when future system upgrades occurred.

In the 2010s, and of primary interest to this review paper, the philosophy and associated framework of fit-for-purpose land administration (i.e., FFPLA), also appeared [70]. The ideals behind the concept had long been understood, and can be said to have essentially grown out of pragmatic thinking (i.e., that associated with engineering disciplines), systems quality management (i.e., out of management disciplines), and evolutionary understandings of development. These philosophies now found more recognition in the land sector from the likes of FIG, the World Bank, and the Global Land Tool Network (GLTN). This saw the development of the 'continuum of land rights' metaphor, where the basic notion of land tenure was extended beyond conventional private and public typologies to include customary and communal forms of land holding, among others. In turn, as a technical response, the idea of FFPLA was that any land administration system must be designed and implemented taking into account the local circumstances and tenures, and should be open to improvement over time. Fit-for-purpose land administration consists of seven primary elements or desirable qualities, with one being upgradability [70]. That is, a system and its data should be constructed with a 'minimum viable' mindset in initial stages, but that over time the system itself and the data held within it could be improved or upgraded. Methods for delivering this upgradability were provided under three frameworks: institutional, legal, and spatial [71]. In this regard, the FFPLA approach and guidelines provided a tangible means by which to deal with the maintenance issue, starting from initial design, moving into day-to-day operations and system evaluation, and was applied as an analytical framework in numerous contexts [2,72–74]. However, like earlier land administration theories and methods, the emphasis tended to remain focused on establishment over maintenance.

The 2010s also saw the spawning of other land administration technology innovations that while not necessarily targeted at it, could further support responses to the maintenance challenge. Developments in high-resolution satellite imagery (HRSI) the decade before [75] started to find application in the land administration space [76–78], with novel methods combining data from older colonial cadastral maps with recent HRSI to develop upgraded digital cadastral fabrics [79]. Notwithstanding issues of cloud cover, the systematic repeated capture of images from the same location, at high resolution, again lent itself to being another data source to support maintenance.

Additionally, unmanned aerial vehicles (UAVs) or drones emerged as a lightweight and more flexible alternative to conventional aerial photography [80,81]. An issue with conventional photography was that over time, base imagery becomes outdated, and re-flying areas is relatively costly. This is particularly an issue in urban or peri-urban areas where changes to land use and tenure can be quite rapid. With UAVs, imagery could be captured at a much lower price, with less expertise, and within a smaller desired area—almost an ideal solution for the land administration data maintenance issue. As with other technology innovations, UAVs face regulatory blockers, issues with lack of trained capacity, and social concerns [13,82]; however, the solution is increasingly a standardized part of the land administration toolkit.

The proliferation of smartphones equipped with GNSS receivers and mapping capability also received much attention in the 2010s, also being mooted as a solution to data maintenance issues [83]. The democratization of mapping would enable citizen surveyors to map, record, and update their own land rights—sending them on to government— the so-called crowdsourced cadastre [84–87]. This idealized vision was somewhat stymied due to the trusted intermediary issue: it was not so much the mapping expertise that was essential for legalizing land rights, but more the need for a trusted agent of state to verify or reverify them. A mobile device could not do this on its own; nonetheless the seed was sown for a more democratized citizen-centric land administration system.

Blockchain or digital ledger technology (DLT) picked up on the democratization or libertarian trend, across multiple sectors, including land administration, and more specifically land registration. Conceptual works explored potential models of application in the sector, ranging from fully public permissionless chains, to more government-controlled private permission chains [88,89], the latter more closely replicating the existing technology and database systems of conventional land administration systems [14]. A range of experimentation and piloting work followed [90]. With regards to maintenance, blockchain solutions potentially distributed the responsibility and risks associated with maintaining land record information. In principle, this would seem a positive proposition; however, the approach could also significantly disrupt the business models of existing actors, including government land agencies, and as such, evidence of scaled application was limited by the close of the decade. It should also be noted that most of these discussions did not move past the demonstration or piloting phase, and typically issues of capacity and competencies needed to maintain and control a scaled system were not countenanced.

Towards the latter part of the 2010s, spurred on by developments in artificial intelligence (AI) and machine learning, the idea of automating updates (and even initial data capture) via automated feature extraction techniques gained substantial research and development (R&D) focus [91–96]. Experimentation and piloting demonstrated the potential of the approach, including boundary change detection. Like other innovations, the conversion from R&D 'cadastrobots' [97] to scaled implementation faces numerous institutional barriers, not to mention issues around quality, performance, and the cost-benefit equation around needing to incorporate human actors into the process. The use of automated feature extraction techniques formed part of a broader examination of the visual nature of cadastral boundaries. While some boundaries are invisible to machine sensors, residing in agreement in the minds of people, many are not. In this regard, more attention was placed on increasing the emphasis on using natural or physical boundaries in cadastral updating—and thereby enabling automatic updating of those, via remote sensing technologies [98]. At the time of writing, these developments largely remain at the level of pilot or demonstration, not yet being considered robust enough to build production-level processes around.

Despite the technical innovations of the 2010s, and suggestions that processes for maintenance hardly differed from processes for the initial creation of data, and that therefore the same processes and tools could be applied [3], in practice the problems of ensuring effective and efficient land administration persisted, particularly in developing contexts. Even in countries where development and renewal programs had been considered a success, for example Rwanda and Turkey, the issue of keeping records up-to-date continued to demand research attention [7,99]. Indeed, the impetus was such that Zevenbergen et al. [2] even framed future research directions in the field entirely around better understanding five (5) levels of 'change': (i) changes in people-to-land relations; (ii) changes in technological possibilities; (iii) land use changes (due to urbanization); (iv) measuring changes; and (v) change agents. In this vein, there remains a strong argument to continue to explore potential options.

Perhaps giving a boost to recognition on the need to better tend to the issues of maintenance, the UN-backed 2030 Agenda and accompanying Sustainable Development Goals (SDGs) provided emphasis on dealing with multiple land issues, particularly in developing contexts, and particularly with regards to securing and recognizing land rights. In this vein, the UNGGIM Expert Group on Land Administration and Management developed the FELA [100,101], which openly sought to encapsulate maintenance approaches throughout:

> *"A reference for developing, **reforming, renewing, strengthening, modernizing**, and monitoring land administration."* [102] (p. 1)

In addition, the nine Strategic Pathways making up the substantive or actionable parts of FELA generally all pay attention to aspects of maintenance. For example, in Strategic Pathway 4 on data:

> *"To support the creation and **maintenance** of land data, data custodianship, acquisition, management, supply chain, curation, metadata concerns, delivery strategies and mechanisms are needed. These should consider cross-sector and multidisciplinary issues—along with privacy and security concerns—and ultimately enable inclusive access and better ordering, integration and searching of information relating to property rights, restrictions, and responsibilities."* (p. 23)

Additionally, for example, in relation to development and use of standards in Pathway 6 on standards:

> *"Standards also assist cost reduction and support removal of duplication and maintenance efforts."* (p. 24)

Therefore, considering all the developments around FFPLA and FELA, perhaps at no earlier time has the emphasis on improving and ensuring day-to-day maintenance of land administration been given more attention: member state-endorsed UN documents make specific mentions and calls for action.

That said, it cannot be denied that implementing these ideals and technological innovations remains challenging in practice: it appears there exists no silver bullet or short term fix: a portfolio approach or suite of tools appears necessary. Therefore, having concluded reporting of the review results, this work makes an attempt to cluster or group the findings and cast forward to future developments.

4. Synthesis and Discussion

Having reviewed both pre- and post-FFPLA literature from the perspective of maintenance, this section seeks to take the synthesis further by (i) constructing a holistic overview of the terminologies, challenges, solutions, and relations with respect to maintenance; and (ii) presenting the maintenance solutions to those key concerns using a contemporary land administration framework: the Framework for Effective Land Administration (FELA) [100]. The aim here is not to conclusively 'solve' the issue land administration maintenance—the scope of that endeavor is outside a single academic paper, and at any rate as mentioned, there appear to be no silver bullets—rather, it is to provide a useful snapshot view of the key underlying problems, contemporary solutions, and areas still requiring further research endeavors.

4.1. Consolidated Analytical Model of Land Administration Maintenance

Figure 1 seeks to summarize the story of the maintenance concept in land administration—in terms of key terminology, typologies, approaches, aspects and options—as it has developed over the previous decades, including the FFPLA era. It was constructed based on the results of the review and subsequent modeling/reasoning activities. For each of the parts of the diagram (i.e., rounded boxes) there are specific challenges to overcome when it comes to maintenance, and in many country contexts these issues remain unresolved.

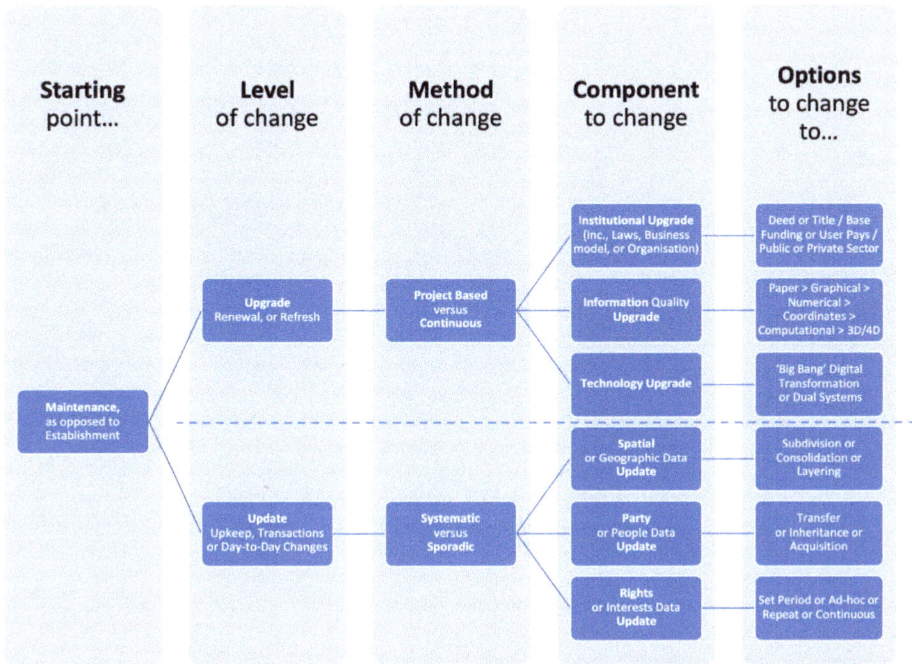

Figure 1. A model for analyzing maintenance in land administration.

Important to note is that the model only demonstrates ways of describing or understanding land administration maintenance; it does not provide judgement on most appropriate solutions to the challenges. The diagram is useful in that it helps to show the differentiation between terms in a crisp manner, as they are represented in the literature. Granted, some of the literature is less precise, but the key hierarchical breakdown is seen as useful for assisting theoreticians and practitioners in ensuring they are communicating on the same issue with the same terminology.

Unpacking the diagram further, first, it starts to become clear that, at least in terms of the existing literature, that beyond the breakdown of 'update' versus 'upgrade', there are essentially six central components that change: institutions, system quality, technology, spatial data, party data, and rights data. It should be noted that these could be possible grouped in any number of ways. For example, we could group the three types of data (i.e., spatial, part, right) into a single category. However, the six categories are decided upon as they are backed by the literature base. That is, each of these categories has a discrete body of research. At a minimum, the approach suggests that any jurisdiction or related land agency should have baseline knowledge, and preferably already enabled supportive processes, for ensuring appropriate maintenance of these components. This would already represent two indicators for measuring and monitoring maintenance within the jurisdiction.

Going further, the options for 'what to change to' or 'how to change' can also be systematically understood. In terms of institutions, this involves unpacking business models, subscription approaches, and even considerations of 'deeds' vs. 'title' approaches. For the component of system quality, this involves understanding where the jurisdiction is placed in terms of paper vs. graphical vs. numerical vs. coordinated vs. computational, and determining if and how to change, and whether leapfrogging between levels is necessary. For the technology architecture component, it involves making assessments as to whether parallel systems are needed and whether 'big bang' change implementations are most efficient or necessary. For the data components, it involves understanding and describing existing processes, assessing them for efficiency and effectiveness, from a range of stakeholder perspectives, and necessarily optimizing them. Of course, this process may have all kinds of change implications back at the previous levels: for institutional arrangements, system quality requirements and procedures, and/or technology architecture.

Whatever the case, the model (Figure 1) offers the ability to take a holistic perspective on maintenance, and to systematically work through existing approaches (and limitations), and potential strategies for improvements. Following on from Figure 1, in the next section, consideration is given to provide more depth and detail on the range of maintenance solutions by repacking the review results against the nine UNGGIM FELA Strategic Pathways. This not only supports providing a holistic view on solutions to maintenance, but also supports the operationalization of the relatively new FELA.

4.2. Fit-for-Purpose Solutions for Land Administration Maintenance

While Figure 1 provides a consolidated depiction of the key terms and issues surrounding land administration maintenance, from an historical point of view, as they have emerged in the literature, Table 1 takes more of a forward-looking perspective. It does this by utilizing the recently endorsed UNGGIM FELA, which is intended to support land administration renewal, reform, and monitoring alongside the UN SDGs, towards 2030. Table 1 provides a general overview of the major maintenance challenges, as identified in the literature, but now categorized by those nine FELA strategic pathways, alongside developed and emerging approaches, as also identified in the literature.

Table 1. Overview of land administration maintenance challenges, developed solutions, and the case examples from which those solutions are derived.

FELA Strategic Pathway	Maintenance Problems	Available Solutions	Specific References and Recent Case Examples
1. **Governance and Institutions**	- Land agencies have project-focus rather than continuous improvement-focus - Land agencies only have a mandate for establishment exists (i.e., no clear mandate for updates) - Land agency organisaitonal resistance to upgrades from within, and external stakeholders - Conventions and traditions guide processes, not FFPLA thinking	- Organisational restructures in line with FFPLA institutional guidelines - Re-allocation of mandates to other land sector stakeholders, in line with FFPLA guidelines - Creation of One-Stop-Shops at local/community levels or online - Use of PPPs or outsourcing	[12,16,20,50,56,69,74,84]; e.g., Rwanda; Namibia; Ethiopia; Eastern European Countries; Indonesia; Vietnam
2. **Law and Policy**	- No developed adopted policy on updates or upgrades. - Failure to create laws for updating and/or upgrading; - Regulations for data capture are outdated or prescriptive - Implementation and enforcement of laws is not in place	- Whole-of-jurisidction FFPLA policy development process - Specific legal reforms on outdated or inadequate legistration/regulations - De-regulation of maintenance processes in terms of requirements and actosr involved - Prioritise moving towards digital data over paper data, in policy/law	[3,4,13,17,23,32,43,55,98,100]; e.g., EU Inspire; Nepal; Colombia; Rwanda; Mekong Delta Countries
3. **Finances**	- Funding dependencies on allocated government budget (i.e., not self sustaining) - Existing business models result in government losses - Land agency rent seeking behaviours - Petty and/or grand corruption	- Free or subsidised registration (subsidies) for post-establishment land transactions - More sustainable donor funding models (less-project-oriented); - Pay-per-Use - Yearly fees or subscripton models	[20,36,49,51,56,57,74,81]; e.g., Eastern European Countries; Turkey; Indonesia

Table 1. *Cont.*

FELA Strategic Pathway	Maintenance Problems	Available Solutions	Specific References and Recent Case Examples
4. Data and Processes	- Analogue data persists across spatial and party data - Transactions remain paper based/manual - New transactions are not recorded - Spatial updates are not made at all - Lack of quality of control over data processes	- Scanning/digitisation of paper records and digital archive creation - Data model development (e.g., LADM) - Business process redesign ('as is' and 'to be') - Increased use of imagery, as per FFPLA - Move from graphical, to numerical, to coordinated, to computational cadastre - Explore automatic-feature extraction/update - Explore 3D/4D upgrade	[7,14,17,18,48,51,56,67–72,93–103]; e.g., Pakistan, South Africa, Nigeria, Malaysia, Indonesia, Singapore, Australia/New Zealand, Armenia,
5. Innovation	- No innovation processes embedded to promote and enable change within land agencies - No promotion of entrepreneurship and/or innovation in the land sector - No existing IT infrastructure and/or technology blueprint; - Legacy IT infrastructure no longer supported	- Embed and promote innovation processes into land agencies - Foster entrepreneurial acumen amongst land sector stakeholders, beyon land agencies - Embed IT system renewal into land administration processes - Replace or maintain parallel IT system	[2,14,51,59,62,73,82,84,87,88,90,103]; e.g., South Korea, Singapore, Netherlands, New Zealand/Australia, its4land Project (Ethiopia, Rwanda, Kenya), Slovenia, Dubai
6. Standards	- Lack of standards on initial capture and maintenance - Quality control and enforcement issues, where standards do exist	- Consdier utilization of ISO for organisational management, archive administration, and land data model standards - Consider also OGC on web services standards - Embed and mandate standards uptake and adaptation processes at national level	[13,49,52,56,57,64–69,100,101]; EU Inspire; LADM/STDM pilots and implementation; OGC web service application examples
7. Partnerships	- Failure to create and maintain partnership networks (local, national, international) - Lack of inter-organisational processes at business, semantic, information, or technology levels - Dependencies on other data providers, - Prevalence of data silos amongst land agencies; - Poorly constructed or enforced PPPs	- Invest in partnership building programs - Undertake inter-organisation process mapping exercises with a view to process redesign - Institute regular and more meaningful land sector forums, conferences and meetings, with incentives (i.e., professional development and accreditation)	[8,15,17,18,50,55,83–87,90,100]; Eastern African Land Administration Network (EALAN); Network of Excellence in Land Governance (NELGA); Global Land Tool Network (GLTN)
8. Capacity and Education	- Staff skills outdated or beneath required levels - Educaitonal curricula outdated in terms of theories, methods, and technologies - Staff composition too static or too frequently changed - Cross-border or cross-disciplinary brain-drain in terms of IT/technical capacities	- Maintain staff skills and retrain where necessary - Ensure staff renewal programs and capacity creation programs are instituted - Ensure close relationship between ducational providers and land agencies - Explore relevance and appropriateness of out-sourcing or off-shoring	[16–18,21,52,55,82,97]; Rwanda, Ethiopia, Namibia, Ghana, Kenya, Uganda, Tanzania, South Africa, Vietnam, Indonesia, Nepal, China, South Korea
9. Communications and Awareness	- Trust and awareness levels in public are low - No engagement with processes and public services - No formalisedcommunication plan or channels	- Establishment or refirm of PR and media units within land agencies - Regular and funded public relations campaigns including grass-roots programs - Establishment of local one-stop-shops, where Internet connectivity and remains limited	[3,17,21,55,83,100]; Netherlands; Mozambique; Vietnam

As can be seen in Table 1, each of the nine pathways (column 1) is encumbered with challenges linked to maintenance (column 2), but in each case solutions are also available (column 3), with varying levels of maturity and application (column 4). In this regard, Table 1 acts as a consolidation of the broad suite of tools developed to combat the land administration maintenance challenge: while there may be no silver bullet, viewed holistically, the body of land administration literature focused on maintenance does already provide much guidance on what changes may be needed in land administration systems, and also how to make those changes.

For any jurisdiction grappling with land administration maintenance challenges, combined use of Figure 1 and Table 1 can help to assess the 'as is' situation in terms of performance or gaps in the existing system, and also to help chart the development of the 'to be' situation, via solution identification and selection from the range of developed and available approaches. For example, are issues primarily linked to law and policy, or data quality and standards, or communications and awareness, or a combination?

From the perspectives of the authors, and with numerous country experiences extending over at least four decades of professional practice, it is suggested that in many cases, it is the combination of unclear legal mandate, inadequate business models, and poor community communication and awareness, that often feeds into inadequacies around data quality and regular updating. Consequently, the solutions linked to those specific maintenance issues are perhaps of most utility and worthy of examination and potential application in many country contexts.

4.3. Emerging Solutions and Issues

In this final part of the discussion, focus is afforded several aspects, identified by the authors through the review process, as either (i) not identified in the literature review, but understood at least historically and/or anecdotally to be important in the context of land administration maintenance; and (ii) being innovations under development, but which are yet to appear in the academic literature.

First, the need to take a perspective beyond land administration processes alone appears important. Land administration establishment and maintenance does not exist in a vacuum: it interacts with many other government and community concerns. In this regard, it appears very necessary to ensure initiatives aimed at improving maintenance of land administration data are integrated, and interoperable with, related administrative areas such as spatial planning (including 3D), land valuation, and marine/coastal management. Already promising on this track are the developments taking place around the update of ISO 19152 LADM (i.e., LADM II).

Second, not covered in the review, but becoming increasingly apparent as a solution being employed in more developed contexts, is the concept of reconstructing cadastral fabrics based on original survey measurements, found on survey plans or field sketches (i.e., 'back capture' projects). This harks back to the evolutionary improvement process espoused in [29]. First-wave digital cadastral databases or layers, generally developed in the 1980s or early 1990s, made use of heads-up digitization of cadastral index maps, not the survey measurements themselves. This was generally seen as a faster and more appropriate way to develop digital cadastral layers. Technology limitations with regards to database size and the ability to effectively store large amounts of field computations also influenced the approach. Those original digital cadastral layers could only ever be of worse accuracy than the cadastral index maps they were digitized from, and in some cases, those were more topologically correct representations rather than anything approximating survey accuracy. For example, digitized boundary representations might be many meters displaced from actual locations on the ground, relative to other features. While there has been ongoing accuracy improvements, via rubber sheeting for example, as new data was brought into those cadastral layers over the decades, the maps are still generally considered as indexes to the original survey plans, rather than carrying any legal weight. Meanwhile, as database technologies and data exchange approaches have improved, and computer processing abilities have also, there now exists the possibility to go back to the original survey plans to rebuild or create from scratch a digital cadastral layer from the original survey measurements. Generally speaking, these measurements, while still containing errors, will be of far greater accuracy than the original index maps. These 'back capture' projects are now being actively explored and piloted in jurisdictions across Australia and the Netherlands, to name a few places. The processes suggested make use of a combination of manual and digital processes to capture data from those original plans, and even make use of artificial intelligence to identify and resolve errors (or disputes) within or between the original survey data. The approach, if successful, is likely to gain traction in other jurisdictions in the coming decade, as a way of leapfrogging cadastral layer quality upgrades, rather than the continuous improvement model from the previous decades.

Third, in the context of FFPLA, it needs to be noted that many of the emerging approaches, such as the 'back capture' approach mentioned above, or those listed in Table 1, even from the era of FFPLA (e.g., automatic feature extraction, blockchain), in reality remain quite distant from those countries or communities (i.e., developing contexts) in most need of sustainable approaches to maintenance. They are arguably too technology-intensive, demanding high levels of IT acumen, and large quantities of that expertise that simply does not exist in those jurisdictions. While many of those developments are driven and introduced with the idea that the innovations will eventually 'trickle down' to all stakeholders, which may or may not be realistic, it is worth considering how grassroots IT support can be provided and embedded into local communities. This would be in a similar vein to grassroots legal support and advice on land issues or the grassroots/barefoot surveyor concept. On this, the organization known as 'Namati' and the range of grassroots legal services they provide, acts as an example. This is not to say that such ICT4D (ICT for development) initiatives do not exist (c.f., [103]) with supportive methods under development, rather it is to say that in the context of a whole-of-jurisdiction land data infrastructure, providing more funding and support could be a way to ensure more sustainable maintenance of capture land data and transaction information.

Fourth, and finally, a gap in the land administration maintenance literature, especially around upgrades to digital data and processes, and solutions around outsourcing or use of cloud and webservices, appears to be that relating to cybersecurity and data privacy. The importance of securing land rights and transaction information has long been recognized [29], even in paper-based systems; however, cyber risks are increasingly recognized as a key business-continuity concern, not just IT concern, facing land agencies, and other government databases more generally. As mentioned, these issues could do with more attention in the land administration domain, particularly with regards to better understanding the sources and reasons of the threats, understanding whether existing or conventional approaches are satisfactory, and identifying and developing emerging cybersecurity and data privacy solutions that could be utilized in the land administration domain.

5. Conclusions and Future Prospects

This paper sought to provide a contemporary and consolidated update of the land administration maintenance challenge, both in terms of problems and solutions, and by including a particular emphasis on FFPLA, and what that has meant for maintenance more generally.

In this regard, a research synthesis was undertaken of land administration literature, using reputable databases and sources, from the 1950s to the contemporary era. The review identified that the challenges of maintaining land administration systems and the data held within them has been long recognized. The emergence of the digital era in the 1970s–1980s gave the issue more impetus as data and processes moved from paper-based and manual, towards digital and automated. The 1990s saw the more formal arrival of the domain of land administration; however, maintenance, while recognized, tended be a secondary concern as opposed to system establishment or complete rebuild, particularly as many post-Communist countries sought to establish cadastres. The 2000s placed more emphasis on more holistic, less technically-centric land administration designs, with sociotechnical systems thinking influencing developments, and with a special focus placed on building systems in developing contexts. Again, in general, maintenance was generally a secondary concern. The post-2010 FFPLA era has delivered a vast range of new social and technological innovations, and even countenanced the issue of maintenance, update, and upgrade more directly. However, while the underlying philosophy and supportive tools can be considered well formed, scaled implementations that are sustainable in terms of maintenance remain a challenge.

Following on, this paper consolidated those findings in terms of problems and solutions into a conceptual model, one enabling use at jurisdictional level, to unpack in more detail which specific aspects of maintenance may be problematic. This included breaking down the issue of maintenance into the level of change, methods for changing, components to change, and options for what to change to. Going further, it also utilized the recently endorsed UNGGIM FELA to explore more specific challenges to the nine strategic pathways in that framework, to illustrate the already available and emerging suite of solutions available to practitioners when it comes to resolving land administration maintenance issues, with a FFPLA mindset. It is suggested that due to the scope and size of what can be considered maintenance issues, there exist no silver bullets—instead a meaningful appraisal of Figure 1 and Table 1, for example, can help to identify the persistent maintenance problems within a country context, and the most appropriate solution set from the suite of available options. We encourage the use of and further reports on the approach outlined here, at a country level, and to have results of the application reported back to the academic community

Finally, while the review carried forward to the present era, it is also identified that new solutions, as yet unpublished, and newly identified challenges, are also emerging. These too could be the focus of further research effort, and relate to, for example, creating integrated maintenance processes of other spatial datasets, such as those relating to spatial planning, valuation, and the marine environment, transferring the lessons of the emerging 'back capture' cadastral projects in more developed jurisdictions, support for grassroots IT development (alongside grassroots lawyers and surveyors), and cybersecurity and data privacy concerns.

Author Contributions: Conceptualization, R.M.B. and C.L.; methodology, R.M.B.; validation, P.D., E.-M.U., and C.L.; formal analysis, R.M.B.; investigation, R.M.B.; data curation, R.M.B.; writing—original draft preparation, R.M.B.; writing—review and editing, E.-M.U. and P.D.; visualization, R.M.B. All authors have read and agreed to the published version of the manuscript.

Funding: This research received no external funding.

Institutional Review Board Statement: Not applicable.

Informed Consent Statement: Not applicable.

Data Availability Statement: Not applicable.

Acknowledgments: Many thanks to the School for Land Administration at the University of Twente, Netherlands for supporting the publishing of this study.

Conflicts of Interest: The authors declare no conflict of interest.

References

1. Van der Molen, P. The dynamic aspect of land administration: An often-forgotten component in system design. *Comput. Environ. Urban Syst.* **2002**, *26*, 361–381. [CrossRef]
2. Zevenbergen, J.; De Vries, W.; Bennett, R.M. *Advances in Responsible Land Administration*; Zevenbergen, J., De Vries, W., Bennett, R.M., Eds.; CRC Press: Boca Raton, FL, USA, 2015.
3. Henssen, J. *Land Registration and Cadastre Systems: Principles and Related Issues*; Technische Universität München: München, Germany, 2010.

4. Zevenbergen, J. Systems of land registration aspects and effects. In *Publications on Geodesy, 51*; TU Delft: Delft, The Netherlands, 2002; Volume 51.
5. Effenberg, W. Spatial Cadastral Information Systems: The Maintenance of Digital Cadastral Maps. Ph.D. Dissertation, University of Melbourne, Melbourne, Australia, 2001.
6. Enemark, S.; McLaren, R.; Lemmen, C. *Fit-for-Purpose Land Administration Guiding Principles*; Global Land Tool Network (GLTN): Copenhagen, Denmark, 2015.
7. Biraro, M.; Bennett, R.; Lemmen, C. Accelerated land administration updates. In *Advances in Responsible Land Administration*; Zevenbergen, J., De Vries, W., Bennett, R.M., Eds.; CRC Press: Boca Raton, FL, USA, 2015; pp. 145–162.
8. Barry, M. Fit-for-purpose land administration—Administration that suits local circumstances or management bumper sticker? *Surv. Rev.* **2018**, *50*, 383–385. [CrossRef]
9. Rossiter, D.G. *Research Concepts & Skills Volume 1: Concepts*; ITC Facuty, The University of Twente: Enschede, The Netherlands, 2011; p. 35.
10. Rockson, G.; Bennett, R.; Groenendijk, L. Land administration for food security: A research synthesis. *Land Use Policy* **2013**, *32*, 337–342. [CrossRef]
11. Simbizi, M.C.; Bennett, R.M.; Zevenbergen, J. Land tenure security: Revisiting and refining the concept for Sub-Saharan Africa's rural poor. *Land Use Policy* **2014**, *36*, 231–238. [CrossRef]
12. Alemie, B.K.; Bennett, R.M.; Zevenbergen, J. Evolving urban cadastres in Ethiopia: The impacts on urban land governance. *Land Use Policy* **2015**, *42*, 695–705. [CrossRef]
13. Stöcker, C.; Bennett, R.; Nex, F.; Gerke, M.; Zevenbergen, J. Review of the Current State of UAV Regulations. *Remote Sens.* **2017**, *9*, 459. [CrossRef]
14. Bennett, R.M.; Pickering, M.; Sargent, J. Transformations, transitions, or tall tales? A global review of the uptake and impact of NoSQL, blockchain, and big data analytics on the land administration sector. *Land Use Policy* **2019**, *83*, 435–448. [CrossRef]
15. Asiama, K.; Bennett, R.; Zevenbergen, J. Towards Responsible Consolidation of Customary Lands: A Research Synthesis. *Land* **2019**, *8*, 161. [CrossRef]
16. Williamson, I.; Enemark, S.; Wallace, J.; Rajabifard, A. *Land Administration for Sustainable Development*; ESRI Press Academic: Redlands, CA, USA, 2010.
17. Zevenbergen, J.; Augustinus, C.; Antonio, D.; Bennett, R. Pro-poor land administration: Principles for recording the land rights of the underrepresented. *Land Use Policy* **2013**, *31*, 595–604. [CrossRef]
18. Hendriks, B.; Zevenbergen, J.; Bennett, R.; Antonio, D. Pro-poor land administration: Towards practical, coordinated, and scalable recording systems for all. *Land Use Policy* **2019**, *81*, 21–38. [CrossRef]
19. Badampudi, D.; Wohlin, C.; Petersen, K. Experiences from using snowballing and database searches in systematic literature studies. In Proceedings of the 19th International Conference on Evaluation and Assessment in Software Engineering, Nanjing, China, 27 April 2015; pp. 1–10.
20. Dale, P.; McLaughlin, J. *Land Administration*; Oxford University Press: Oxford, UK, 2000.
21. Zevenbergen, J.; De Vries, W.; Bennett, R.M. Future directions in responsible land administration. In *Advances in Responsible Land Administration*; Zevenbergen, J., De Vries, W., Bennett, R.M., Eds.; CRC Press: Boca Raton, FL, USA, 2015.
22. Singh, A.K. The preparation and maintenance of the village land records—A case study of the office of the patwari in the north western provinces, 1834–1900. *Proc. Indian Hist. Congr.* **1987**, *48*, 491–501.
23. Hanstad, T. Designing land registration systems for developing countries. *Am. U. Int'l L. Rev.* **1997**, *13*, 647.
24. Dowson, E.; Sheppard, V.L. *Land Registration*; HM Stationary: London, UK, 1956.
25. Williamson, I.P. Cadastres and land information systems in common law jurisdictions. *Surv. Rev.* **1985**, *28*, 114–129. [CrossRef]
26. Dahlberg, R.E. The public land survey system: The American rural cadastre. *Comput. Environ. Urban Syst.* **1984**, *9*, 145–153. [CrossRef]
27. Kauter, B. The swiss cadastre and the district surveyor. *Can. Surv.* **1987**, *41*, 541–553. [CrossRef]
28. Flemming, N.L. Continuous updating—An essential requirement of a modern cadastre. *Can. Surv.* **1975**, *29*, 64–74. [CrossRef]
29. Braasch, H.W. The arrangement of numerical cadastral data in a modern cadastre of land holdings. *Can. Surv.* **1975**, *29*, 39–48. [CrossRef]
30. Williamson, I.P. The role of the Cadastre in a statewide land information system. *Aust. Surv.* **1983**, *31*, 567–581. [CrossRef]
31. Smith, G. Implementation of a co-ordinated Cadastre for South Australia. *Aust. Surv.* **1988**, *34*, 251–277. [CrossRef]
32. Kaufmann, J.; Steudler, D. *Cadastre 2014: A Vision for a Future Cadastral System*; FIG Commission 7: Berne, Switzerland, 1998.
33. International Federation of Surveyors. *The FIG Statement on the Cadastre*; International Federation of Surveyors: Canberra, Australia, 1995.
34. International Federation of Surveyors. *Congress. The Bathurst Declaration on Land Administration for Sustainable Development*; International Federation of Surveyors (FIG): Bathurst, Australia, 1999.
35. Larsson, G. *Land Registration and Cadastral Systems; Tools for Land Information and Management*; Longman Scientific and Technical: Harlow, Essex, UK, 1991.
36. Van Der Molen, P. Land registers and cadastre as prerequisites for land and building taxation: A case study of The Netherlands. In Proceedings of the XXVIII Incontro di Studio: Valori Immobilari Catasto e Fiscalita, Rome, Italy, 19 October 1998.
37. Polley, I.; Williamson, I. GIS, the Internet and the cadastre: Coming together. *Aust. Surv.* **1999**, *44*, 183–188. [CrossRef]

38. Lemmen, C.; van Oosterom, P. Efficient and automatic production of periodic updates of cadastral maps. *JEC* **1995**, *95*, 137–142.
39. Dale, P.F.; McLaren, R.A. GIS in land administration. *Geogr. Inf. Syst.* **1999**, *2*, 859–875.
40. Grant, D.B. A dynamic datum for a dynamic cadastre. *Aust. Surv.* **1995**, *40*, 22–28. [CrossRef]
41. Blick, G.; Grant, D. Possibility of a dynamic cadastre for a dynamic nation. In *Advances in Positioning and Reference Frames*; Springer: Berlin/Heidelberg, Germany, 1998; pp. 107–113.
42. Effenberg, W.; Williamson, I.P. Digital cadastral databases: The Australian experience. In Proceedings of the AGI' 97 Conference, Birmingham, UK, 1997.
43. Enemark, S. Building land information policies. In Proceedings of the Special Forum on Building Land Information Policies in the Americas, Aguascalientes, Mexico, 26 October 2004; Volume 26, p. 2004.
44. Bogaerts, T.; Zevenbergen, J. Cadastral systems—alternatives. *Comput. Environ. Urban Syst.* **2001**, *25*, 325–337. [CrossRef]
45. Stoter, J.; Salzmann, M. Towards a 3D cadastre: Where do cadastral needs and technical possibilities meet? *Comput. Environ. Urban Syst.* **2003**, *27*, 395–410. [CrossRef]
46. Lauhkonen, H. Cadastral renewal in Finland—The challenges of implementing LIS. *GIM Int.* **2007**, *21*, 4.
47. Dawe, P.; Gariepy, D.; Hall, B. Integrated parcel maintenance: Trends and directions. *Geomatica* **2007**, *61*, 410.
48. Van Oosterom, P.J.; Lemmen, C.H. Spatial data management on a very large cadastral database. *Comput. Environ. Urban Syst.* **2001**, *25*, 509–528. [CrossRef]
49. Zevenbergen, J.; Frank, A.; Stubkjær, E. *Real Property Transactions—Procedures, Transaction Costs and Models*; IOS Press: Amsterdam, The Netherlands, 2007.
50. Fourie, C. Land management, land administration and geospatial data: Exploring the conceptual linkages in the developing world. *Geomatica* **2002**, *56*, 351–361.
51. Enemark, S.; Williamson, I.; Wallace, J. Building modern land administration systems in developed economies. *J. Spat. Sci.* **2005**, *50*, 51–68. [CrossRef]
52. Williamson, I.; Ting, L. Land administration and cadastral trends—A framework for re-engineering. *Comput. Environ. Urban Syst.* **2001**, *25*, 339–366. [CrossRef]
53. Bennett, R.; Wallace, J.; Williamson, I. Organising land information for sustainable land administration. *Land Use Policy* **2008**, *25*, 126–138. [CrossRef]
54. Navratil, G.; Frank, A.U. Processes in a cadastre. *Comput. Environ. Urban Syst.* **2004**, *28*, 471–486. [CrossRef]
55. Williamson, I.P. Land administration "best practice" providing the infrastructure for land policy implementation. *Land Use Policy* **2001**, *18*, 297–307. [CrossRef]
56. Burns, T.; Grant, C.; Nettle, K.; Brits, A.M.; Dalrymple, K. Land administration reform: Indicators of success and future challenges. *Agric. Rural Dev. Discuss. Pap.* **2007**, *37*, 1–227.
57. Steudler, D.; Rajabifard, A.; Williamson, I.P. Evaluation of land administration systems. *Land Use Policy* **2004**, *21*, 371–380. [CrossRef]
58. Vučić, N.; Roić, M.; Mađer, M.; Vranić, S.; Van Oosterom, P. Overview of the croatian land administration system and the possibilities for its upgrade to 3D by existing data. *ISPRS Int. J. Geo-Inf.* **2017**, *6*, 223. [CrossRef]
59. Drobež, P.; Fras, M.K.; Ferlan, M.; Lisec, A. Transition from 2D to 3D real property cadastre: The case of the Slovenian cadastre. *Comput. Environ. Urban Syst.* **2017**, *62*, 125–135. [CrossRef]
60. Guo, R.; Li, L.; Ying, S.; Luo, P.; He, B.; Jiang, R. Developing a 3D cadastre for the administration of urban land use: A case study of Shenzhen, China. *Comput. Environ. Urban Syst.* **2013**, *40*, 46–55. [CrossRef]
61. Vandysheva, N.; Sapelnikov, S.; Van Oosterom, P.J.; De Vries, M.E.; Spiering, B.; Wouters, R.; Hoogeveen, A.; Penkov, V. The 3D cadastre prototype and pilot in the Russian Federation. In Proceedings of the FIG Working Week 2012, Territory, Environment and Cultural Heritage, Rome, Italy, 6–10 May 2012.
62. Stoter, J.; Ploeger, H.; van Oosterom, P. 3D cadastre in The Netherlands: Developments and international applicability. *Comput. Environ. Urban Syst.* **2013**, *40*, 56–67. [CrossRef]
63. Oldfield, J.; Van Oosterom, P.; Beetz, J.; Krijnen, T.F. Working with open BIM standards to source legal spaces for a 3D cadastre. *ISPRS Int. J. Geo-Inf.* **2017**, *6*, 351. [CrossRef]
64. Lemmen, C.; Van Oosterom, P.; Bennett, R. The land administration domain model. *Land Use Policy* **2015**, *49*, 535–545. [CrossRef]
65. Elia, E.A.; Zevenbergen, J.A.; Lemmen, C.H.; Van Oosterom, P.J. The land administration domain model (LADM) as the reference model for the Cyprus land information system (CLIS). *Surv. Rev.* **2013**, *45*, 100–110.
66. Babalola, S.O.; Rahman, A.A.; Choon, L.T.; Van Oosterom, P.J.M. Possibilities of land administration domain model (ladm) implementation in Nigeria. *ISPRS Ann. Photogramm. Remote Sens. Spat. Inf. Sci.* **2015**, *II-2/W2*, 155–163. [CrossRef]
67. Tjia, D.; Coetzee, S. Application of the land administration domain model to the city of johannesburg land information system. *S. Afr. J. Geomat.* **2013**, *2*, 260–279.
68. Zulkifli, N.A.; Rahman, A.A.; Van Oosterom, P.; Tan, L.C.; Jamil, H.; Teng, C.H.; Looi, K.S.; Chan, K.L. The importance of Malaysian land administration domain model country profile in land policy. *Land Use Policy* **2015**, *49*, 649–659. [CrossRef]
69. Antonio, D.; Zevenbergen, J.; Augustinus, C. Social tenure domain model: An emerging land governance tool. In *Advances in Responsible Land Administration*; CRC Press: Boca Raton, FL, USA, 2015; Volume 14, pp. 251–268.
70. Enemark, S.; Bell, K.C.; Lemmen, C.; McLaren, R. *Fit-for-Purpose Land Administration*; International Federation of Surveyors: Copenhagen, Denmark, 2014.

71. Habitat, U.N. *GLTN/Kadaster (2016) Fit-for-Purpose Land Administration: Guiding Principles for Country Implementation*; E-book; United Nations Human Settlements Programme (UN-HABITAT): Nairobi, Kenya, 2016.
72. Musinguzi, M.; Enemark, S. A fit-for-purpose approach to land administration in Africa: Supporting the global agenda. *Int. J. Technosci. Dev.* **2019**, *4*, 69–89.
73. Moreri, K.; Fairbairn, D.; James, P. Issues in developing a fit for purpose system for incorporating VGI in land administration in Botswana. *Land Use Policy* **2018**, *77*, 402–411. [CrossRef]
74. Bennett, R.M.; Alemie, B.K. Fit-for-purpose land administration: Lessons from urban and rural Ethiopia. *Surv. Rev.* **2016**, *48*, 11–20. [CrossRef]
75. Alexandrov, A.; Hristova, T.; Ivanova, K.; Koeva, M.; Madzharova, T.; Petrova, V. Application of QuickBird satellite imagery for updating cadastral information. In Proceedings of the XX Congress of ISPRS, Istanbul, Turkey, 12–23 July 2004; pp. 386–391.
76. Ali, Z.; Tuladhar, A.; Zevenbergen, J. An integrated approach for updating cadastral maps in Pakistan using satellite remote sensing data. *Int. J. Appl. Earth Obs. Geoinf.* **2012**, *18*, 386–398. [CrossRef]
77. Bennett, R.; Oosterom, P.; Lemmen, C.; Koeva, M. Remote sensing for land administration. *Remote Sens.* **2020**, *12*, 2497. [CrossRef]
78. Rao, S.S.; Sharma, J.R.; Rajasekhar, S.S.; Rao, D.S.; Arepalli, A.; Arora, V.; Singh, R.P.; Kanaparthi, M. Assessing usefulness of high-resolution satellite imagery (HRSI) for re-survey of cadastral maps. In Proceedings of the ISPRS Annals of Photogrammetry, Remote Sensing & Spatial Information Sciences, Hyderabad, India, 9–12 December 2014; Volume 2.
79. Sengupta, A.; Lemmen, C.; Devos, W.; Bandyopadhyay, D.; Van der Veen, A. Constructing a seamless digital cadastral database using colonial cadastral maps and VHR imagery–an Indian perspective. *Surv. Rev.* **2016**, *48*, 258–268. [CrossRef]
80. Puniach, E.; Bieda, A.; Ćwiąkała, P.; Kwartnik-Pruc, A.; Parzych, P. Use of unmanned aerial vehicles (UAVs) for updating farmland cadastral data in areas subject to landslides. *ISPRS Int. J. Geo-Inf.* **2018**, *7*, 331. [CrossRef]
81. Ramadhani, S.A.; Bennett, R.M.; Nex, F. Exploring UAV in Indonesian cadastral boundary data acquisition. *Earth Sci. Inform.* **2018**, *11*, 129–146. [CrossRef]
82. Koeva, M.; Stöcker, C.; Crommelinck, S.; Ho, S.; Chipofya, M.; Sahib, J.; Bennett, R.; Zevenbergen, J.; Vosselman, G.; Lemmen, C.; et al. Innovative remote sensing methodologies for kenyan land tenure mapping. *Remote Sens.* **2020**, *12*, 273. [CrossRef]
83. McLaren, R. Crowdsourcing support of land administration–A partnership approach. In Proceedings of the FIG Working Week 2012: Territory, Environment, and Cultural Heritage, Rome, Italy, 6–10 May 2012.
84. de Vries, W.T.; Bennett, R.M.; Zevenbergen, J.A. Neo-cadastres: Innovative solution for land users without state based land rights, or just reflections of institutional isomorphism? *Surv. Rev.* **2015**, *47*, 220–229.
85. Asiama, K.O.; Bennett, R.M.; Zevenbergen, J.A. Participatory Land Administration on Customary Lands: A Practical VGI Experiment in Nanton, Ghana. *ISPRS Int. J. Geo-Inf.* **2017**, *6*, 186.
86. Gkeli, M.; Potsiou, C.; Ioannidis, C. A technical solution for 3D crowdsourced cadastral surveys. *Land Use Policy* **2020**, *98*, 104419. [CrossRef]
87. Basiouka, S.; Potsiou, C. The volunteered geographic information in cadastre: Perspectives and citizens' motivations over potential participation in mapping. *GeoJournal* **2014**, *79*, 343–355.
88. Vos, J.A.; Lemmen, C.H.; Beentjes, B.E. Blockchain based land administration feasible, illusory or a panacea. In *Netherlands Cadastre, Land Registry and Mapping Agency. Paper Prepared for Presentation at the 2017 World Bank Conference on Land and Povertry, Washington, DC, USA, 20–24 March 2017*; The World Bank: Washington, DC, USA, 2017.
89. Lemieux, V.L. Evaluating the use of blockchain in land transactions: An archival science perspective. *Eur. Prop. Law J.* **2017**, *6*, 392–440. [CrossRef]
90. Bennett, R.; Miller, T.; Pickering, M.; Kara, A.-K. Hybrid approaches for smart contracts in land administration: Lessons from three blockchain proofs-of-concept. *Land* **2021**, *10*, 220. [CrossRef]
91. Wassie, Y.A.; Koeva, M.N.; Bennett, R.M.; Lemmen, C.H.J. A procedure for semi-automated cadastral boundary feature extraction from high-resolution satellite imagery. *J. Spat. Sci.* **2017**, *63*, 75–92. [CrossRef]
92. Crommelinck, S.; Bennett, R.; Gerke, M.; Nex, F.; Yang, M.Y.; Vosselman, G. Review of automatic feature extraction from high-resolution optical sensor data for UAV-based cadastral mapping. *Remote Sens.* **2016**, *8*, 689. [CrossRef]
93. Wierzbicki, D.; Matuk, O.; Bielecka, E. Polish cadastre modernization with remotely extracted buildings from high-resolution aerial orthoimagery and airborne LiDAR. *Remote Sens.* **2021**, *13*, 611. [CrossRef]
94. Xia, X.; Koeva, M.; Persello, C. Extracting cadastral boundaries from UAV images using fully convolutional networks. In Proceedings of the IGARSS 2019-2019 IEEE International Geoscience and Remote Sensing Symposium, Yokohama, Japan, 28 July–2 August 2019; pp. 2455–2458.
95. Kumar, P.; Rahman, A.A.; Buyuksalih, G. Automated extraction of buildings from aerial lidar point cloud and digital imaging datasets for 3D cadastre–preliminary results. In *Cadastre: Geo-Information Innovations in Land Administration*; Springer: Cham, Switzerland, 2017; pp. 159–165.
96. Luo, X.; Bennett, R.; Koeva, M.; Lemmen, C.; Quadros, N. Quantifying the overlap between cadastral and visual boundaries: A case study from vanuatu. *Urban Sci.* **2017**, *1*, 32. [CrossRef]
97. Bennett, R.; Asiama, K.; Zevenbergen, J. The intelligent cadastre. In Proceedings of the FIG Commission, St. Juliens, Malta, 16 November 2015; Volume 7.
98. Bennett, R.; Kitchingman, A.; Leach, J. On the nature and utility of natural boundaries for land and marine administration. *Land Use Policy* **2010**, *27*, 772–779. [CrossRef]

99. Yildiz, O.; Erden, Ç. Cadastral updating: The case of Turkey. *Surv. Rev.* **2020**, *7*, 1–4. [CrossRef]
100. Unger, E.M.; Bennett, R.; Lemmen, C.; de Zeeuw, K.; Zevenbergen, J.; Teo, C.; Crompvoets, J. Global policy transfer for land administration and disaster risk management. *Land Use Policy* **2020**, *99*, 104834. [CrossRef]
101. de Zeeuw, K.; Benn, T.; Unger, E.M.; Bennett, R.; Teo, C. The proposed United Nations framework for effective land administration (FELA): Progress, pathways and prospects. In Proceedings of the Land and Poverty Conference 2020: Institutions for Equity and Resilience, Washington, DC, USA, 16–20 March 2020.
102. UNGGIM. Framework for Effective Land Administration, August 2021, New York, United States. Available online: https://ggim.un.org/meetings/GGIM-committee/10th-Session/documents/E-C.20-2020-29-Add_2-Framework-for-Effective-Land-Administration.pdf (accessed on 10 May 2021).
103. Kaur, R.; Ahamed, A.; Sargent, J.; Stockdale, R. Interactions in ICT4D projects: A sociotechnical perspective. In Proceedings of the 24th Pacific Asia Conference on Information Systems (PACIS 2020), Dubai, United Arab Emirates, 20–24 June 2020.

land

MDPI

Article

Initial Insights on Land Adjudication in a Fit-for-Purpose Land Administration

Monica Lengoiboni [1,*], Christine Richter [2], Paul van Asperen [3] and Jaap Zevenbergen [1]

[1] Faculty ITC, University of Twente, 7514 AE Enschede, The Netherlands; j.a.zevenbergen@utwente.nl
[2] Fraunhofer, 04109 Leipzig, Germany; christine.richter@imw.fraunhofer.de
[3] Rijkswaterstaat, 3500 GE Utrecht, The Netherlands; paul.van.asperen@rws.nl
* Correspondence: m.n.lengoiboni@utwente.nl

Abstract: Land adjudication constitute a series of sequential steps that if followed carefully and correctly, can lead to a sufficient determination of the varied interests in land including whether, and where they overlap, complement, conflict or compete with each other. This is a preliminary study aiming to find out how the adjudication process as it is conducted in the context of a fit-for-purpose land administration (FFPLA). A framework of components for adjudication in the FFPLA context is first developed. Further, the steps involved in accomplishing the adjudication components are compiled, assessed, and discussed from the perspective of the theory of collaborative governance. The study poses questions for consideration by implementers of land tenure documentation activities on how to identify the interests in land as they exist in their undocumented form. An understanding of the interaction between different types of interests in land in undocumented form as defined from the perspective of the communities themselves rather than from the law, could help assess which tenures and their attributes—can overlap or complement each other, or inform how they equate to specific rights in the legal perspective with minimal conflicts.

Keywords: community-based land adjudication; components of adjudication; fit-for-purpose land administration; land tenure; land rights

Citation: Lengoiboni, M.; Richter, C.; Asperen, P.v.; Zevenbergen, J. Initial Insights on Land Adjudication in a Fit-for-Purpose Land Administration. *Land* **2021**, *10*, 414. https://doi.org/10.3390/land10040414

Academic Editor: Robin McLaren

Received: 23 March 2021
Accepted: 9 April 2021
Published: 14 April 2021

1. Introduction

In the formal land administration (LA), adjudication is a prerequisite for intentions to register land rights [1,2]. In itself land adjudication does not create rights, it only establishes existing rights [2–4]. Processes involved in the determination and definition of legal rights for registration must be understood because the rights become legally binding when registered [5]. Almost universally, however, legal frameworks guiding the adjudication process remain flawed because the determination of rights for registration is almost always restricted to land tenure that is based on individual/Western parcel-based statutory tenures and has been observed to have difficulties with catering for other forms of land tenure [6]. This is because land rights in many countries are often complementary (e.g., when different parties share the same interest in the same land parcel), or overlapping (e.g., when several parties have different rights to the same parcel of land) [7]. A typical customary setting has a complex set of tenure arrangements/land rights, which may coexist and also seasonally shift [8]. In such a setting, private tenure forms can evolve over time in response to external pressures and as land gains value and as population grows [9]. Okoth-Ogendo and Van der Molen [3,4,6] observe that the adjudication process modifies tenure arrangements and that the 'traditional overlaps' are 'deleted' by individualized Western form of tenures—and thereby how land rights are organized. Interests conferred upon formal registration are often incompatible with how people use land in a customary tenure setting, which subsequently can lead to disputes and conflicts. Kabalamu [10] observes that despite international support and tenure reforms undertaken by successive governments in sub-

Sahara Africa, inequalities, skewed distribution of land resources and land disputes and conflicts in form of disagreements and contestations over interests in land persist.

To embrace traditional overlaps in land tenure (hence the notion of 'inclusivity'), Okoth-Ogendo and Van der Molen [3,4,6] recognize that overlapping and complementary land rights could be preserved using an adjudication process that codifies the recognized rules within the customary tenure setting.

But how does the land adjudication process 'modify' or 'delete' overlapping or complementary rights, even though it is not supposed to alter existing rights? Laws can define land tenures—whether primary (hereon 'ownership rights') or secondary forms of land rights (hereon 'non-ownership rights'). In this study, 'ownership' is associated with the bundle of rights that give more, if not complete and exclusive control of a piece of land [11,12]—while acknowledging that different countries/contexts may have different definitions or conceptualizations of what arrangements within the bundle of rights constitute full or less control of a piece of land. Non-ownership interests, in this study, refer to those associated with limited rights that do not amount to ownership. Non-ownership interests are often overlapping with ownership interests and as such, in this study, the non-ownership rights are the overlapping rights to land. Further, complementary rights can be in the form of ownership or non-ownership interests. Barrows and Roth [13] state that difficulties of sorting out and recording the complex set of land rights associated with an area (e.g., in a customary setting) resulted in the exclusion of certain right holders in practice. Those difficulties may pertain to the way in which framing the concept/idea of land rights is done during the adjudication process.

Ways in which stakeholders frame issues and the process of their interaction may explain collaborative success or failure [14]. According to Gray [14], frames refer to the lenses that stakeholders use to make sense of an issue or situation, and that the principal frames adapted by the stakeholders subsequently influence both the process and the outcomes. Besides difficulties in framing, a number of circumstances also lead to rightful claimants losing their land rights during the adjudication process. These include the people who are absent during the adjudication process, and when adjudication is conducted incorrectly, for example adjudication of land rights for pastoralists being at odds with their strategies of seasonal migrations based on their traditional land use patterns [15], and alterations into practices such as shifting cultivation and hunting and gathering. More recent studies show that such problems persist [11,16,17]. Therefore, the framing of land rights, as well as procedural factors during adjudication can influence the outcome of land rights registration.

New approaches for documenting land tenure have emerged in recent years, with emphasis on using participatory approaches for recording legal and social tenures using quick and affordable approaches understood as "pro-poor land administration" and "fit-for-purpose land administration" (FFPLA) [18–21]. Three core components of the FFPLA concept are the spatial, the legal and the institutional frameworks. The spatial framework is about adopting technologies and methodologies for collecting spatial data fast and affordably using general boundary approaches; the legal framework is about recognition of and recording social, legitimate, tenures as may exist in an area [21]; and the institutional framework concerns the governance aspect of land administration [18,19].

Non-governmental organizations (NGOs), international non-governmental organizations (INGOs), community-based organizations (CBOs) and others have emerged and collaborate with local communities in documenting tenure information using the FFPLA approach. In this sense the relevance of and need for a framework describing and clarifying the processes of adjudication in the context of a FFPLA is clear. So, what is needed for effective adjudication of both ownership and non-ownership tenure rights? Organizations facilitating community-based tenure documentation activities—knowingly or not—engage in some form of adjudication, where land rights are identified, mapped and documented. Unique is that collaborative approaches are employed, where different actors are brought together to collaborate in the adjudication process. Collaboration describes "a process through which parties who see different aspects of a problem can constructively explore their differences and search for solutions that go beyond their own limited vision of what is possible" [22]. Benefits of collaborative processes are a greater responsiveness and deliberations that can lead to effective, efficient and flexible solutions—often with a greater public acceptability [23].

It is unclear how the agencies facilitating a community-based adjudication activity capture social tenures. This paper reports on a preliminary study which aimed to (i) develop a framework for land adjudication in FFP land tenure documentation setting; (ii) gain insights into how the adjudication processes of those rights are being carried out in projects/programs designed to secure tenure; and (iii) develop initial ideas on how the concept of land tenure is framed (ownership and non-ownership) and how this influences what is eventually recorded. Factors that enable or hinder the framing or the adjudication process and their implications are discussed.

The paper is organized as follows: Section 2 presents the methodology, where a framework for land adjudication in a fit-for-purpose land tenure documentation is developed; Section 3 presents the results, where the processes of adjudication components—carried out in projects/programs designed to secure land tenure—are described; Section 4 builds on the results whereupon observations and questions for attention—in relation to the theory of collaborative governance—are raised by the researchers. This is followed by conclusions and recommendations. Recommendations are in the form of questions for future consideration by implementers of tenure documentation activities.

2. Research Methodology

This is an exploratory study focusing on the identification of adjudication components that yield to the documentation of both ownership and non-ownership tenure forms and land rights associated with them. An exploratory approach was preferred as it concerns the initial examination of an issue to gain insights and ideas about its underlying nature; or information gathering which can form the basis of an explanation for, or for immediate application to an administrative problem [24].

The study was undertaken in three phases: (i) establishing and defining the components of land adjudication; (ii) describing the processes used by the organizations to achieve these components; and (iii) gain insights on how land tenure and land rights are framed and how this influences the outcomes of what is eventually recorded.

Components of adjudication identified for this study serve as an initial analytical framework in order to summarize the findings from the questionnaire. As such, rather than an empirically fully substantiated theory for the case of FFPLA approaches, the frame provides entry points for understanding the interaction between different types of land tenures and land rights (e.g., when overlapping, complementing each other) as defined from the community perspective and as defined from perspective of statutory law and administrative procedure—depending on how this interaction plays out for each organization.

2.1. Phase 1: Establishing and Defining the Components of Adjudication

This part was guided by a review of two key documents: one, on the idea of adjudication in the context of formal land administration, and one, on the idea of adjudication in the context of FFPLA, which then led to the identification of a preliminary set of components of a community centered adjudication process.

Firstly, a set of components and their strategies—derived from the formal land administration context were identified. In the context of formal land administration, adjudication is anchored in the definition of Simpson [1], who is widely cited on the topic. The term adjudication is used for the "process by which all existing rights in any particular parcel or area of land are finally and authoritatively ascertained". According to Simpson [1], the purpose of adjudication is to establish finality on the existing land rights. Thus prior to the initial compilation of a register a legally defined procedure for official adjudication and legal registration of rights to land require three steps to be performed by formal adjudication committees. Step 1 identifies the persons and any qualification of his or her interests. This involves: (i) identification of owner (or holder of interests) and any qualification of his or her interest. Here, particulars of primary interests (comparable to "ownership" forms of tenures—in the context of this study) including the rights, restrictions and responsibilities (RRRs) attached to those tenures are identified; and (ii) identification of any particulars and claims to subordinate or derivative interests (comparable to "non-ownership" forms of tenures—in the context of this study) and persons attached to them so that relevant info is not overlooked. The name of the owner and particulars of any interest affecting the ownership must be ascertained at one and the same time; claims to ownership and claims to non-ownership interests must be considered together for anything relevant not to be overlooked [1]. Step 1 results in qualitative information about the land right holders. Step 2 identifies the boundaries and results in a map in three steps/strategies [1]: (i) demarcation, which is the physical marking of the area/parcel to be mapped; (ii) indication of the parcel by pointing out to the official concerned its position as demarcated by physical features or by visible monuments or, if it is invisible, where it lies; and (iii) survey of the boundaries and the preparation of a map or plan illustrating them. The order of how these steps can be followed is context dependent. Step 3 registers the rights in the land register and/or cadaster: a database of who-has-what-rights-where.

Secondly, a set of components and their strategies—derived from the FFPLA context were also identified. Here, the term adjudication is derived from the FFPLA guiding principles for country implementation [18]. These guidelines imply that the adjudication procedures are similar to the formal adjudication procedures i.e., identify, delineate, and survey the visible land parcel/area boundaries, and the rights determined are entered into a register [18]. In addition to formal land administration, the FFPLA places emphasis on (i) applying participatory adjudication methods to support the identification of tenures for documentation and (ii) monitoring and evaluation of tenure documentation activities [18,25]. According to Enemark et al. [18], participatory approaches to adjudication are expected to reveal a range of land parcels; all types of right holders (i.e., individuals, couples, households, etc.); and the relationships between land and people for documentation.

Thirdly, these components were combined to form the basic components of a framework for modelling the adjudication process in a FFPLA setting. By merging the key adjudication themes from the formal and FFPLA domains, the following emerged as fundamental items to inform the adjudication process:

2.2. Phase 2: Describing the Processes Used by the Organizations to Achieve These Components

After identifying the set of components of adjudication for a FFPLA setting, the next step was to assess the processes used to achieve these components in the real world. This study drew lessons from the organizations implementing tenure documentation activities in their various contexts. Criteria for selecting the organizations responding to this study are: (i) they are non-state actors e.g., NGOs, INGOs, CBOs or other; (ii) they support or conduct tenure documentation activities with communities; and (iii) they use participatory approaches when conducting land documentation activities. The organizations work in various countries and contexts. Two sets of documents were sent: (i) a background to the topic, and a need to gather the experiences on the practice of adjudication in the context of FFPLA; and (ii) a questionnaire, asking to provide systematic procedures or strategies used by them to achieve the components. The questionnaire was structured along the five components in Table 1 and were sent by email to the 11 organizations in January 2019. The questionnaire consisted of Likert statements formatted to scale (1 = never and 6 = always) and open-ended questions. Eight questionnaires were returned—mostly partially completed. For confidentiality reasons, identities of the responding organizations are not disclosed.

Table 1. Essential components of a community-based adjudication process.

Component	Is About
1. Mobilization	Using participatory adjudication procedures, awareness creation
2. Adjudication of rights to land	Unpacking the varieties of interests to land (RRRs) and their right holders
3. Adjudication of boundaries	Identification, delimitation, demarcation and mapping boundaries where land rights begin/end. Unpacking the boundaries of varied types of land rights that exist in the area
4. Documentation	Documentation of the adjudicated tenures and rights corresponding to those tenures and their spatial areas in a cadastral database, including validation and dispute resolution
5. Monitoring	Following up on the impacts of the documentation activities on tenures and their land rights holders

2.3. *Analysis*

Results from the eight organizations responding to the questionnaires were synthesized. Content analysis was used to derive the categories of procedures used to accomplish each land adjudication component. The process of each adjudication component was addressed individually. The theory of collaborative governance was used as a benchmark for discussing the results. This theory describes the collaborations between public actors and non-state, semi-state or other state stakeholders, defined as "a process where one or more public agencies directly engage non-state, semi-state or other state stakeholders in a collective decision-making that is formal, consensus-oriented, and deliberative and that aims to make or implement new public policy or manage public programs or assets differently" [26]. In describing the kinds of power held by participants in collaborative processes, Purdy [27] reveals that power can be exercised structurally and relationally and suggests that "three arenas" of power come into play, which influence process design for collaboration: (i) participants arena: describes who is involved in a collaborative process and who leads it; (ii) process for collaboration arena: describes the where, when, and how of collaborative governance, influencing the nature of interaction and the modes that are used for communication and decision making; and (iii) content of collaboration arena: this describes the setting the agenda and establishing expectations regarding outcome of the process. Agreements in this phase will reflect in the outcomes, as the interpretations that people use to identify issues and understand alternatives are closely linked to the success of the process [14]. Framing the issue can result in stakeholders having different interpretations of the same issue and consequently, procedural factors—when acting on the issue—can explain outcomes [14]. In this study, the content arena is about how the topic of 'adjudicating land rights' or 'determining ownership rights' or 'determining non-ownership land rights'.

3. Results: Processes Used to Accomplish the Adjudication Components in a FFPLA Setting

Results presented here are based on the general strategies and experiences by the organizations when documenting tenure in a FFPLA approach. Processes within each land adjudication component are presented.

3.1. *Component: Mobilization Process*

This section describes the activities and experiences of the organizations during the mobilization process. Some of the countries where the organizations have conducted land tenure documentation activities (at the time of the interviews) include Africa (Burundi, Democratic Republic of Congo, Mozambique, Kenya, Nigeria, Uganda; Sudan, Namibia, Zambia, Benin, Gambia), South America (Bolivia, Colombia), and Asia (India, Iraq, Nepal, Philippines, Laos, Indonesia).

All organizations engage in community mobilization events prior to tenure documentation. The approaches used to convey information on tenure documentation activities to communities are diverse, including more standard approaches such as informing community leaders, holding community meetings and sharing of project promotional materials (written documents) such as flyers, banners, artistical drawings of the community delimitation process. Focus groups, theatre plays, on-the-job training and learning events are also a means to promote the tools and the participatory enumeration techniques, with the aim of getting feedback on applicability for the given context in the community. Moreover, approaches employing locally relevant media and means, including newspapers, radio and TV spots, quizzes, fairs on land tenure documentation and "door-to-door socializing" are also used, all the while attention is given to attendance and participation of women and vulnerable groups.

Various factors determine the number of community awareness sessions that can take place. These factors can be grouped into themes: (i) geographical factors, for example accessibility to target communities in remote rural areas; (ii) logistical factors, for example meetings are held via video in case pilot areas are located in very remote areas (iii), communal factors, for instance presence of different groups of the community at different times of the day, mechanisms to reach consensus in the community, absenteeism and composition of the community for participation process and (iv) interference of the process i.e., through internal or external factors. Sub processes as determining ownership and non-ownership interests and the mapping of ownership and non-ownership boundaries requires awareness creation. All organizations report that they always aim to understand the local ownership rights situation prior to the tenure documentation activities. There are those who lead the communities in the adjudication processes, as 'facilitators'. Six organizations responded to the questions on how the facilitators are selected, which follows two approaches: (i) in accordance with the law, where five organizations report government involvement in the adjudication process and (ii) through open recruitment, where community members are appointed as facilitators. All organizations always create awareness of the mapping process by training communities on approaches that enable them to indicate where the boundaries begin/end. Methods used to train on boundary mapping are diverse.

Regarding the role of technology developers in the mobilization, the study finds that they participate in community mobilization occasionally and in the case of two organizations very frequently.

Barriers to community mobilization can be categorized into the following types: logistical and infrastructural, for instance time limitations and transportation costs, barriers relating to the communities, for example community members' lack of understanding the benefits of tenure documentation, and cultural and political interferences, especially mistrust in the government, and dynamics of social exclusion, including gender biases.

To address these issues and reduce barriers, organizations focus on accommodating meeting places and times that are a best possible fit for the community in terms of work schedules, and using existing meetings, such as church meetings for the mobilization and awareness campaigns; and also, by directly supporting transport. Those who still cannot attend are notified in writing. Local coordinators or grassroot surveyors visit the homes of those who could not attend the meetings. In order to address barriers related to community's mistrust and fear the focus rests on spending sufficient time on socializing as the main goal of activities, to highlight the benefits of tenure documentation, and having multiple meetings to build trust and understanding.

Five organizations report positive results of addressing the barriers, in the sense that it contributed to a broadening of participation to more communities who are being involved in the documentation processes and more acceptance of the initiatives.

3.2. Component: Adjudication of Interests in Land

This section describes the organizations' activities and experiences on adjudication of both the ownership and non-ownership interests in land.

Most organizations train the facilitators and communities on the procedures for adjudicating and establishing the interests and their accompanying RRRs. Six organizations gave their methods on procedures, which follow one of three possible pathways: (i) establish the interests and RRRs during the actual practice of adjudication and rely on publication of results for communities to validate/verify the interests in land; (ii) begin by inquiring about the existing situation of land rights before embarking on any adjudication activities, sought from political, religious, and cultural leaders, existing documents on land ownership and tenure, including existing titles, and communities or neighbors themselves as witnesses and to verify interests in land; and (iii) a combination of the other two. In the case of missing formal administrative tenure records, a field survey may fill the gap.

Four sets of methods can be identified deployed by organizations to ensure a complete inventory of (non)-ownership interests in land: (i) baseline studies through means such as focus groups, interviews, key informant interviews and discussions with communities and community leaders; (ii) consultation meetings and workshops with community leaders, community members, government institutions; (iii) trainings given to both enumerators/facilitators and community members; and (iv) test runs and pilots, including the utilization of participatory mapping activities. Three organizations believe that non-ownership rights are recorded, and other responses are more cautious on this.

All organizations indicate that some barriers exist that inhibit participating communities in adjudicating the ownership interests. Barriers include those related to: (i) community attributes i.e., conflicts at family level such us those linked to inheritance, men hijacking women's land rights, disputes with neighbors and conflicts at a larger level such as those arising between herders and farmers as well as due to ethnic division. Lack of knowledge and understanding on non-ownership rights and some community members partaking in "forum shopping" during adjudication activities are also in this category; (ii) attributes from the legality or legitimacy of interests in land such as unclear legal settings, bureaucracy and lack of capacity is mentioned as barriers, while lack of documentation on property lead to uncertainties on legitimacy of land rights; and (iii) attributes relating to mapping of boundaries—i.e., concerning the fuzziness and uncertainty of boundary locations.

Organizations use several approaches to address the barriers e.g., through training, lobbying, mediation, issuance of documents on land rights to right holders and using lessons from previous experiences in future planning. Resolution of barriers relating to community attributes have resulted to progress such as reduction in conflicts; willingness of communities to cooperate; better social inclusion of women and vulnerable groups.

3.3. Component: Adjudication of Boundaries

During the deployment of boundary mapping activities participating communities are able to indicate where boundaries begin/end. Communities can therefore map the boundaries relating to ownership interests. On non-ownership interests, seven organizations responded, with mixed responses on whether communities are able to map boundaries of non-ownership rights.

Categories of barriers to the mapping of boundaries relate to: (i) community attributes, including boundary disputes at family level, with neighbors, or two people claiming the same spatial unit, or a group of people disagreeing on the ownership of the same spatial unit are reported, but also ethnic divisions, violent conflicts, migration and displacement, encroachment of farming activities on pastoralist paths; (ii) legal attributes and/or questions of legitimacy of interests in land, including unclear legal settings, bureaucracy; lack of capacity, and that government does not recognize rights of poor, vulnerable and marginalized people and women; and (iii) the process of mapping of boundaries itself, for example difficulties in perceiving/conceptualizing non-ownership interests.

Organizations use several approaches to address the barriers e.g., through trainings, lobbying and mediation. Where needed, "Community development mechanisms" with the government as a stakeholder are established. Dialogues could lead to the "fair" relocation of people if necessary, for example through construction of settlements to protect the target population. On mediation, dedicated teams exist to handle cases and extents of gains and losses to parties involved are explained until an understanding is reached. Lobbying involves organizing multi-stakeholder meetings and awareness creation. Organizations mention that most of the conflicts are solved during the mediation process, which lead to 'better social environment' within the community members and also between the community and land administration. Increased participation during mapping activities creates consensus on the location of boundaries.

3.4. Component: Documentation Process

With the variety of interests in land, their accompanying RRRs and their boundaries determined, the next step is confirmation by documenting them. This part of the process is address in the following sub-sections.

The following types of interests are being recorded according to the respondents: (i) customary family based, (ii) customary clan based, (iii) occupancy, (iv) tenancy, including where the tenant owns crops but not the land (v) ownership, (vi) informal, (vii) possession, including rotating possession (viii) individual (freehold), leasehold, (ix) easements and rights of way: paths, roads, (x) state land, (xi) community lands allocated for specific purposes (e.g., churches, schools, sacred forests, community forest access), (xii) sacred places (in general) (xiii) conflict, (xiv), community lands reserved for future allocation, (xv) encumbrances, (xvi) pastoralist paths, (xvii) access to water, (xviii) use, (xix) extraction, (xx) community forest access, (xxi) storage places, (xxii) buffer along roads or other infrastructure as consequence of planning regulations Depending on country context, interests e.g., leaseholds can be categorized as ownership while in others as non-ownership. This contextualization blurs a strict distinction of 'ownership' and 'non-ownership' interests, making it difficult to generalize.

Upon documentation, the organizations report that they display tenure information for public view in the community, for example at community resource centres, village boards, or local leaders' offices; and/or at government offices at sub-county and local level, and, depending on context, at district level and in land administration agencies. The timeframe within which the objections and corrections can be voiced range between: 15–30 days; 45 days; 6 months (in accordance with the law) or is left open until the project closes. All organizations report that objections have been successfully addressed in the past.

Examples of barriers encountered that inhibit the documentation of ownership and non-ownership interests in the cadastral database relate to four attributes: (i) communities, e.g., social issues within the community or between communities or between a community and land administration agencies, (ii) legal, e.g., unclear legal settings and disagreements over the legitimacy of interests in land, (iii) the mapping process, e.g., difficulties to identify the boundaries and descriptions of RRRs and (iv) technical attributes of the process, e.g., minor omissions or additions done unintentionally by the field teams, this has potential to create confusion or conflict.

Trainings, lobbying and mediation are used to address the barriers. Training address technical issues including addressing unclear technical requirements/processes, while lobbying relates to awareness raising with political actors. Mediation is mentioned without further explanation. Meetings and workshops are also mentioned where examples of similar settings, region and topology are shared.

3.5. Component: Monitoring Process

Seven organizations responded to the questions on monitoring. About half the organizations return to the communities for monitoring. In terms of overall focus areas for evaluation two arenas can be distinguished: (i) those related to community impacts, including potential conflicts and level of community agreement, transactions and retrieval of documents, impact on communities in general and specifically what land interests have been overlooked (in most cases); and (ii), those elements relating to the process of documentation, including speed and quality of the process, the identification of challenges and adaptation, as well as maintenance of equipment.

Two organizations reported that barriers to monitoring exist. Barriers relate to (i) community factors such as conflict, and (ii) the monitoring process, e.g., lack of skills and short-term given more importance over long-term monitoring. These barriers are addressed through multi-stakeholder meetings, lobbying with political actors, mediation, awareness raising, but also technical capacity building.

4. Discussion

The five components of a FFP adjudication process are addressed in this section. The theory of collaborative governance provides the analytical framework to discuss the results and raise questions for consideration about the results. Observations from the results are discussed against the theory of collaborative governance's three arenas of collaboration, whereupon the authors of this study raise questions for attention by implementers of tenure documentation activities. The three arenas are:

(i) participation arena (who is involved and who leads the process);
(ii) process for collaboration arena (where, when and how of collaborative process); and
(iii) content arena (setting the agenda and establishing expectations regarding the outcome of the process).

4.1. Mobilization Process

This phase emphasizes the use of participatory approaches in tenure documentation, thereby harnessing the potential for efficiency, inclusiveness, empowerment and sustainability [28–30].

(i) Participation arena: with a variety of state and no-state actors engaged, questions for consideration include:
    - To what degree can the various stakeholders influence the perspective of, or how the concepts of land tenure and land rights are framed in the awareness raising session?
    - Which participants have the authority to make decisions on the issue of land tenure and land rights to be recorded, and what are the consequences of this?

(ii) Process for collaboration arena: modes of communication during the mobilization phase were varied. Determinants of "where", "when", and "how" collaboration in the tenure documentation process are communicated in the community mobilization and awareness sessions through various means. Meanwhile, factors influencing the nature of interaction between the organizations and the communities are determined by a number of factors as listed in Section 3.1. Questions for consideration include:
    - How do the various types of communications and interactions e.g., via TV, radio etc. influence participation in awareness raising sessions?
    - What alternatives exist for community members who cannot be reached via TV, radio etc. to enable them to participate in tenure documentation processes?
    - What are the consequences of community members being present and/or absent in the next phases of the tenure documentation processes?
    - How do the various determinants of community awareness creation sessions—i.e., feedback, accessibility, community social structure, knowledge gap, distractions, logistics, budget or urgency—singly or combined—influence the subsequent phases of tenure documentation processes?

(iii) Content arena: setting the agenda of land tenure documentation may depend on the way in which the facilitators of adjudication activities frame the topic of land tenure adjudication, land rights determination and mapping to the communities during the mobilization phase. Communities may form their own interpretations of what they believe is expected of them during the tenure documentation process. It is worth noting that interests in land are different in nature and, depending on the perspective taken to identify and classify them, a number of key attributes (bundles of rights) make up a specific interest in land [31–33]. For every right an individual holds, rules exist that authorize what can or cannot be done with the right [32,33]. Hence land tenure is typically associated with a bundle of rights that define the conditions attached to them [32,33]. Bundles of rights therefore differ for different types of tenures, with "ownership" forms of tenure perceived as holding the complete bundle of rights over a particular resource compared to "non-ownership" forms of tenures [34].

It is important to consider the role of the developers of technologies for recording the tenure information. The processes of information sharing and translating between community and the technology developers, with emphasis on iterative feedback loops for the design of databases and applications in order to ensure that the technology reflects and allows for incorporation of land rights diversity. The questionnaire investigated to what extent the technology developers are informed about the local ownership and non-ownership rights for inclusion in the data model. From the responses it appears that this is done, though not in all cases. Having feedback loops is reported less. This is a point of concern, because the community may check how, or if their land rights have been modelled according to the existing situation. It could also be a matter of design: where only the developer can set-up a model of land rights, or the system is made flexible to the extent that the facilitators or community themselves may design and implement their own data model.

Overall, the results here do not allow drawing generalized conclusions about the exact role played in the documentation of statutory legal vis-à-vis existing tenure and tenure rights. There are indications that statutory legal land rights over-write socially legitimate and locally specific, overlapping land rights. However, the questionnaire findings do not provide conclusive insights into the exact translation of rights according to local context into statutory frameworks. Also, the involvement of land administration officials in the process of land tenure identification indicates that land tenure becomes anchored in statutory and administrative legal foundations.

Of interest for this study is the idea of overlapping and complementary land rights. Two organizations reported the criteria being used to determine the existence of multiple types of rights on a parcel, or sorting out the different types of rights existing on a parcel: (i) baseline studies; and (ii) relying on community members to provide an overview of the rights.

The above approaches—singly or in combination—may reveal the types of land rights in an area, including whether they are overlapping or not. Subsequently, mapping using participatory, faster and affordable approaches serve to confirm the boundaries of the tenures.

Questions for consideration:

- How does the framing of the terms such as land adjudication, land rights, mapping etc. in the community awareness sessions reflect in subsequent processes (determination of land rights), and manifest as the types of land rights that are eventually recorded?
- In how far are the land rights from the community perspective translated to resemble a certain type of land right from the legal perspective, and subsequently, to which tenures will packaging of different bundles of land rights equate to?
- With just two out of eight organizations responding on the criteria for determining whether multiple types of rights exist, or for sorting out different types of rights existing on land, does this mean that most do indeed lack cues or criteria to categorize the variety of interests in land for documentation?
- What is the role of translation from one language to another (e.g., of terms such as land adjudication, land rights, mapping etc. and through the use of different terminologies in procedures, instructions, user interfaces, and on certificates) have on the outcome of the tenure documentation process?

4.2. Adjudication Process—Adjudication of Land Rights

A variety of interests in land have been identified as given in 3.1. How then, having principal frames around the identification of "interests in land"—adapted in the community mobilization phase—manifest in the implementation phase? Observations are drawn from the types of barriers inhibiting the determination of the land rights process, whose elements are hereby contextualized within the three arenas of collaborative theory:

(i) Participants arena: categories of participants in the mobilization phase will likely participate in the implementation phase as well, unless hindered by certain factors. According to Agarwal [28], participatory institutions/organizations operate on principles of cooperation but excluding members of the community can occur due to systemic factors—which in turn unfavorably affect both equity and institutional efficiency. Purdy [27] adds that as participation in a collaborative activity is often voluntary, it is biased towards those who have the resources to enable them to participate. On the other hand, beneficiaries of the land rights documentation project may also choose not to participate. Land rights holders' lack of participation due to various factors, as well as due to refusal to take part, is cited in this study as one of the obstacles during the identification of land rights activities. This has implications on the next phase. Questions for consideration:
    - Depending on the proportion of beneficiaries participating, who determines the quorum for the community documentation process to proceed? And what alternatives exist for those absent?
    - To what degree is the lack of participation due to systemic factors, voluntary decision rather than of circumstances that hinder participation, or lack of understanding—among other challenges listed as barriers in the identification of interests phase?

(ii) Process for collaboration: from the community perspective, those who were reached through the various modes of communication (during mobilization and implementation phases) are likely to be present during the rolling out of the adjudication of land rights and mapping activities. According to Ansell and Gash [26], "incentives to participate in collaborative governance will increase if stakeholders perceive achievement of their goals to be dependent on cooperating from other stakeholders", which requires stakeholders to build trust and interdependencies in working together towards achieving the goal. In the case of this study, where, when and how organizations leading the process convey the information, facilitate interactions and build trust with the communities can influence both the quality of the process and the outcomes of the documentation process. Questions for consideration:
    - What proportion of anticipated beneficiaries of the tenure documentation activities are reached through the various modes of communication and how does this influence the tenure documentation process?
    - What proportion of the anticipated beneficiaries of tenure documentation activities participate in the determination of land rights?

(iii) Content arena: an array of conceptions of forms of interests in land have been realized—as given in Section 3.4. Criteria for assessing and converting the communities' perspectives of land rights to their equivalent in the recorded or legal perspective did not feature in the results of this study. An understanding of the processes leading to differentiating the natures of, and classifying the types of traditional/community or informal land rights into their new forms e.g., "use", "extraction", "possession" or other in the recorded and legal context need to be further studied and understood. The questions that lead the adjudicators and communities to determine the types of land rights existing in an area, and then packaging the various land rights into their corresponding interest in the recorded or legal perspective are unclear. Content of the questions and methods that guide the unpacking and classifying of the different types of land rights from their undocumented form into the recorded or legal form, also taking into consideration the effects that language translation may have in the process need to be understood. Zevenbergen et al. [21] suggest the importance of ensuring that land documentation design lays a foundation for movement along the continuum of land rights, without having to jump out of one system (community perspectives of tenure and land rights) into another. As such, it is unclear how the baseline studies can determine and translate community land rights into those in Section 3.4. Further, criteria used to determine the existence of multiple land rights existing in a specific area and criteria for sorting out multiple types of land rights in order to determine that single or overlapping types of tenures exist also need to be understood. This would provide knowledge on how arrangements of bundles of rights in different types of interests in land enable them to coexist as "overlapping" in the same space. Questions for consideration:

- What criteria need to be framed and or adopted to ensure that communities comprehend the nature of and differences between overlapping and non-overlapping interests in land? Results, as is, suggest that community perspectives of land rights are in some cases being modified in the determination of land rights phase into some form of land rights in the legal context. The way in which this process is conducted did not feature in the results of this study, and this raises the question, what attributes, or sticks in the bundle of rights in the community perspective qualify to become a specific type of right in the legal context?
- Where interests are framed around the legal perspective, (e.g., freehold, leasehold, etc.) could this be a cascading effect arising from the lack of criteria that guide the notion of land tenure and land rights—from the perspective of the communities? And as such is this direct replacement with the de jure interests perhaps the easy way out of the complexities of describing interests in land from the perspective of the communities?

4.3. *Adjudication of Boundaries*

Sequentially, the identification of interests in land precedes the boundary mapping. The role of the boundaries therefore would be to confirm the limits within which those interests (already identified) in land apply. The identification of interests in land is therefore of fundamental importance for the preservation of traditional overlaps in land rights prior to the mapping. Equating adjudication of land rights to adjudication of boundaries can lead to individuals or groups of people losing their rights to land and this has been a long-standing discussion. For example, the original tenure reform statutes or laws in Kenya framed adjudication in customary areas around i.e., cultivation and residential occupation rights, and subsequently "the adjudication of land rights came to mean little more than the identification of homesteads plus the area of land to which household heads had cultivation rights" [3]. This resulted in "altering the structure of access to land in the family economy by vesting the attributes of ownership in adult male heads of households without at the same time giving adequate protection to the de facto or potential rights of women and children or others who had general access rights not accompanied by the power of ultimate control and further, the categories of land rights were unable to accommodate those who had other forms of rights in the adjudication area, nor describe land rights in those parts of the country where no permanent settlements nor cultivation existed e.g., in pastoralist areas" [3]. Coldham and Migot-Adholla et al. [15,35] observed that due to the problems involved in the adequate definition and protection of customary land tenure, land adjudication had the effect of depriving some people of their rights while conferring on others greater rights than they are entitled under customary law. Moreover, the "very narrow view on land rights in the statutes make it virtually impossible to bring to the adjudication register all the multiple rights claimable under customary tenures, and as such rightful claimants do lose their land rights as part of land adjudication" [36]. With the implementation of tenure documentation activities in the context of FFPLA, early indications highlight that the inclination towards recording single rather than the multiple types of tenures that may exist in an area persist [19,25].

Cutting across the three arenas of collaborative governance, questions for consideration include:

- To what degree do the communities understand what is expected of them in terms of the adjudication process, and that the adjudication of land rights is different from the adjudication of boundaries?
- To what degree are the communities able to capture the boundaries of overlapping and non-overlapping interests in land?

4.4. Documentation Process

How have processes of identifying and mapping interests in land manifested in the documentation phase? Tenure types identified and mapped are subsequently forwarded to the documentation unit, where they are entered in the cadastral database. Thereafter a public display of the tenure information is placed to give room for objection or correction. Questions for consideration include:

(i) Participants arena: This phase may have implications on completeness in terms of representation on the types of interests in land recorded, and the right holders recorded. This is an important phase because the interests become legally binding once they are formally registered [1,5,6], with potential for conflict when interests are misrepresented [10]. Public display of the tenure information therefore confirms the aim of the entire process and opens opportunities for objections and corrections. Omission and mistakes that appear during the public display of the outcomes could lead to frustrations and disputes. Questions for consideration:
   - Whose interests in land are recorded? Whose is left out? Whose interest is altered?
   - How to deal with future claims in case of omissions, or conflicts between community perspectives of land rights and the legal perspectives?

(ii) Observation on the process for collaboration arena and for content arena: In this study, it was mentioned that the process of documentation is not a problem as it is about confirming what has resulted from both the processes of mobilization and implementation of adjudication. For this reason, questions for consideration here are similar to those raised in the participants arena—under the documentation process.

(iii) Same as (ii) above.

4.5. Monitoring

Monitoring and evaluation focus in this study is specifically on which land interests are affected/overlooked by documentation and how documentation is conducted. It will be interesting to find out, in how far the nature as well as the extent of barriers to these processes (e.g., conflict, ethnic division) differ from barriers to the documentation process itself, where similar issues seem to play a role (e.g., mistrust, exclusionary trends in community). Results showed that barriers such as conflicts are addressed through training, lobbying and mediation—with positive outcomes. This confirms the view that conflict can actually create a powerful incentive for collaborative governance for conflict resolution [26]. Consensual approaches such as mediation try to find a harmony through intensive discussions and negotiations, during which all sides learn to understand the other party's concerns, conflict issues, fears and desires, with the aim to re-establish a positively functioning relationship. Also known as Alternative Dispute Resolution (ADR) approaches, they are cheaper and faster therefore often a preferable alternative to overloaded courts [37]. ADRs thus provide an avenue to deliver justice, a key element of the Sustainable Development Goal 16 [38].

The questionnaire for this study did not cover the process side and usability of system development, while this might contain challenges as well. First of all, the end user is not clearly defined yet. Is it the facilitator? Is it the youth from the community trained to capture the data? Or is it the community at large (do they capture the land rights and the boundaries themselves or does it relate to a system where every person can query the database and eventually transfer land or use other services provided)? In other words, the scope and the user group for each and every implementation may vary and may be given more attention in future research.

Questions by the researchers regarding participants, process and content for collaboration arenas:

- How is the match between "community composition" before/during/after established?
- What means exist/can be developed to address structural challenges to longer term monitoring and evaluation, e.g., financial, political stability, migration/displacement?
- How to ensure consideration of and adjustment to discovering "the unexpected?"
- How do barriers and problems in evaluation and monitoring differ specifically from those encountered during documentation process itself?
- How to consider contextual complexity of environmental factors that influence outcomes at different points in time and over longer periods?

5. Conclusions

This study explores how ownership and non-ownership interests in land are being recorded in a FFPLA context—with a special focus on overlapping and complementary land rights. An exploratory research method was employed and results discussed against the theory of collaborative governance. Results from the eight organizations show that when discussing the topic of "adjudication of land rights", the legal perspective of land rights intersects with the perspective of communities regarding legitimate rights and interests in various ways during the documentation process. During adjudication of land rights alterations in the nature of land tenure and rights to legal forms may happen. Hence, the benefits as well as the limitations associated with the legal tenures, such as their exclusive nature are also introduced. As such, the study raised questions for consideration for future tenure documentation activities anchored on the three arenas of the collaborative governance theory i.e., the participation, process and content arenas. These questions may aid in consensus-oriented understandings on ownership rights, non-ownership rights and of course mapping and documentation of those land rights. Understanding complementarity and variability between interests in land in both the community and the legal perspectives will offer multiple lines or strategies for integration or conciliation when recording land interests. Without this understanding, opportunities for recording certain types of rights to land, including overlapping ones risk to fail also in the context of pro-poor land administration. Further, in the process of documenting the interests in land, future research and evaluation can look more carefully at the processes of information sharing and translation between community and technology developers, with emphasis on iterative feedback loops for the design of databases and applications.

This study offers preliminary insights to the questions posed at the outset. As mentioned in the methodology section, the components of adjudication identified serve as an initial analytical framework in order to summarize the findings from the questionnaire. As such, limitations of the study surround: (i) the many responses being contextual, which make it difficult to generalize; (ii) while results allow building a methodological skeleton for future inquiry, key themes and categories that emerged for each component can be used to fine-tune future questionnaires in future studies; and (iii) it may be possible that terminologies used in the questionnaire were interpreted differently by different respondents.

Author Contributions: Conceptualization, M.L., C.R., P.v.A.; methodology, M.L., C.R., P.v.A.; validation, M.L., C.R., P.v.A.; formal analysis, M.L., C.R., P.v.A.; investigation, M.L., C.R., P.v.A.; data curation, M.L., C.R., P.v.A.; writing—original draft preparation, M.L., C.R., P.v.A.; writing—review and editing, M.L., C.R., P.v.A., J.Z., funding acquisition, M.L. All authors have read and agreed to the published version of the manuscript.

Funding: This research was funded by the Dutch programme "Land Administration for National Development" (LAND) a partnership of the Ministry of Foreign Affairs and Kadaster, grant number 9002110. Funding of the publication costs for this article has kindly been provided by the School of Land Administration Studies, University of Twente, in combination with Kadaster International, The Netherlands.

Institutional Review Board Statement: Not Applicable.

Informed Consent Statement: Not Applicable.

Data Availability Statement: Not Applicable.

Acknowledgments: Authors gratefully thank the organizations implementing participatory tenure documentation approaches for contributing to this study by providing data. Authors also acknowledge Christiaan Lemmen's invaluable support on editing the manuscript.

Conflicts of Interest: The authors declare no conflict of interest. The funders had no role in the design of the study; in the collection, analyses, or interpretation of data; in the writing of the manuscript, or in the decision to publish the results.

References

1. Simpson, S.R. *Land Law and Registration*; Cambridge University Press: Cambridge, UK, 1976.
2. Lawrance, J.D. Land adjudication. In *Proceedings World Bank Seminar on LIS*; World Bank: Washington, DC, USA, 1985; pp. 1–39.
3. Okoth-Ogendo, H.W.O. *The Perils of Land Tenure Reform: The Case of Kenya*; University of Nairobi: Nairobi, Kenya, 1986.
4. Van der Molen, P. The dynamic aspect of land administration: An often-forgotten component in system design. *Comput. Environ. Urban Syst.* **2002**, *26*, 361–381. [CrossRef]
5. Kaufmann, J.; Steudler, D. *CADASTRE 2014—A Vision for a Future Cadastral System*; International Federation of Surveyors: Copenhagen, Denmark, 1998.
6. Van der Molen, P. Future Cadastres. In *Fair Land Governance: How to Legalise Land Rights for Rural Development*; Otto, J.M., Hoekema, A., Eds.; Leiden University Press: Leiden, The Netherlands, 2012.
7. FAO. *Land Tenure and Rural Development*; FAO: Rome, Italy, 2002.
8. Tanner, C. *Law Making in an African Context: The 1997 Mozambican Land Law*; FAO Legal Papers Online 26; FAO: Rome, Italy, 2002.
9. Lipton, M. *Land Reform in Developing Countries*; Routledge: London, UK, 2009.
10. Kalabamu, F.T. Land tenure reforms and persistence of land conflicts in Sub-Saharan Africa—The case of Botswana. *Land Use Policy* **2019**, *81*, 337–345. [CrossRef]
11. Meinzen-Dick, R.; Mwangi, E. Cutting the web of interests: Pitfalls of formalizing property rights. *Land Use Policy* **2009**, *26*, 36–43. [CrossRef]
12. Davy, B. After form. The credibility thesis meets property theory. *Land Use Policy* **2018**, *79*, 854–862. [CrossRef]
13. Barrows, R.; Roth, M. Land Tenure and Investment in African Agriculture: Theory and Evidence. *J. Mod. Afr. Stud.* **1990**, *28*, 265–297. [CrossRef]
14. Gray, B. Strong opposition: Frame-based resistance to collaboration. *J. Community Appl. Soc. Psychol.* **2004**, *14*, 166–176. [CrossRef]
15. Coldham, S. The Effect of Registration of Title Upon Customary Land Rights in Kenya. *J. Afr. Law* **1978**, *22*, 91–111. [CrossRef]
16. Lengoiboni, M.; Bregt, A.K.; van der Molen, P. Pastoralism within land administration in Kenya—The missing link. *Land Use Policy* **2010**, *27*, 579–588. [CrossRef]
17. Broegaard, R.B.; Vongvisouk, T.; Mertz, O. Contradictory land use plans and policies in Laos: Tenure security and the threat of exclusion. *World Dev.* **2017**, *89*, 170–183. [CrossRef]
18. Enemark, S.; McLaren, R.; Lemmen, C. *Fit-for-Purpose Land Administration Guiding Principles for Country Implementation*; UN-793 Habitat/GLTN: Nairobi, Kenya, 2016.
19. Hendriks, B.; Zevenbergen, J.; Bennett, R.; Antonio, D. Pro-poor land administration: Towards practical, coordinated, and scalable recording systems for all. *Land Use Policy* **2019**, *81*, 21–38. [CrossRef]
20. Lengoiboni, M.; Richter, C.; Zevenbergen, J. Cross-cutting challenges to innovation in land tenure documentation. *Land Use Policy* **2019**, *85*, 21–32. [CrossRef]
21. Zevenbergen, J.; Augustinus, C.; Antonio, D.; Bennett, R. Pro-poor land administration: Principles for recording the land rights of the underrepresented. *Land Use Policy* **2013**, *31*, 595–604. [CrossRef]
22. Gray, B. *Collaborating: Finding Common Ground for Multiparty Problems*; Jossey-Bass: San Francisco, CA, USA, 1989.
23. Sousa, D.J.; Klyza, C.M. New directions in environmental policy making: An emerging collaborative regime or reinventing interest group liberalism. *Nat. Resour. J.* **2007**, *47*, 377.
24. McNabb, D.E. *Research Methods for Political Science: Quantitative and Qualitative Methods*; Routledge: London, UK, 2015.
25. Barry, M. Fit-for- purpose land administration—Administration that suits local circumstances or management bumper sticker? *Surv. Rev.* **2018**, *50*, 383–385. [CrossRef]
26. Ansell, C.; Gash, A. Collaborative Governance in Theory and Practice. *J. Public Adm. Res. Theory* **2008**, *18*, 543–571. [CrossRef]
27. Purdy, J.M. A framework for assessing power in collaborative governance processes. *Public Adm. Rev.* **2012**, *72*, 409–417. [CrossRef]
28. Agarwal, B. Participatory Exclusions, Community Forestry, and Gender: An Analysis for South Asia and a Conceptual Framework. *World Dev.* **2001**, *29*, 1623–1648. [CrossRef]
29. Uphoff, N. *Fitting projects to people. Putting people first: Sociological variables in rural development*; World Bank: Washington, DC, USA, 1985.
30. Knight, R.; Brinkhurst, M.; Vogelsang, J. *Community Land Protection: Facilitators Guide*; Namati: Washington, DC, USA, 2016.

31. Bennett, R. Property Rights, Restrictions and Responsibilities: Their Nature, Design and Management. Ph.D. Thesis, The University of Melbourne, Melbourne, Australia, 2007. Available online: http://www.csdila.unimelb.edu.au/publication/thesis/Rohan_Bennett_PhD_Thesis.pdf (accessed on 30 March 2021).
32. Bruce, J.W. *Review of Tenure Terminology*; Land Tenure Center, University of Wisconsin-Madison: Madison, WI, USA, 1998.
33. Schlager, E.; Ostrom, E. Property-rights regimes and natural resources: A conceptual analysis. *Land Econ.* **1992**, *68*, 249–262. [CrossRef]
34. Barry, D.; Meinzen-Dick, R. The invisible map: Community tenure rights. In Proceedings of the 12th Conference of the International Association for the Study of the Commons, (IASC), Cheltenham, UK, 14–18 July 2008.
35. Migot-Adholla, S.; Hazell, P.; Blarel, B.; Place, F. Indigenous land rights systems in sub-Saharan Africa: A constraint on productivity? *World Bank Econ. Rev.* **1991**, *5*, 155–175. [CrossRef]
36. Okoth-Ogendo, H.W.O. African Land Tenure Reform. In *Agricultural Development in Kenya: An Economic Assessment*; Oxford University Press: Nairobi, Kenya, 1976.
37. Wehrmann, B. *Land Conflicts—A Practical Guide to Dealing with Land Disputes*; GTZ: Eschborn, Germany, 2008.
38. UN General Assembly. *Transforming Our World: The 2030 Agenda for Sustainable Development*; Report No. A/RES/70/1; United Nations General Assembly: New York, NY, USA, 2015.

land

MDPI

Article

Fit-for-Purpose Land Administration in Violent Conflict Settings

Clarissa Augustinus [1,*] and Ombretta Tempra [2]

1 Independent Consultant, P17 V448 Kinsale, Ireland
2 UN-Habitat, Giza 12611, Egypt; ombretta.tempra@un.org
* Correspondence: augustinusc@paulaugustinus.com

Abstract: According to the United Nations (UN) Refugee Agency, there were 79.5 million forcibly displaced people worldwide by the end of 2019. Evictions from homes and land are often linked to protracted violent conflict. Land administration (LA) can be a small part of UN peace-building programs addressing these conflicts. Through the lens of the UN and seven country cases, the problem being addressed is: what are the key features of fit-for-purpose land administration (FFP LA) in violent conflict contexts? FFP LA involves the same LA elements found in conventional LA and FFP LA, and LA in post conflict contexts, as it supports peace building and conflict resolution. However, in the contexts being examined, FFP LA also has novel features as well, such as extra-legal transitional justice mechanisms to protect people and their land rights and to address historical injustices and the politics of exclusion that are the root causes of conflict. In addition, there are land governance and power relations' implications, as FFP LA is part of larger UN peace-building programs. This impacts the FFP LA design. The cases discussed are from Darfur/Sudan, Democratic Republic of Congo, Honduras, Iraq, Jubaland/Somalia, Peru and South Sudan.

Keywords: fit-for-purpose land administration (FFP LA); violent conflict; United Nations; extra-legal; transitional justice; peace building; land governance; power relations

Citation: Augustinus, C.; Tempra, O. Fit-for-Purpose Land Administration in Violent Conflict Settings. *Land* **2021**, *10*, 139. https://doi.org/10.3390/land10020139

Received: 28 December 2020
Accepted: 27 January 2021
Published: 1 February 2021

1. Introduction

According to the United Nations Refugee Agency (UNHCR), there were 79.5 million forcibly displaced people worldwide by the end of 2019, of which 26 million were refugees and 45.7 million were internally displaced within their own countries [1]. The United Nations (UN) plays a major role in peace and security in these contexts as "saving future generations from the scourge of war was the main motivation for creating the UN ... Since its creation, the UN has often been called upon to prevent disputes from escalating into war, or to help restore peace following the outbreak of armed conflict, and to promote lasting peace in societies emerging from wars [2]." Land is a major root cause of violent conflict [3] (p. 9). "The most common land-related human-rights abuses include violent disputes over land and territory, forced evictions from houses and land, the loss of access to livelihoods and natural resources, and the dispossession of land and houses [4] (p. 9)."

This paper focuses very narrowly on the land administration (LA) aspects of UN work in violent conflict contexts, which are often protracted. The seven country cases undertaken by the UN on land in violent conflict contexts reviewed in this paper overturn the conventional view that it is not possible to undertake LA interventions in these settings [5] (p. 5). The information comes from cases that UN-Habitat has been involved with, either through the process of supporting other UN agencies to document their work, or where UN-Habitat was directly involved. UN-Habitat facilitates the Global Land Tool Network (GLTN) [6] and much of this work is undertaken through this network.

The UN as a system started to sharpen its position on LA in these settings beginning in 2014 under the guidance of the Secretary-General's Office, led by UN-Habitat and involving many other UN entities working in violent conflict settings. The UN undertook a scoping and status assessment of the land-related work it was undertaking in these settings. One of its recommendations was to "adopt the continuum of land rights and fit for purpose land administration approaches for a sustained and coherent engagement on land and conflict [7] (p. 33)." A Secretary-General's Guidance Note on land and conflict was then developed that gives guidance for the UN system as a whole. This work was led by UN-Habitat and involved numerous UN entities working across the conflict cycle. The Note identifies 15 critical root causes of violent conflict one of which is weak LA. The Note describes this weakness as involving "weak state, land policies, laws and institutions, land administration, land management and land use planning systems, land governance structures and land dispute resolution capacity [3] (p. 9)." The Note recommends the use of "the 'fit-for-purpose' approach to support land administration, developed by the World Bank and UN [3] (p. 7)." This paper examines some of the UN's work in the violent conflict contexts that are the focus of the Note. This paper shows how it accords with FFP LA approaches, but at the same time has its own specific characteristics because of the UN's power and programming approaches.

Through the lens of the UN and seven country cases, the problem being addressed in this paper is: what are the key features of FFP LA in violent conflict contexts? Through a review of some of the LA and FFP LA, post conflict LA and land governance literature, we identify some of the major characteristics of UN-related FFP LA in these contexts. We demonstrate that the approach to transitional justice, (land) governance and power relations accompanying UN work in these settings contributes to the way the FFP LA is designed. This will also speak to the framing of a definition of 'FFP LA' and 'land governance' in these contexts.

Section 2 provides a review of the literature and gives key definitions. Section 3 describes the research methods used to acquire information that is largely based on the re-framing and re-analysis of seven UN case studies published in the grey literature of UN reports such as the UNHCR [1]. Section 4 presents the findings of the research with reference to the case studies, which in turn informs the novel definitions of FFP LA and land governance for violent conflict contexts, described in the discussion, Section 5. This latter section also draws the threads together, identifying the key characteristics of FFP LA in these settings. The conclusions are presented in Section 6. This paper can increase the understanding of how to undertake FFP LA in these settings in general, by the UN, as well as by national governments and other actors.

2. Literature Review and Definitions

The Secretary-General's review of UN peace operations is used to describe the term 'violent conflict contexts.' The term 'UN peace operations' embraces a broad suite of tools managed by the UN Secretariat from special envoys and mediators, political missions (including peace building missions), regional preventative diplomacy offices, to small technical specialist missions, and multidisciplinary operations" [8] (p. 4). The Report of the High-Level Independent Panel on UN Peace Operations to the UN Security Council in 2015 outlines what this paper identifies as the key characteristics of violent conflict contexts through a UN lens. The report states that "UN peace operations meet the expectations of those whose lives are ravaged by armed conflict"; "the UN is today the largest provider of international peace operations"; "since the 1990s, the number and intensity of armed conflicts has declined"; however "the number of civil wars has increased in the last few years and attacks perpetrated by governments and armed groups against civilians have risen for the first time in a decade. This increase is compounded by the rise in violent extremism"; "in addition to indiscriminate killing, appalling abuses are perpetrated against civilians. Sexual violence remains a pervasive tactic of modern warfare."; "many of today's armed conflicts are more intractable and less conducive to political resolution."; "notions of inter-state and intra-state conflict have blurred"; "at the same time, many conflicts are caused by bad governance, where the state is captured by elites who monopolize its levers for power and enrichment in these contexts, efforts to sustain peace have foundered partly due to a failure to establish a fair sharing of resources"; "more often than not, root causes and conflict drivers are not effectively dressed." [8] (pp. 1–3). Violent conflict settings are usually termed by the UN 'emergency' or 'humanitarian' contexts. As indicated above, the Secretary-General has noted that land needs to be analyzed as a root cause of conflict, including weak land administration [3] (p. 7). FFP LA is needed to build peace in conflict contexts where land is identified as one of the root cause of the conflict. For example, in one of the case studies used here, from Jubaland/Somalia, land is specifically identified and analyzed as a root cause of its conflict [9].

The UN has defined transitional justice as "the full range of processes and mechanisms associated with a society's attempt to come to terms with a legacy of large-scale past abuses, in order to ensure accountability, serve justice and achieve reconciliation [10] (p. 3)." Drawing on Collins Dictionary definition of extra-legal, in terms of FFP LA, extra-legal transitional justice mechanisms are defined here as "not governed or regulated by law [11]," not governed by formal regulations and administrative procedures relating to the LAS, including the land use planning system.

FAO and UN-Habitat have defined land governance—it relates to power and political economy issues. There is a direct link between the power structure of society, governance, the way power is distributed in society and land tenure. Land tenure rules can entrench power relations within social groups. "(L)and governance concerns the rules, processes and structures through which decisions are made about access to land and its use, the manner in which the decisions are implemented and enforced, the way that competing interests in land are managed [12] (pp. 1, 2, 9)." In violent conflict contexts, the UN "has unmatched convening power and an ability to bring together disparate interests for common purposes, to elevate issues above sometimes paralyzing regional agendas and to identify and implement impartial strategies that can lead to political solutions [8] (p. 23)." The UN Security Council notes that in regard to peacekeeping operations, the "UN is a powerful presence [8] (p. 3)." One aspect of the way this power is used is through what is termed the 'good offices' of the UN. Good offices are third-party assistance given to conflicting parties to help find a solution to their problems [4] (p. 7)." The seven case studies [4,9,13] show that the UN's power, including good offices, is also used for land governance, and the cases demonstrate its key characteristics in violent conflict settings. The cases show that the definition of land governance given above can be adapted for these settings. This accords with Van der Haar and van Leeuwen, who argue that land governance issues are central in times of displacement during war and violence [5] (p. 8).

There is a paucity of literature, particularly journal articles, on FFP LA in violent conflict contexts through the lens of the UN. The Scopus database of journal articles was reviewed and gives evidence of this paucity. Different searches were applied with a range of results. "Land administration" AND "violent conflict" AND "United Nations" produced 0 results, with "United Nations" AND "violent conflict" AND "land" produced 6 results. "Land administration" AND "fit-for-purpose" AND "conflict" produced 3 results, with 0 results if "violent" was added to that string. "Land administration" AND "violent conflict" gave 5 results. "Land registration" AND "violent conflict" produced 1 result, with 47 results for "land registration" AND "conflict." "Land information" AND "violent conflict" produced 0 results, whereas there were 47 results if the word "violent" was excluded. Finally, there were 20 results on "United Nations" AND "land administration" but none of them were on violence or on conflict [14]. These search results demonstrate that FFP LA in violent conflict contexts as undertaken by the UN is an under researched journal topic. To address the paucity, the literature from conventional and post conflict settings is adapted to violent conflict settings, through an examination of the seven cases. The identification of the LA elements, including novel characteristics, in the cases is intended to help to fill this gap.

There are a number of reasons for this lack of UN-related literature examining these environments. Firstly, UN agencies are focused on dealing with the crisis and there is little staff time dedicated to documenting conflict-related activities. Where documentation exists, it tends to be grey literature reports or information on the web [1]. This was recognized at the first Global Land Tool Network-hosted forum of the Land and Conflict Coalition [15,16]. At the forum, a road map for future work for the Coalition was developed to address this gap [15] (p. 22). A publication documenting the work of different UN entities on land in violent conflict settings was subsequently developed by a number of Coalition members, including UN entities, through a UN-Habitat/GLTN-led writeshop [4] (p. 9). Many of the case studies in this paper come from these documented case studies [4]. The second reason for paucity is that the literature on LA and conflict, including peer-reviewed literature, tends to focus on the post conflict phase rather than on the violent conflict phase [17–21].

Through a review of the literature and using evidence from the case studies (described below), an understanding and definition of FFP LA in violent conflict contexts is developed. This is performed through: (1) the identification of key elements of conventional LASs [22] (pp. 27–28); (2) a review of FFP LA [23–26]; (3) the examination of what happens to LASs during violent conflict adapted from the post conflict literature [18–21]; (4) a review of literature linking land governance to LA [27], and an adaptation of the FAO–UN-Habitat land governance definition to these contexts [12] (p. 9); and (5) adapting and expanding the gender-responsive land governance approach developed for FFP LA by Paradza et al. to violent conflict contexts [28] (p. 1). Finally, this approach frames two proposed novel definitions, one of FFP LA and the other of land governance in violent conflict settings.

3. Research Methodology and Methods

This paper's ontological frame is based on conventional LA elements [22] (pp. 27–28), their FFP forms [23–25] and the way LA manifests in post conflict contexts [18–21,29]. This paper's epistemological framework is based on the authors' of the seven cases authoritative knowledge, as well the authors of this paper's knowledge of these seven case studies [9,13,30–36]. This knowledge emanates from their role as staff in the particular UN entity, or through their support to the entity as a consultant. This paper is based on their knowledge and experience in the field in that country or in that particular project. This paper is also based on the knowledge that these authors had gathered from people in the know, books (UN reports), leaders of organizations" [37] (p. 27), and workshop presentations. The authors of this paper were involved with overseeing the documentation of the case studies, sometimes writing them, and were also present at the writeshop [4] (pp. viii, 9). This paper uses a seven-country case study series approach to gather its findings (see Table 1 below). The cases were chosen because: (1) the authors of this paper were involved in guiding and/or writing the documentation of all seven cases; (2) the documentation of six of the cases was undertaken after a global conference on land and conflict identified the lack of published knowledge on land-related activities in peace-building programs in violent conflict contexts, including through the lens of the UN. Six different UN entities (as well as 3 non UN entities not addressed here) then agreed to write up their knowledge on the subject, through a country program study, to support global capacity development; (3) the knowledge has only been published in grey literature [4] and needed re-framing and vigorous analysis using peer-reviewed literature, here FFP LA, to contribute to filling the gap identified in peer-reviewed literature (see above).

Under UN-Habitat staff guidance, UN staff at country level documented these six cases while working in their organizations' peace-building programs. This documentation included obtaining preparatory material, attending a writeshop to write it up and finalizing it afterwards, including getting approval from their organization to publish the material. UN-Habitat's work in Jubaland/Somalia, undertaken by the second author, is also used as an additional case study [9]. UN-Habitat has updated two of these case studies, namely Iraq [33] and Darfur/Sudan—where an extensive case study was undertaken of the land administration system (LAS) [13] by consultants working for UN-Habitat and UN staff, including authors of this paper. All these cases have been published in the grey literature. None of the case studies are exclusively focused on LA, except for the update of Darfur/Sudan, which assesses the LA in terms of designing for the voluntary returns of displaced people. The LA FFP aspects have been extracted from the published grey literature, in this instance UN published reports, for the purposes of this paper. Here this material is examined to identify the major features of FFP LA in violent conflict contexts and to develop definitions for these contexts. Given space constraints, the full case studies are not summarized here and readers are referred to the source literature (see Table 1 below) for further information, such as geographic information (including maps), descriptions of the larger program, and details of the intervention by the UN entity.

Finally, the FFP LA features identified in these contexts are based on the analysis and comparison of multiple sites where the UN has been working in violent conflict. Hasse argues that working across multiple sites improves research because it shows that a (technical) design that works across several sites is better and will work "consistently across cases, sites, and variations [38] (p. 219)". This paper examines seven countries across 3 global regions to identify common patterns in FFP LA in violent conflict contexts. Each country and 'site' includes the constant variables of violent conflict and a central role of the UN in peace building in these contexts. Cross-country comparison underpins the analysis, typology creation, definitions and the conclusions that show what FFP LA are likely to be found in other cases, sites and variations in similar violent conflict contexts in other countries.

FFP LA as Part of Larger UN Programs

None of the seven cases being examined here (see above) were standalone LAS programs intended to modernize the LAS, often found in post conflict settings [17]. All the LA cases were instead embedded in some form of larger UN peace-building program. A number of the cases concerned the voluntary returns of refugees and IDP to their countries or areas of origin from where they had been displaced. In Iraq, this was central to the program [33] (p. 61). Honduras was about preparation for future returns [32] (p. 45), along the lines proposed by Unruh when recommending instituting "techniques for deriving, protecting and using forms of evidence attesting to (housing, land and property) HLP claims early in a conflict, as opposed to subsequent to a conflict" [39] (p. 111). Some of the returnees were in IDP camps (Iraq [33] (p. 64), Darfur/Sudan [30]) (p. 83) and others were displaced and living outside of camps (Honduras [32] (p. 43)). The Darfur/Sudan activities involved both looking after the current 2.6 million IDP, some of them living in 174 IDP camps in Darfur [30] (p. 80), as well as assessing the current LAS and its ability to respond to voluntary returns [13]. The Iraq program was also focused on the rehabilitation of 17 Yazidi villages to facilitate people returning to the homes that the Islamic State of Iraq and Levant (ISIL/ISIS) had occupied and destroyed [33] (p. 62), [36] (pp. 71–74).

Three of the cases were UN peacekeeping missions involving UN peace keepers set up by the UN Security Council (Darfur/Sudan—UNAMID [30] (pp. 81–82), DR Congo—MONUSCO [31] (pp. 52–53), South Sudan—UNMISS [35] (p. 88)). The focus of these missions was on keeping the peace between different parties in violent conflict, including armed groups. Governments were often party to the conflict. The Peru case was about conflict between mining companies and indigenous peoples. The mining sector contributed 19% of total income tax during the period in question. UNDP supported the government to lead a wide-ranging set of dialogues that led to land-related agreements between the parties in conflict and the transfer of USD 3.3 billion to regional and local governments to fund local infrastructure.

Table 1. Seven country cases [4] (pp. 13, 71), [9,30–35,40,41].

Country	UN Entity	Conflicting Parties	Internally Displaced Persons/Conflicts	Approach
Darfur/Sudan	UN-African Union Hybrid Operation in Darfur (UNAMID)-peacekeeping mission UN-Habitat	Host community Internally Displaced People (IDP)	2.6 million	Good offices agreements for access to land, technical assistance
Democratic Republic of Congo (DR Congo)	UN Organization Stabilization mission in the DR Congo (MONUSCO)-peacekeeping mission	Big farmers Customary owners	4.49 million; Masisi territory case study 4600	Good offices, human rights, protection of civilians
Honduras	UN High Commissioner for Refugees (UNHCR)	Criminal gangs Displaced people	174,000 20 urban municipalities (2004–2014)	Documenting forcibly abandoned houses
Iraq	UN-Habitat	Different ethnic/religious groups	3.3 million; 250,000 Yazidi from case study	Land certificates, GIS, house rehabilitation
Peru	UN Development Program (UNDP)	Extractive industry indigenous people	200 conflicts a year (2006–2016), 70% linked to extractives	Territorial development agreement
Somalia	UN-Habitat	Pastoralists, farmers, urban residents, owners and occupants	1.1 million for the whole of Somalia; at least 44,000 for case study area	Land policy process, LA recommendations
South Sudan	UN Mission in South Sudan (UNMISS)-peacekeeping mission	Cattle owners, farmers, cross border	1.94 million South Sudan; 250,406 IDP Upper Nile case study	Territorial agreement

(rural electrification, hospitals, schools, etc.) and poverty reduction programs [34] (p. 95). The UN mission in Somalia (UNSOM) commissioned the Jubaland/Somalia conflict assessment. The purpose was to better understand how to address the land-related root causes of conflict and mitigate their consequences, including through improving LA [42]. That is, in all seven case studies, LA was only a small part of a larger peace-building program, even for the specific study performed on LA for voluntary returns in Darfur/Sudan.

4. Findings

Through a review of some of the LA, FFP LA, post conflict, and land governance literature, the seven case studies from Darfur/Sudan [13,30], DR Congo [31], Honduras [32], Iraq [33,36], Jubaland/Somalia [9], South Sudan [35], and Peru [34] are examined to identify their FFP LA elements in violent conflict contexts. This section shows, through this examination, that FFP LA in these contexts accords with the LA elements found across conventional LASs and with the FFP approaches associated with such LASs. FFP LA also corresponds with post conflict LAS assessment and design, and the more recent approaches of incorporating land governance as part of LA, including the gender-responsive aspects. However, this section will also argue that the work of the UN in these contexts introduces novel transitional justice FFP LA features associated with its peace-building role and the power dynamics associated with it.

4.1. Building on Land Administration Frameworks

Williamson et al. [22] (p. 27) (quoting the UN Economic Commission for Europe (UNECE) Land Administration guidelines [43]) describe what constitutes LA. They state that "land administration (is): the processes of recording and disseminating information about ownership, value, and use of land when implementing land management policies." The use of the term 'ownership' has been outdated by a move to "the continuum of land rights" [44] (p. 13) that takes into account all forms of "legitimate land rights" [45] (p. 3), not just ownership rights. Williamson et al. then go on to outline ways of seeing the constituent parts of LA, such as "procedures by which land rights are allocated and recognized; the definition and delimitation of boundaries between parcels; the recording of information about land rights, rights holders, and parcels; procedures governing transactions in land, including sales; the resolution of uncertainty of adjudication of disputes concerning land rights and boundaries; institutions and processes for planning, controlling, and monitoring of land use [22] (pp. 27–28)." These constituent parts resonate with the FFFP LA features in the seven cases.

All the cases involved some form of recognition and protection of land rights by the UN, in most cases of legitimate land rights (South Sudan-customary [35] (p. 94), Peru—indigenous [34] (p. 96), Jubaland/Somalia—customary and IDP/informal settlement [9] (p. 48) rather than ownership rights. All the designs to strengthen security of tenure were part of extra-legal transitional justice mechanisms, as were the administrative procedures, the land information creation and recordation, and the land use planning. In DR Congo, peacekeepers intervened in a dispute that could have led to the eviction of customary occupants by security forces [31] (pp. 53–55). In Peru and South Sudan, territorial planning, not based on land parcels, plots or sites, was undertaken to manage conflict between identity groups, mining companies and indigenous peoples, respectively; and pastoralists and settled farmers [34] (p. 96), [35] (pp. 88–90). In Peru, territorial agreements were intended to address the geographical gaps in the way government funding, generated by mining tax dollars in indigenous areas, was allocated for development in the areas of indigenous groups [34] (p. 13). In South Sudan, the territorial agreement involved agreements across international borders between Sudan and South Sudan; and between pastoralists and their livestock migration routes and areas of settled farmers in South Sudan [35] (see map in Figure 1 below). In Iraq, the land information on the rights and boundaries in the villages was recorded on a GIS run by UN-Habitat [33] (p. 62). This was because there was a lack of capacity of the local government and the rehabilitation of the villages and the houses was performed outside of the national legal framework [33] (pp. 62–65). These examples also show that the institutions involved in the processes of planning, controlling and monitoring land use were often UN institutions, sometimes working alongside government bodies other than the LAS entities.

4.2. Building on Fit-for-Purpose Land Administration Frameworks

Enemark et al. state that FFP LA is about directly aligning with country specific needs, affordability, and flexibility in accommodating different tenure types and "for shaping the legal and institutional frameworks." It is also about the ability to upgrade when opportunities arise and "it is highly participatory, can be implemented quickly and provides security of tenure" [23] (pp. vii–viii). The Framework for Effective Land Administration (FELA) states in regard to fit-for purpose LA, that "an effective LA and management system in conflict contexts prevents land-related conflict, stabilizes situations and fosters peaceful, just and inclusive societies [46] (p. 11)." We show that FFP LA in violent conflict contexts shares the same characteristics of FFP LA in conventional contexts. In Honduras, gangs have evicted people from their properties. The UNHCR, together with the land registry, developed procedures and land information using the knowledge of local Catholic parishes about the land rights of the people who have been evicted. Information will be placed on the title deeds of registered properties so that they cannot be sold [32] (pp. 43, 44, 46, 50). This implies a new procedure for land rights recognition and a new way of recording information on title deeds about evicted land rights holders' rights. This is also a new way of governing transactions in land, including sales and a way of adjudicating future disputes over rights.

In Iraq, ISIL/ISIS evicted 250,000 Yazidi people. UN-Habitat, working with local communities and with the local government, created new extra-legal LA planning, documentation and land information management procedures for 17 villages for returnees [33] (pp. 60–65). These cases show the type of country contexts that need to be aligned to in violent conflict settings. They show what flexibility means in terms of different types of extra-legal tenures, non-standard procedures in a land registry, and the shaping of non-statutory legal and institutional arrangements that can evolve with political processes. They also demonstrate the features of security of tenure in these contexts and the role of the UN in terms of good offices.

Groenendijk et al. [25] (p. 29) and Bennett and Alemie [26] have identified capacity development as a key FFP LA element. FAO also notes that there is likely to be limited government capacity in post conflict environments [18] (p. 14). This resonates with all of the seven cases [4] (pp. 2–27), [9,13]. Often areas that have been in protracted conflict for long periods have very weak LAS and capacity [19] (pp. 27–30). Darfur/Sudan has less than 1 percent registered land rights and there is insufficient capacity to support planning for voluntary returns. There is only one computer in each Land Department in each of the five states. Land cannot be planned for development without first being surveyed. There is no master plan covering the region or the cities. In regard to customary tenure areas, currently outside of the LAS, the witness system is used and involves the customary authorities, neighbors and relatives. Land information is in the memories of elders and traditional leaders [13] (pp. 9, 53–56). Yet the Doha peace agreement (2011) for the region states that, "individuals in the local communities may register their customarily owned land as their own lands [13] (p. 68)." Against this background, to support the Darfur Land Commission to plan for the return of 2.6 million people, the UN developed a number of FFP LA recommendations to build the capacity required. The recommendations were based on "a fit-for-purpose approach applicable to humanitarian settings focus(ed) on a particular purpose (e.g., records for returnees, regional planning for returns), flexible and capable of incremental improvement [13] (p. 81)." The FFP LA recommendations to support the voluntary returns of millions of people include

- Incremental building on existing capacity and legal frameworks. Focusing on adjusting regulations as much as possible as new laws can take years.
- Rapid identification and use of government-owned land for rapid delivery of security of tenure to returnees.
- Reducing planning standards and undertaking one stop planning for rapid planning approvals. This is vital for upgrading informal settlements and IDP camps.
- Reducing surveying and registration requirements. This includes allocating unplanned non-surveyed land, called Grade IV in the state-level statutory system, thereby excluding the final registration step in the Judiciary at national level. Using group areas instead of individual plots. For example, the allocation by customary authorities of land for villages for returnees can be registered as a community based right (Plot 1), where instead only the outside boundary of the village is registered.
- Legalizing the customary land management system.
- Delivering serviced land rapidly in urban areas by increasing the amount of serviced land by minimizing the time taken to deliver the services, while ensuring minimum standards.
- Improving information management to track large-scale returns and developing strategic plans for the overall management of returns, including through regional planning, land management and for conflict analysis.
- Identifying a range of dispute resolution mechanisms based on the assessment of possible return areas and conflict hot spots, and the tenure and land document types found in these areas. Using land-dispute resolution tools such as mediation, territory-

wide land use agreements, and dispute resolution at the plot and territory levels [13] (pp. 9, 19, 30, 52–55, 79–84).

While targeted capacity development of major actors linked to the LAS is important, the Darfur/Sudan examples given above, of potential FFP LA interventions for voluntary returns, go beyond this. They demonstrate that FFP LA in these contexts is also about filling capacity gaps by using and adapting existing systems and options, including through altering norms and standards. That is, capacity development is not just about training and knowledge development but also about adjusting the LAS system, using FFP LA approaches, to achieve tenure security results and scale for the most vulnerable by matching approaches and procedures better with existing capacity.

Another example of weak capacity is South Sudan. Ten million cattle cross from Sudan annually and there has been protracted violent conflict between pastoralists and farmers. The UN has built conflict management capacity in the parties to the conflict and facilitated territorial agreements that cross the international boundary with Sudan, using land use zoning agreements between different identity groups to address and prevent conflict. This involved inclusive decision-making around the mapping and designation of territories [35] (pp. 88–92). The areas involved and cross-border cattle migration routes are shown in Figure 1 below. In DR Congo, the UN has built capacity in local communities. Its Civil Affairs Section "monitors areas affected by eviction through its extensive community alert network in order to prevent the escalation of the situation and to respond to incidents." Local people contact the UN mission when threatened [31] (pp. 52, 55).

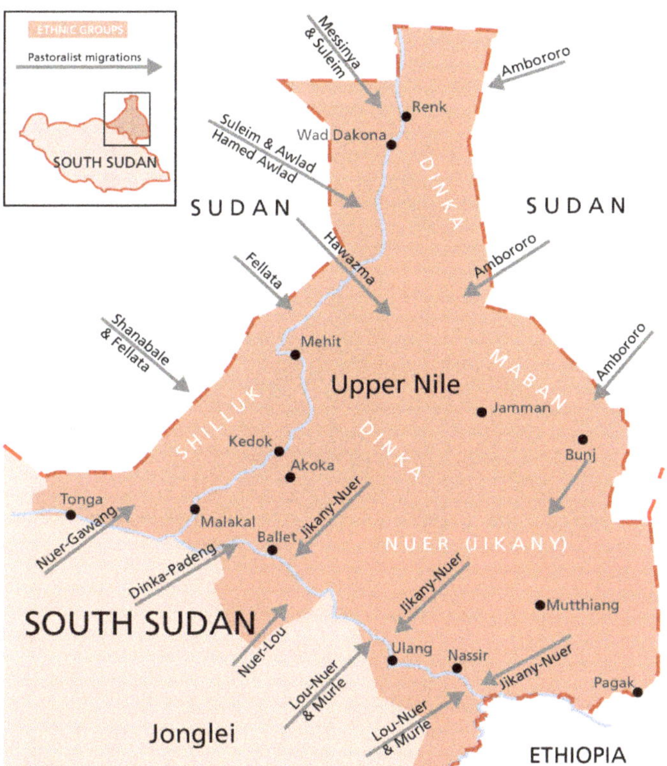

Figure 1. Migration of pastoralists in the former Upper Nile State of South Sudan [35], (p. 89); copyright UNMISS.

4.3. Building on Post Conflict Land Administration Approaches

Todorovski et al. [20], Jossam et al. [21], UN-Habitat [19], and Augustinus and Barry [29] describe LASs during post conflict. We show that these frameworks also have explanatory power when applied to violent conflict contexts. The key characteristics they identify are firstly, the land management and LAS is likely to be largely dysfunctional as it has either been destroyed [20] (pp. 76–77), [21] (p. 237), [19] (p. 9), or did not exist in the first place as it is a customary area. This latter can be found in DR Congo [31] (pp. 53–54) and Darfur/Sudan [30] (p. 83) [13] (pp. 8, 47, 50, 68) cases where the LAS was very weak in the areas of conflict.

Secondly, the rule of law has broken down. This can mean land dealings outside of a legal framework or routine technical process and powerful actors grabbing both public and private land [20] (p. 73), [21] (p. 238), [19] (p. 7). In Honduras, gangs grabbed houses [32] (p. 44). In DR Congo, "land documents are not properly registered but can be fraudulently acquired and traded. Private interests repeatedly use security institutions and sideline or co-opt the justice system to protect and further their own goals [31] (p. 51)." The cases show that a very "common rule of law issue is the eviction of people from their homes and land" [4] (p. ix). In Peru, indigenous groups' land rights were threatened by investors [34] (p. 96). Eviction is often also linked to the destruction of housing, such as in Iraq [33] (p. 61), [18] (p. 14). Another rule of law issue relates to overlapping rights and claims, or secondary rights, on the same piece of land, also identified by Todorovski et al. [20] (p. 76), Jossam et al. [21] (pp. 233, 241), and Van der Haar and van Leeuwen [5] (p. 3). In Darfur/Sudan, pastoralists have claimed the land of customary farming communities [30] (p. 80), [47]. In Kismayo, in Jubaland/Somalia, there are overlapping rights and claims between pastoralists, farmers and urban residents vying for rights over the same land; and between absentee owners of registered land and occupants, between government and urban residents and between IDP and host communities (see Figure 2 below) [9] (p. 48).

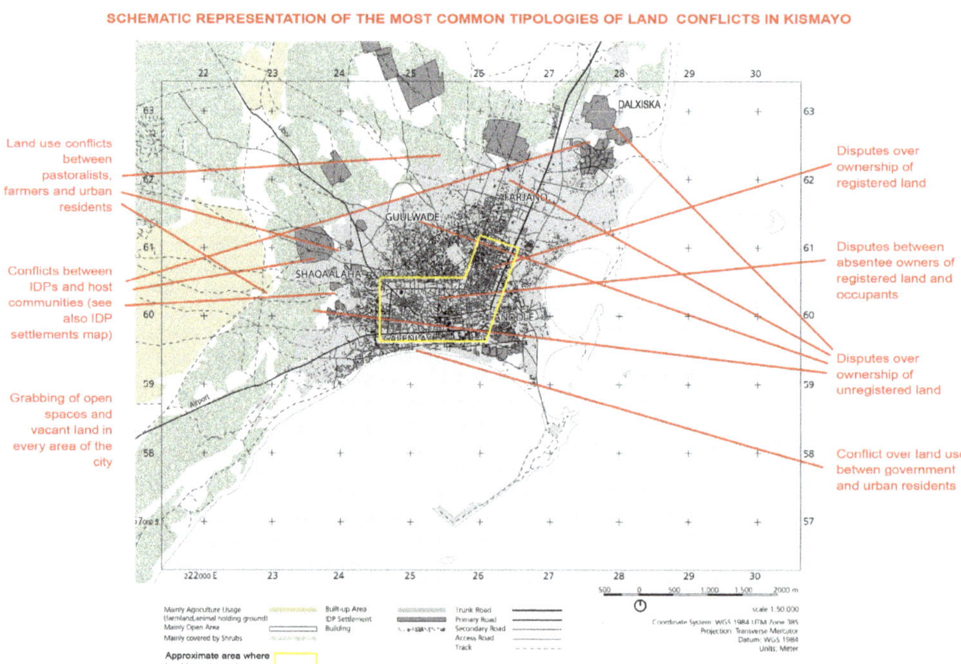

Figure 2. Analysis of the land-related root causes of conflict in Jubaland [9], (p. 48); copyright UN-Habitat.

Thirdly, there is likely to be large-scale ambiguity and gaps in the land-related regulatory framework [29] (p. 671), [21] (p. 246), [19] (p. 7). Some of the cases demonstrate this ambiguity and fluidity of the land-related institutional landscape. In Iraq, the extra-legal LAS created by UN-Habitat became politicized when the Iraq Federal Government army took back the area from the Kurdish state government control in 2018. For a while, it was unclear as to whether the extra-legal LAS and its land certificates (see Figure 3 below) that UN-Habitat and the local government had put in place for Yazidi returnees were going to be accepted by the national government. They were accepted by the Prime Minister's office and is in the process of being extended to the whole of the Governorate of Ninewah [36] (pp. 73–74). The ambiguity and gaps in the land-related regulatory framework, be it through the statutory, customary or religious systems, undermine the weak land right claims of women [19] (p. 71). In Jubaland/Somalia, there has been forty years of civil war. The LAS was only embryonically developed when the state collapsed and the decades-long civil war has added further challenges for the formal, legal and institutional LAS. LA functions are undertaken by a mixture of practices inspired by religious, customary and statutory laws, with the last based on laws that existed before the current federal nature of Somalia was set up [9] (pp. 26, 32–37).

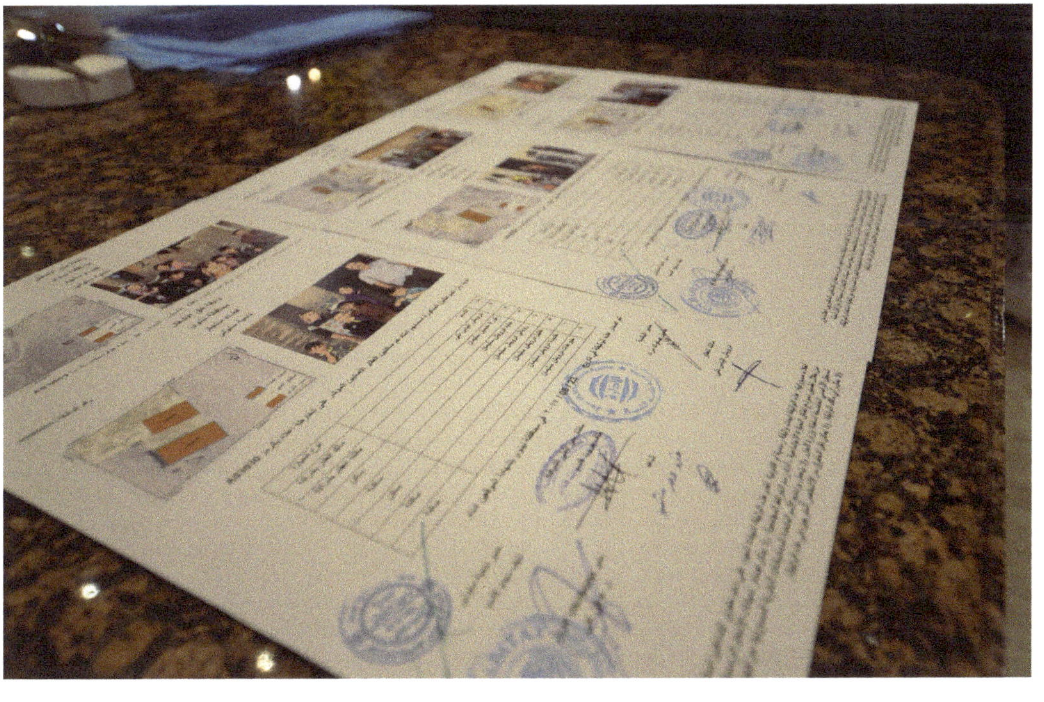

Figure 3. Certificate of occupancy for Yazidi village [33], (p. 62); copyright UN-Habitat.

Augustinus and Barry, in their analysis of post conflict contexts, argue that LAS design for these contexts needs to include a number of specific elements. These elements are also found in the design of the LASs in the cases. The LAS elements or "constituent parts" prioritized for action should contribute to the higher macro-environmental objectives of the system [29] (p. 676), such as dispute resolution. Dispute resolution is central to violent conflict settings [19] (p. 22). Territorial dispute resolution is found as a primary objective in the case of Peru [34] (p. 95) and South Sudan [35] (p. 88). DR Congo demonstrates the use of security assets by the UN, in the form of peacekeeping patrols, to address disputes [31] (p. 53). Another goal for LAS design can be political or peace building. van der Haar and van Leeuwen argue that "technical solutions need to be connected to a strong moral compass [5] (p. 9)." All three of the UN missions in Darfur/Sudan, DR Congo and South Sudan were peacekeeping missions mandated by the UN Security Council, with all three involved in some form of LAS (see above).

Augustinus and Barry also argue that the LAS elements prioritized for strengthening should assist with conflict resolution and reconciliation [29] (p. 680). Iraq [33] is an example of where the UN supports people who have been evicted to return to their homes and land. DR Congo [31] (pp. 53–55) is an example of where the UN prevents people from being evicted from their homes or land. The adjudication of rights, including land records with overlapping rights and claims, is vital for conflict resolution [21] (p. 239), [18] (pp. 27–28), [19] (pp. 50, 52, 53). The cases show this to be true for parcels and territories. In Iraq, UN-Habitat supported village communities to demarcate their land parcels during the rehabilitation of the village and their homes [33] (pp. 62–65). The UN supported identity groups to adjudicate or demarcate territorial boundaries in Peru, through territorial development plans [34] (p. 98); and in South Sudan, through migration route planning [35] (pp. 89–93).

It is likely that new institutions will be set up and/or existing ones re-purposed [19] (pp. 58, 59). These situations are extremely fluid "with a lack of clarity about where land functions are placed in government with gaps, ambiguities around the law and policy, and large scale opportunistic behavior it is necessary to position the land administration functions within this fluid environment [29] (p. 679)." This accords with the findings of FAO [18] (p. 28). This can be seen in the Iraq case, where UN-Habitat and the local government set up the village LAS, even though LAS was a national function; and its endurance was in doubt when the national government reasserted control over the area [33]. In South Sudan, extra-legal territorial planning was used to manage conflict and, while the UN wanted the approach to be formalized, it was unclear whether and when this would happen because of the weakness of the South Sudan state [35] (p. 91). Jubaland/Somalia support is being provided to the newly established Land Commission [42].

4.4. Land Governance and the UN in Violent Conflict Contexts

LA is not just made up of the constituent parts identified by Williamson et al. [22] (pp. 27–28). Land governance has also increasingly come to be seen as core to LA, particularly with the development of the Voluntary Guidelines on the Responsible Governance of Tenure of Land, Fisheries and Forests in the Context of National Food Security [45]. The connection between them has been strengthened by the World Bank's Land Governance Assessment Framework (LGAF). LGAF helps countries to examine their land governance status against global good practice in regard to "land rights recognition, land use planning, management and taxation, expropriation, public provision of land information, and conflict resolution" [27] (p. 76). That is, land governance is closely associated with the LA constituent parts originally identified by Williamson et al. [22], which as we have shown match LA elements also found in violent conflict contexts (see above). Alemie et al. also demonstrate this connection between LA and land governance and unpack some of its characteristics further by identifying the key roles of the cadastre in supporting urban land governance. They identify these as: improving tenure security; improving transparency and participation; providing easy access to information; improving the governments' and citizens' decision making and efficiency; reducing corruption; and improving equity [24] (p. 57). The case studies show that the UN plays many land governance roles in violent conflict settings, both in urban and rural areas, and that they have distinct UN-related features (see below).

The case studies describe how the UN uses its convening power in violent conflict contexts, including good offices to bring parties in conflict together, to strengthen land governance, as defined by FAO and UN-Habitat [12], based on the parameters described by Deininger et al. [27] and Alemie et al. [24]. In South Sudan [35] (pp. 88–92), Jubaland/Somalia [9] (pp. 70–71), Darfur/Sudan [30] (pp. 81–85), Peru [34] (p. 97) and DR

Congo [31] (pp. 54–55), because of the good offices of the UN and its power to engage in these settings, new "rules, processes and structures" were put in place "through which decisions are made about access to land and its use [12] (p. 9)." The UN used its power, including its security assets, good offices and convening power, to introduce extra-legal tenure types (Iraq [33] (pp. 64–65)) and territorial/land use planning (South Sudan [35] (pp. 89–90)), negotiate access to land by getting agreement between IDP and areas controlled by armed groups (Darfur/Sudan [30] (p. 83)), and get agreement between (sub-) national government, mining companies and indigenous people about territorial land uses and access (Peru [34] (pp. 97–99)), and enforced customary rights against the security establishment (DR Congo [31] (pp. 52–53)). The UN also used its technical assistance capacity such as in Darfur/Sudan—re-thinking the LAS system [13]; in Iraq—development of a GIS which recorded the extra-legal tenures [33] (pp. 62–3); in Honduras—development of a land information system based on evidence supplied by local churches [32] (p. 46); and in DR Congo—re-establishment of three surveying and land registration teams (cadastral brigades) [31] (p. 56). In Peru, the UN supported the government to lead a wide-ranging set of dialogues that led to territorial agreements between the parties in conflict and the transfer of USD 3.3 billion to regional and local governments to fund local infrastructure and poverty reduction programs [34] (p. 95).

The in-depth case study of LAS for voluntary returns in Darfur/Sudan [13] gives further details of the range of land governance instruments that are useful in these contexts. These include

- The targeted clarifying and coordinating of LAS government functions across national, state and local levels for a specific humanitarian purpose such as voluntary returns.
- Amending regulations rather than developing new laws.
- Ensuring due diligence and minimum standards to protect people and their land rights.
- By using land information management, tracking trends of people returning to their areas of origin or other areas, conflict and dispute resolution patterns, and to identify areas that need additional land-related peace-building interventions. This also requires early warning systems at the community level and along migratory routes.
- Strengthening the Land Commission to "perform its function and mandate to arbitrate between willing contending parties on land claims, (and) assess appropriate land compensation" [13] (pp. 82–85).

Gender-responsive activities are a vital aspect of land governance in these environments. "Conflict worsens discrimination against women regarding land. Women are particularly vulnerable to losing their land rights, and they are more likely than men to be forcibly displaced or evicted [4] (p. 28)." This resonates with the findings of FAO [18] (p. 26) and the World Bank [48]. "Conflict should be used as an opportunity to empower women. Women should be involved in land interventions both as beneficiaries and as partners [4] (pp. 28–29)." The UN often promotes gender responsiveness, because of its adherence to human rights principles and practices, and a number of the cases demonstrate this.

Paradza et al. [28] outline crucial features of land governance and women's land rights in customary settings. A number of the features that they identify are found in the Darfur/Sudan [13] and Jubaland/Somalia [9] extended case studies. This indicates that Paradza et al.'s approach can also be adapted to other areas in violent conflict contexts. Paradza et al. argue that "mapping initiatives generate opportunities, innovations, and novel spaces for securing women's access to land in customary areas which include increasing awareness of women's interests, providing opportunities for women to participate in decision-making forums [28] (p. 1)". In Darfur/Sudan, a range of land-related activities were recommended to support women's land rights during voluntary returns. These include awareness raising regarding women-headed household land rights under statutory law, Islamic law (dower or *mahar*, inheritance or *mirath*) and strengthening of local norms; civil society organizations being encouraged to increase their "knowledge and understanding of women's land rights, their importance and how to promote them in practical and culture/context specific manner," with special attention being given to empowering women leaders in playing a bigger role in land-related discussions; and the creation of information support centers to support women to understand what options are available to them [13] (pp. 73, 85, 86). The Puntland/Somalia and Iraq case studies also demonstrate a similar accord with Paradza et al. [28] findings. In Puntland/Somalia, dialogues and training are provided to increase the understanding of the importance of protecting women's housing, land and property rights and how to achieve it [9] (pp. xi, 10). The Iraq case study demonstrates that innovative mapping creates a novel space for securing women's land rights as "the project placed a strong emphasis on gender, including during the selection of beneficiaries. Female-headed households, including widows and households with young pregnant women, were given priority [33] (p. 66)."

Additional gender-responsive land governance features for violent conflict settings are identified by Abukashawa et al. [13] and Tempra [9]. These include

- Ensuring the processes for addressing housing, land and property challenges in a humanitarian setting do not discriminate against women [9] (p. 70).
- Supporting women's access to land and tenure security across the full range of tenure types and the identification of the most viable tenure options that can reach the greatest number of women in the shortest time [13] (pp. 68, 73), as well as support to their access to justice mechanisms [9] (p. 70).
- Providing special provisions for women, particularly widows and women-headed families of returnees, in customary areas where their legal status needs to be protected. Women-headed households and widows should be considered as heads of households in the customary system with a right to access land [13] (pp. 75, 79, 84). This may also involve support for the provision of civil documentation [9] (p. 70).

5. Discussion

The findings, using the case study material, show that LA in violent conflict settings involves the same key elements found generally in LASs, not dissimilar elements. They serve the same purposes, even while they seem unfamiliar. All the cases involved some form of protection of land rights, in most cases legitimate land rights rather than ownership rights. The designs to strengthen security of tenure were part of extra-legal transitional justice mechanisms outside of the formal LAS, as were the administrative procedures, the land information and the land use planning, control and enforcement. The land use planning had conflict management between identity groups at its core, using territorial approaches not based on land parcels, plots or sites. Extra-legal transitional justice procedures even governed sales transactions procedures in the land registry. That is, extra-legal transitional justice LA rule making and keeping, which was FFP, was central to the UN's role in these contexts. However, capacity development for FFP LA in these settings is also about achieving security of tenure objectives rapidly through altering LAS norms, standards and objectives such as the rapid identification and use of government-owned (state) land for the settlement of displaced people; the delivery of the maximum amount of land in the shortest possible time, with due diligence and minimum standards, and services in urban areas and villages; lowering planning standards, particularly relating to site size, and using innovative territorial planning both statutory and extra-legal; moving away from registered land ownership as the only option to using other forms of tenure along the continuum innovatively, such as statutory group rights, unregistered rights, customary tenure, including through the reduction in surveying requirements and registered outside boundaries for villages, as well as extra-legal land certificates; and using land information management for large-scale overall human settlement management for humanitarian purposes.

When comparing the post conflict impact on LASs with that of violent conflict settings found in the case studies, we find that they correspond. The shared characteristics include a breakdown in the rule of law, fluid institutional environments and ambiguity, dysfunctional LAS, overlapping (secondary) rights and claims. Further, the types of LA designs proposed for post conflict settings match those used by the UN in violent conflict contexts. These include political or peace-building goals rather than purely technical goals, prioritization of dispute and conflict resolution and reconciliation, mechanisms for the adjudication of 'suspect' land rights.

The cases show that land is part of larger UN peace-building programs. In some cases, land is part of a voluntary returns program, such as a housing and/or settlement rehabilitation program, encouraging people to return after being evicted, or is part of a program to protect their future right to return. In other cases, it is part of conflict management by the UN between parties in conflict at the territorial level rather than the level of individual land rights, something not commonly dealt with by LASs. The cases show how the UN sets up FFP LA extra-legal transitional justice mechanisms to protect people's tenure security to address historical injustices and the politics of exclusion, as part of its larger peace-building programs. These extra-legal mechanism include tenure types, creation and recordation of land information, LAS administrative procedures, land use and territorial planning, and land governance, including around women's land rights.

Sometimes these mechanisms are not part of any discussion with national governments, as these governments are parties to the conflict [30] or too fragile [36]. Sometimes these mechanisms are part of government [34] or there is a process of trying to formalize them into government [35]. Based on this, we can conclude that in discussing FFP LA as used by the UN in violent conflict settings, and as argued by Lengoiboni et al., we also need to "ask not only fit-for-what purpose, but for whose purposes and at what point in time?" Whose responsibility is it to upgrade the resulting documents to official recognized tenure certificates? (Or, as in the case of South Sudan, the territorial plans [35]). What procedures should be followed? [49] (p. 29). The cases show that upgrading the resulting documents is not guaranteed even when the UN is intentionally trying to support such an upgrade, such as in Iraq with the land certificates [33]. The cases also show that if the UN supports the re-purposing of formal territorial planning or an operating land registry from the outset like in Peru [34] and Honduras [32], formalization of the new conflict management features is more likely. However, where there is no state capacity to manage violent conflict, like in South Sudan with the territorial planning [35], and the UN fills the role for peace-building purposes, LAS state building is required prior to the handover of the function from the UN to the newly reconstituted state.

Just as in conventional settings, land governance is a key part of LA in violent conflict settings, including through improving tenure security, equity, transparency and participation, and the governments' and citizens decision making and efficiency; providing easy access to information; and reducing corruption. However, in these settings, it has novel traits. The UN uses its power to address multilevel power dynamics through its good offices, convening power and security assets to strengthen land governance and manage power relations. The UN does this by introducing a wide range of extra-legal FFP LA transitional justice mechanisms to protect and defend vulnerable and displaced people's land rights. These mechanisms include norms and standards that fit humanitarian purposes; ensuring land-related human rights practices; using land information management to track displacement, conflict and dispute resolution patterns and trends; and to identify areas that need additional land-related peace-building interventions. In regard to human rights, this is particularly important in regard to women's land rights because conflict worsens discrimination for women. The UN uses its power to open new spaces for women's land rights through awareness raising, mapping, rapid engagement and special land-related provisions for women and transitional justice mechanisms that address discrimination. Some aspects of these are the same as those found in conventional settings. Some are specific to violent conflict environments.

Through a UN lens, we have shown that FFP LA in violent conflict contexts involves the same LA elements found in conventional LA and FFP LA contexts. At the same time, it shares the same features as LA in post conflict contexts, as it supports peace building and conflict resolution. FFP LA also has other novel characteristics that shape its design. FFP LA has distinctive characteristics that relate to the role of the UN in these contexts. FFP LA in violent conflict contexts often (1) involves extra-legal transitional justice mechanisms to protect people (particularly women) and their land rights to start addressing some of the historical injustices and the politics of exclusion that are some of the root causes of conflict, as part of the UN's peacekeeping and peace-building programs; and (2) it is only a small part of these much larger programs and this has important power relations implications for land governance in these contexts. These power relations are linked to the UN's convening power, good offices, and in some situations security assets, and in others its technical assistance. We also adapt the FAO and UN-Habitat definition of land governance [12] (p. 9) to these contexts. This paper has argued that the role of the UN in land governance in violent conflict contexts concerns: the FFP LA extra-legal rules, processes and new and re-purposed structures through which decisions are made by the UN and other parties about access to land and its use; the manner in which the decisions are implemented and enforced (including through good offices); and the way that competing interests in land are managed (including through negotiating between parties in conflict and security assets). Just like in other settings, land governance is key, but it has novel features because of the power dynamics associated with the UN in these settings.

6. Conclusions

The UN was created from the outset to build peace in violent conflict contexts. As acknowledged by the Secretary-General's Guidance Note on land and conflict [3], land is a major root cause of conflict and hence of interest to the UN in its peace building. We have examined the types of FFP LA-related activities that the UN has undertaken in seven different countries. The FFP LA it used is the same as that found in conventional settings, but it has novel characteristics. They range from peacekeeping missions protecting people and their access to land and legitimate land rights, to technical assistance to plan for large-scale voluntary returns of people who have been displaced. The cases demonstrate how the UN has used what we have shown to be FFP LA extra-legal transitional justice mechanisms, as part of larger peace-building programs, to start addressing land-related historical injustices and the politics of exclusion, which are some of the root causes of conflict. In these contexts, land governance is of particular importance because of the way the land-related power dynamics play out across the conflict cycle. The cases demonstrate that the UN uses its power and capacity to strengthen land governance, including around women's land rights. The UN does this by using its good offices and convening power and security assets. This is a critical component of the FFP LA design in these settings. In the seven country cases, the UN's power was used to manage the competing interests in land. Its power was an important contributory factor concerning which FFP LA rules and processes were used, and the manner in which FFP LA was implemented. The cases also give early insights into how the UN's convening power and good offices' roles can be used to move transitional justice FFP LA mechanisms into more formal LAS arrangements.

Finally, the type of FFP LA approaches described here, found across multiple countries and sites, give practical options to support peace-building operations by the UN and other stakeholders. The cases overturn the view that it is not possible to undertake LA interventions in violent conflict settings. On the contrary, FFP LA approaches are a critical component of peace building and are often designed to serve as a basis for future work on LA in the post conflict and development phases.

Author Contributions: C.A. contribution included visualization, conceptualization, methodology, investigation, formal analysis, and writing—original draft preparation. O.T. contribution included visualization, conceptualization, validation, investigation, and writing—review and editing. All authors have read and agreed to the published version of the manuscript.

Funding: The research and writing of this paper received no external funding. Funding of the publication costs for this article has kindly been provided by the School of Land Administration Studies, University of Twente, in combination with Kadaster International, The Netherlands.

Institutional Review Board Statement: Not applicable. All the material used in this study has already been published in United Nations reports and no new studies were undertaken.

Informed Consent Statement: This case study series did not involve humans, as it did not involve any new human data beyond that already published as per the United Nations approval system. The six United Nations entities approved the publication of the country case studies in the grey literature and on the web.

Data Availability Statement: All the case study material are referenced in the manuscript and are available on the web.

Conflicts of Interest: The authors declare no conflict of interest.

References

1. United Nations Refugee Agency (UNHCR). Figures at a Glance. Available online: https://www.unhcr.org/en-ie/figures-at-a-glance.html (accessed on 23 October 2020).
2. United Nations. Peace, Dignity and Equality on a Healthy Planet, Peace and Security. Available online: https://www.un.org/en/sections/issues-depth/peace-and-security/ (accessed on 23 October 2020).
3. United Nations Secretary General. The Guidance Note of the Secretary-General The United Nations and Land and Conflict; Endorsed on the 15th March 2019. Available online: https://gltn.net/2019/03/15/guidance-note-of-the-secretary-general-the-united-nations-and-land-and-conflict-march-2019/ (accessed on 7 June 2020).
4. UN-Habitat; Global Land Tool Network; International Institute of Rural Reconstruction. *Land and Conflict: Lessons Learned from the Field on Conflict Sensitive Land Governance and Peacebuilding*; UNON: Nairobi, Kenya, 2018; pp. 1–126.
5. Van der Haar, G.; van Leeuwen, M. War-Induced Displacement: Hard Choices in Land Governance. *Land* **2019**, *8*, 88. [CrossRef]
6. Global Land Tool Network. Available online: https://gltn.net/ (accessed on 3 October 2020).
7. Decorte, F.; Augustinus, C.; Lind, E.; Brown, M. *Scoping and Status Study on Land and Conflict: Towards UN System-Wide Engagement at Scale*; UN-Habitat/Global Land Tool Network, Working Paper May 2016, Report 5/2016; UNON: Nairobi, Kenya, 2016; pp. 1–49.
8. United Nations; United Nations Security Council. *Report of the High-Level Independent Panel on Peace Operations on Uniting Our Strength for Peace: Politics, Partnership and People*; A/70/95, S/2015/446; United Nations: New York, NY, USA, 2015; pp. 1–104. Available online: https://www.un.org/en/ga/search/view_doc.asp?symbol=S/2015/446 (accessed on 12 November 2020).
9. Tempra, O. *Land and Conflict in Jubaland Root Cause Analysis and Recommendations*; UN-Habitat/Global Land Tool Network and Department of Political and Peacebuilding Affairs; UNON: Nairobi, Kenya, 2018; pp. 1–84.
10. United Nations Office of the High Commission for Human Rights (OHCHR). *Transitional Justice and Economic, Social and Cultural Rights*; 11 April 2014, HR/PUB/13/5; United Nations: New York, NY, USA; Geneva, Switzerland. Available online: https://www.refworld.org/docid/5347cbe74.html (accessed on 18 December 2020).
11. Collins Dictionary. Available online: https://www.collinsdictionary.com/dictionary/english/extralegal (accessed on 27 October 2020).
12. UN-FAO; UN-Habitat. *Towards Improved Land Governance*; FAO: Rome, Italy, 2009; pp. 1–55.
13. Abukashawa, S.; Augustinus, C.; Mustafa, A.; Tempra, O. *Darfur Land Administration Assessment—Analysis and Recommendations*; UN-Habitat Sudan/Global Land Tool Network; UNON: Nairobi, Kenya, 2020; pp. 1–91.
14. Scopus Preview. Available online: https://www.scopus.com/home.uri (accessed on 1 July 2020).
15. Global Land Tool Network. *Land and Conflict Forum*; Developing an Issue-Based Coalition, Final Draft, Internal Document; UN Gigiri Complex, New Office Facility: Nairobi, Kenya, 2015; pp. 1–29.
16. UN-Habitat; Global Land Tool Network. *Annual Report 2015*; UNON: Nairobi, Kenya, 2015; pp. 1–56.
17. Van der Molen, P.; Lemmen, C. (Eds.) *Land Administration in Post Conflict Areas: Proceedings of a Symposium held by FIG Commission 7, Cadastre and land management on 29 and 30 April 2004 at the European Headquarters, Palais des Nations, Geneva, Switzerland*; International Federation of Surveyors: Frederiksberg, Denmark, 2004; pp. 1–216.
18. FAO. *Access to Rural Land and Land Administration after Violent Conflicts*; FAO Land Tenure Studies 8; FAO: Rome, Italy, 2005; pp. 1–74.
19. UN-Habitat. *From Emergency to Reconstruction a Post-Conflict Land Administration and Peace-Building Handbook*; Series 1: Countries with Land Records; UN-Habitat: Nairobi, Kenya, 2007; pp. 1–102.
20. Todorovski, D.; Zevenbergen, J.; van der Molen, P. Land Administration in Post Conflict Contexts. In *Advances in Responsible Land Administration*; Zevenbergen, J., De Vries, W., Bennett, R., Eds.; CRC Press, Taylor and Francis: Boca Raton, FL, USA, 2016; pp. 73–90.
21. Jossam, P.; van der Molen, P.; Boerboom, L.; Todorovski, D.; de Vries, W. Displacement and land administration. In *Advances in Responsible Land Administration*; Zevenbergen, J., De Vries, W., Bennett, R., Eds.; CRC Press, Taylor and Francis: Boca Raton, FL, USA, 2016; pp. 235–249.
22. Williamson, I.; Enemark, S.; Wallace, J.; Rajabifard, A. *Land Administration for Sustainable Development*; ESRI Press Academic: Redlands, CA, USA, 2010; pp. 1–487.

23. Enemark, S.; Lemmen, C.; McLaren, R. *Fit-for-Purpose Land Administration Guiding Principles*; UN-Habitat/Global Land Tool Network; UNON: Nairobi, Kenya, 2015; pp. 1–118.
24. Alemie, B.K.; Bennett, R.M.; Zevenbergen, J. Urbanization, Land Administration, and Good-Enough Governance. In *Advances in Responsible Land Administration*; Zevenbergen, J., De Vries, W., Bennett, R., Eds.; CRC Press, Taylor and Francis: Boca Raton, FL, USA, 2016; pp. 53–57.
25. Groenendijk, L.; Lemmen, C.; Bennett, R. Experiences and lessons from Mozambique. *GIM Int.* **2016**, *30*, 29–31.
26. Bennett, R.M.; Alemie, B.K. Fit-for-purpose land administration: Lessons from urban and rural Ethiopia. *Surv. Rev.* **2016**, *48*, 11–20. [CrossRef]
27. Deininger, K.; Hilhorst, T.; Songwe, V. Identifying and addressing land governance constraints to support intensification and land market operation: Evidence from 10 African countries. *Food Policy* **2014**, *48*, 76–87. [CrossRef]
28. Paradza, G.; Mokwena, L.; Musakwa, W. Could mapping initiatives catalyze the interpretation of customary land rights in ways that secure women's land rights? *Land* **2020**, *9*, 344. [CrossRef]
29. Augustinus, C.; Barry, M. Land management strategy formulation in post conflict societies. *Surv. Rev.* **2006**, *38*, 668–681. [CrossRef]
30. Dzakuma, M. Sudan: Intercommunal reconciliation of land disputes in Darfur. In *Land and Conflict, Lessons Learned from the Field on Conflict Sensitive Land Governance and Peacebuilding*; UN-Habitat, Global Land Tool Network, International Institute for Rural Reconstruction, Eds.; UNON: Nairobi, Kenya, 2018; pp. 80–87.
31. Weyns, Y. Democratic Republic of Congo: Interventions to prevent evictions of subsistence farmers communities. In *Land and Conflict, Lessons Learned from the Field on Conflict Sensitive Land Governance and Peacebuilding*; UN-Habitat, Global Land Tool Network, International Institute for Rural Reconstruction, Eds.; UNON: Nairobi, Kenya, 2018; pp. 51–59.
32. El Abdellaoui, J.; Padilla, L.N. Honduras: Protecting the housing and land of displaced persons. In *Land and Conflict, Lessons Learned from the Field on Conflict Sensitive Land Governance and Peacebuilding*; UN-Habitat, Global Land Tool Network, International Institute for Rural Reconstruction, Eds.; UNON: Nairobi, Kenya, 2018; pp. 43–50.
33. Frioud, I. Iraq: Social tenure and house rehabilitation to support the return of Yazidis in Sinjar. In *Land and Conflict, Lessons Learned from the Field on Conflict Sensitive Land Governance and Peacebuilding*; UN-Habitat, Global Land Tool Network, International Institute for Rural Reconstruction, Eds.; UNON: Nairobi, Kenya, 2018; pp. 60–69.
34. Thais, L.F. Peru: Dialogue to prevent and manage conflicts over the use of natural resources. In *Land and Conflict, Lessons Learned from the Field on Conflict Sensitive Land Governance and Peacebuilding*; UN-Habitat, Global Land Tool Network, International Institute for Rural Reconstruction, Eds.; UNON: Nairobi, Kenya, 2018; pp. 95–104.
35. Okumu, J. South Sudan: Migration dialogues to prevent conflict between host communities and pastoralists. In *Land and Conflict, Lessons Learned from the Field on Conflict Sensitive Land Governance and Peacebuilding*; UN-Habitat, Global Land Tool Network, International Institute for Rural Reconstruction, Eds.; UNON: Nairobi, Kenya, 2018; pp. 88–94.
36. Sylla, O.; Tempra, O.; Decorte, F.; Augustinus, C.; Frioud, I. Land and conflict: Taking steps towards peace. *Forced Migr. Rev.* **2019**, *62*, 71–77.
37. Kivunja, C.; Kuyini, A.B. Understanding and applying research paradigms in educational contexts. *Int. J. High. Educ.* **2017**, *6*, 26–41. [CrossRef]
38. Hasse, C. The multi-variation approach: Cross-case analysis of ethnographic fieldwork. *Paldyn* **2019**, *10*, 219–227.
39. Unruh, J.D. Evidencing the restitution landscape: Pre-emptive and advance techniques for war-torn land and property rights reacquisition. *Land Use Policy* **2014**, *38*, 111–122. [CrossRef]
40. OCHA. Democratic Republic of Congo: Internally Displaced Persons and Returnees (as of December 2017). Available online: https://reliefweb.int/sites/reliefweb.int/files/resources/drc_factsheet_trim4_2017_en_07022018.pdf (accessed on 17 January 2021).
41. OCHA. South Sudan: Humanitarian Shapshot June 2017. Available online: https://reliefweb.int/sites/reliefweb.int/files/resources/SS_170717_Humanitarian_Snapshot_June.pdf (accessed on 17 January 2021).
42. Tempra, O. (UN-Habitat, Nairobi, Kenya). Personal communication, 2020.
43. United Nations Economic Commission for Europe. *Land Administration Guidelines, with Special Reference to Counties in Transition*; United Nations: New York, NY, USA; Geneva, Switzerland, 1996; pp. 1–112.
44. New Urban Agenda. Available online: https://habitat3.org/wp-content/uploads/NUA-English.pdf (accessed on 12 December 2020).
45. FAO. *Voluntary Guidelines on the Responsible Governance of Tenure of Land, Fisheries and Forestry in the Context of National Food Security*; FAO: Rome, Italy, 2012; pp. 1–40.
46. United Nations Global Geospatial Information Management (UN-GGIM). *Framework for Effective Land Administration, a Reference for Developing, Reforming, Renewing, Strengthening, Modernizing, and Monitoring Land Administration*; E/C.20/2020/29/Add.2.; United Nations Department of Economic and Social Affairs: New York, NY, USA, 2020. Available online: http://ggim.un.org/meetings/GGIM-committee/10th-Session/documents/E-C.20-2020-29-Add_2-Framework-for-Effective-Land-Administration.pdf (accessed on 20 December 2020).
47. Abdul-Jalil, M.; Unruh, J. Land rights under stress in Darfur: A volatile dynamic of the conflict. *War Soc.* **2013**, *32*, 156–181. [CrossRef]
48. World Bank. *Violent Conflict and Gender Inequality: An Overview*; World Bank Policy Research Working Paper; World Bank: Washington, DC, USA, 2013; Volume 6371, pp. 1–35.
49. Lengoiboni, M.; Richter, C.; Zevenbergen, J. Cross-cutting challenges to innovation in land tenure documentation. *Land Use Policy* **2019**, *85*, 21–32. [CrossRef]

land

MDPI

Article

Decentralization as a Strategy to Scale Fit-for-Purpose Land Administration: An Indian Perspective on Institutional Challenges

Serene Ho [1,*], Pranab R. Choudhury [2], Nivedita Haran [3] and Rebecca Leshinsky [4]

[1] School of Science, RMIT University, Melbourne, VIC 3001, Australia
[2] NRMC Centre for Land Governance, Bhubaneshwar, Odisha 751003, India; pranabrc@nmrc.co
[3] Ente Bhoomi Trust, Thiruvananthapuram 695005, India; nharan@gmail.com
[4] School of Property, Construction and Project Management, RMIT University, Melbourne, VIC 3001, Australia; rebecca.leshinsky@rmit.edu.au
* Correspondence: serene.ho2@rmit.edu.au; Tel.: +61-3-9925-7654

Abstract: Many countries grapple with the intractable problem of formalizing tenure security. The concept of 'fit-for-purpose land administration' (FFPLA) offers a way forward by advocating a shift towards a more flexible, pragmatic and inclusive approach for land rights recording. Inherently, the process and outcome of implementing FFPLA will have significant socio-political ramifications but these have not received much attention in the literature; additionally, few papers have considered this in the context of decentralization, an endorsed strategy for implementing FFPLA. This paper contributes to this gap by critically analyzing three land formalization initiatives in India which have employed flexible recording approaches and where decentralization is used to scale implementation. The cases show how quickly decentralization can kickstart implementation at scale via collaborations with local governing bodies and partnerships with non-state actors. An institutionalist approach highlights ensuing political contests between new and traditional land actors that inhibit political authority, and the challenges of coordinating a network of public and private actors without clear formal collaborative governance structures to ensure democratic outcomes. In doing so, we contribute to governance knowledge around FFPLA implementation so that it is 'fit-for-people' and better able to support policies and processes to secure land rights at scale.

Keywords: land administration; decentralization; India; fit-for-purpose; institutions; governance; politics

Citation: Ho, S.; Choudhury, P.R.; Haran, N.; Leshinsky, R. Decentralization as a Strategy to Scale Fit-For-Purpose Land Administration: An Indian Perspective on Institutional Challenges. *Land* **2021**, *10*, 199. https://doi.org/10.3390/land10020199

Academic Editor: Stig Enemark

Received: 5 January 2021
Accepted: 12 February 2021
Published: 16 February 2021

1. Introduction

Many countries around the world grapple with the seemingly intractable problem of formalizing land and property rights at scale, which is believed to disadvantage some of the most marginalized groups including Indigenous Peoples, women, and informal settlement communities. Acknowledging that a key barrier in titling lies in adherence to traditional surveying methodologies which are resource-intensive (time, money and capabilities), the publication of the concept of 'fit-for-purpose land administration' (FFPLA) advocate a shift towards more flexible, pragmatic and inclusive approaches for land rights recording that has legal backing, as well corollary shifts in organizational structures governing land administration [1]. More recently, the publication of a guide draws attention to a set of twelve key principles as starting points for implementing and scaling FFPLA that aligns with good governance values [2]. A body of knowledge is starting to accrue around strategic, structural and technical/functional aspects pertaining to implementing FFPLA (e.g., [3–10]). However, the process and outcome of implementing FFPLA is likely to have significant socio-political ramifications, and these have not received much attention in the literature [11].

This paper contributes to this gap by critically analyzing three land formalization initiatives in India: all employ flexible recording methods supported by legislation and receive significant state or federal government support and resources (see Table 1). In design, these cases echo the institutional principles proposed in the FFPLA guide; importantly, these are also cases where decentralization is used as an implementation strategy—also endorsed by the FFPLA guide—which includes partnerships with non-state actors [2]. The intention of this paper is to broadly explore the socio-political aspects of implementing decentralized land administration rather than specific analysis against each FFPLA institutional principle or the impact of new property rights (due to lack of empirical data as two of the three initiatives have only recently commenced).

Table 1. Overview of Indian land formalization initiatives.

Tenure Type Being Formalized	Year	Name of Initiative
Forest tenure (federal reform)	2006–no fixed end line	The Scheduled Tribes and Other Traditional Forest Dwellers (Recognition of Forest Rights) Act 2006 (FRA)
Urban tenure (state reform)	2018–2023	*Odisha Land Rights to Slum Dwellers Act* 2017 and Jaga Mission (OLRSD)
Rural tenure (federal reform)	2020–2024	Survey of Villages Abadi and Mapping with Improvised Technology in Village Areas) (SVAMITVA)

Decentralization generally refers to both a form of governance and organizational structure, as well as an institutional process where power, functions, and/or resources are reallocated from central to local authorities. It can be fiscal, political, and/or administrative and is commonly believed to lead to greater accountability, stability and coordination. There are three broad types [12]:

- *Deconcentration,* where central responsibilities are simply shifted to local branches (i.e., administrative decentralization).
- *Delegation,* where central responsibilities are transferred to local governments who then act on behalf of the central authority.
- *Devolution,* where authority is given to local governments to act, although they can still be held accountable to a central authority (i.e., political decentralization).

These forms of decentralization represent a continuum of central-local relations in terms of shifts in power, authority and resources, and hence, democratic outcomes intended by decentralization (see Figure 1). Thus, it is a popular policy lever for improving the legitimacy of government and enabling democracy [13], especially in developing countries where the emphasis has tended to be on devolution [14–16]. With the promise of delivering political, governance and efficiency values [17], decentralization has also been linked to good governance outcomes and essentially also characterized as principles of good land governance [18]. In the context of land administration, decentralization is increasingly argued as a type of fit-for-purpose strategy for effectively addressing poor land governance at multiple levels [19–22]. Additionally, with a global push towards formalizing diverse social tenures, there has been an increasing turn towards a collaborative (or partnerships) approach in decentralized land governance, such as utilizing existing local institutions that are already well integrated and recognized by communities, or introducing new structures constituted of local representatives and tasked with dealing with local needs [23].

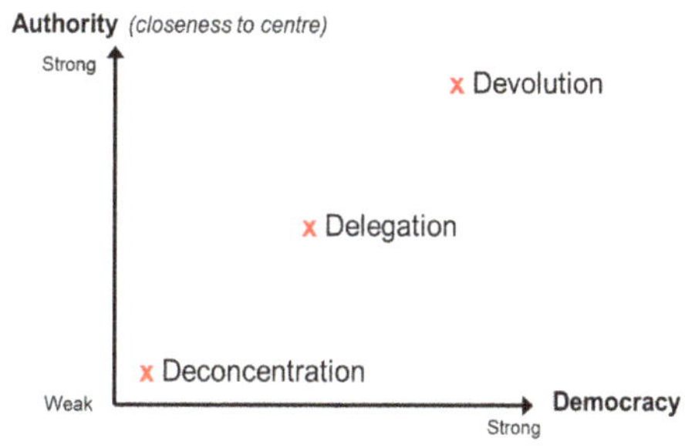

Figure 1. Continuum of decentralized relations and trade-offs between authority and democracy.

However, neoliberal reforms that seek to institutionalize decentralization have, in reality, often imposed new forms of control that tipped the balance of bargaining power back towards the center, e.g., by withholding financial resources [24]. In this way, decentralization has been criticized as a political instrument that does not benefit local communities and, in developing countries, realizing the benefits of decentralization remains predicated on a complex combination of institutional factors including local capacity and autonomy, multi-level accountability mechanisms, and commensurate fiscal decentralization [25]. We operationalize this in the analysis by looking at the consequences of decentralization on new authority structures, local capacity and autonomy related to new land administration processes, and how this is coordinated and governed across multiple levels.

Theoretically, our analysis draws on institutional theory as institutions are strongly implicated in attempts to introduce innovations in land administration [19,26,27]. Additionally, property rights research has shown that similar institutions may not always produce consistent effects due to contextual conditions related to how organizational power is distributed throughout society [28,29]. Institutional theory argues that actors (individuals or organizations) are always embedded in social structures that influence and condition behavior. These social structures, i.e., institutions, are stable regulatory, normative or cognitive elements that provide the 'rules of the game'—those formal or informal prescriptions providing organizing principles for collective behavior [30,31]. In this way, behavior can be 'locked in' or institutionalized as a consequence of longstanding repetition and internalization of rules [32]. Actors are compelled to comply with institutional rules to gain legitimacy and hence, access to resources; therefore, legitimacy exists when an organization is perceived to meet needs thereby gaining cultural, cognitive and normative support and where its goals and performance become unchallenged and unquestioned [33]. This perspective directs analysis to institutional dynamics between (new) local land administration actors and macro-level political and administrative contexts, as well as new and existing land administration structures.

Methodologically, the critique is interpretive, based on secondary data. A broad concurrent search and analysis was conducted across diverse data sources to develop a rich description of the multiple realities of those implementing or being impacted by the schemes. The search criteria were defined by the paper's aim (i.e., new authority structures, local capacity and autonomy) and data sources included publicly available policy documents, reports and media articles[1] regarding the structure and implementation of the schemes, as well as grey and academic literature[2] which provided empirical context and findings. An iterative, critically reflexive analysis (relating theory, history and context of Indian land administration and published experiences) supported identification of common themes across the literature, whereby an institutional lens guided meta-analysis of the consequences of decentralization. Research rigor was maintained through: diverse

1 Using mainstream Indian news media and articles published in English,

2 These are limited peer review studies for *Odisha Land Rights to Slum Dwellers Act 2017* (OLRSD) and the Survey of Villages and Mapping with Improvised Technology in Village Areas (SVAMITVA) due to the schemes being recently implemented,

documentary evidence, which strengthened the representation of 'truth' with validity predicated on triangulating across sources to identify repetition; interpretive awareness, enhanced by critical reflexivity; and a collaborative, multi-researcher approach consisting of a range of disciplinary backgrounds as well as some co-authors having professional experiences with the schemes [34–36].

Although the findings are oriented to the Indian experience, the outcomes of the paper have global resonance in three ways:

- Practically, it is highly likely that decentralization and its focus on improved local service provision will be attractive as a key modality for operationalizing and scaling FFPLA in many countries.
- Ideologically, decentralization reflects the type of exogenous (neoliberal) institutional logics—including good governance—that tend to characterize public and land administration reforms in the Global South, where we see a growing role for non-state actors that prioritizes small government and market-oriented policies for delivering public services [19,21,37–40].
- Theoretically, if successfully used, decentralization can heighten the legitimacy of central government and local actors, as well as the formal land tenure document, something particularly important in land administration reforms given the widely acknowledged issue of corruption in land transactions and limited perceived security or utility of formal land tenure documents [41].

The paper is organized accordingly. First, the institutional context of land administration in India is briefly overviewed, touching also on the long history of decentralization of land administration in the country. Secondly, the three cases are presented describing local land issues and the nature of decentralization, with each case concluding with its own institutional analysis of decentralization efforts. Comparative learnings are then drawn out in the discussion before the paper concludes with a recognition of research limitations and recommendations for further research.

2. The Institutional Context of Land Administration in India

2.1. Class, Caste and Land Relations

In India, societal class and caste relations have historically wielded significant power and influence over land relations, while economic growth aspirations, especially of late, have added competition and contestations around land. In the medieval period, kings paid priests (brahmins) and military leaders (kshatriyas) to deliver state services around land, while the colonial regime, based on these existing land ownership patterns, consolidated and formalized caste-based land ownership distribution [42]. At independence, India's large landowners were typically drawn from the upper castes [43] and the lower castes or dalits were largely landless laborers and servants for, or tenants to, the upper castes [44]. More than 60% of dalits do not own land, and their share of the village population is strongly correlated with local land inequality [42].

Similarly, in the land reform principle, 'Land to the Tiller,' law makers (then and now) continue to see the tiller as male. This attitude causes women's independent identity to be subsumed under the identity of the (male-headed) household. Despite, their extensive involvement with agriculture work, 65% of all Indian agricultural workers are women while only around 14% of all landholders are women [45]. Today, entrenched gender and class systems continue to affect Indian land administration especially at local levels, perpetuating exclusionary behaviors (especially during land formalization, locally known as survey and settlement processes) despite constitutionally enshrined equity and a plethora of legal and institutional reforms to enhance the land rights of vulnerable groups [46,47].

2.2. Decentralization: Old Game, New Name

Decentralization has a long history in Indian land administration. In pre-colonial times, as per the ancient Hindu system, the king was entitled to the share of the produce as the protector of subjects and not as owner of the land. Land belonged to the cultivators and not to the king, and was settled collectively with the communities and administered at village level, with the village headman responsible for collecting land revenue and paying it to the head of the Pargana (a group of villages) [48]. A revenue official (the gopa) was entrusted with the duty to assess and collect land revenue and maintain various registers regarding village boundaries and land use. The Mughals, who, primarily transformed the customary and unwritten accounts of Hindu land administration to systematic management of revenue records through survey, measurement and settlement of revenue, introduced zamindars as intermediaries—non-hereditary, transferable state officials of land who, after the decline of the Mughals, became hereditary and locally indispensable land administrators [48].

To maintain political equilibrium, the British retained existing local land administration norms. The intermediary structure was evident in the form of three systems of land revenue collection introduced: zamindari (permanent settlement, with lands owned by zamindars), ryotwari (peasant cultivation), and mahalwari (combination of zamindari and ryotwari, with land organized into mahals (one or more villages) with ownership vested with peasants) [49]. However, to maximize revenue collection, colonial administrators introduced western-style survey and settlement processes, revenue administration and dispute resolution systems.

2.3. The 'Rules' of the Decentralization Game

Post-independence, land fell under three lists that divide powers between the center and the states (as per the seventh schedule under Article 246 of the Constitution). The union list governs taxes and duties on and related to succession of non-agriculture land and over which Parliament can legislate. The state list governs agricultural land including rights in or over land, land tenures including the relations between landlord and tenant and the collection of rents; transfer and alienation of agricultural land; land revenue, including the assessment and collection of revenue, the maintenance of land records, survey for revenue purposes and records of rights, and alienation of revenues, which are governed by state legislatures. As a state subject, land is administered via a hierarchical system down to the village level, largely reproducing colonial architectures, although tribal areas are governed under local customary laws and special protections. The concurrent list governs land subjects where both Parliament and state legislatures have jurisdiction, including transfer of property other than agricultural land, as well as registration of deeds and documents around land transfers.

Table 2 broadly overviews the structure of Indian land administration. At the state level, responsibility for land is shared by at least four departments and some independent bodies. These have acquired and command legitimacy across state and market institutions. Forest land is managed by the Forest Department; public and private agricultural land as well as common lands are under the Revenue Department. Land use planning sits with the Town and Country Planning Department under the Urban Development Department, while rural planning sits with Gram Panchayats (village local governing bodies), who are granted powers via the Indian constitution to perform functions as mandated by state legislatures. Land for industrial use is often acquired by public industrial development corporations. Religious lands are governed separately under religious boards and endowment departments. Urban and rural local bodies (such as municipalities and Panchayats respectively), as decentralized structures (i.e., the third tier of governance in India), manage urban and rural lands, mostly properties and in some states, also rural public lands (e.g., Karnataka and Rajasthan). Urban and rural land are administered by one department, i.e., the Land Revenue Department in most states, with one registry.

Table 2. Land administration structure in India.

<table>
<tr><th>Land Administration/Use</th><th>Land Survey and Record Management</th><th>Sale, Purchase, Transfer</th><th>Land Use Change, Land Rent/Cess</th><th>Property Tax, Land Use Planning</th><th>Overall Control (Trustee)</th></tr>
<tr><td>Urban</td><td rowspan="4">Land Revenue-Survey and Settlement Department (state)</td><td rowspan="4">Land Revenue-Registration Department (state)</td><td rowspan="4">Land Revenue Department (state)</td><td>Municipalities</td><td rowspan="4">Land Revenue Department with local bodies, Industry Department, Endowment Department (state)</td></tr>
<tr><td>Rural</td><td>Panchayats</td></tr>
<tr><td>Industrial</td><td>Industry Department (state)</td></tr>
<tr><td>Religious Use</td><td>Endowment Department (state)</td></tr>
<tr><td>Forest</td><td colspan="5">Forest Department (state and federal)</td></tr>
</table>

In the 73rd amendment of India's Constitution, land administration was proposed to be devolved to Gram Panchayats and hamlet/habitation level institutions in tribal areas (Palli Sabha). However, except for the state of West Bengal,, the devolution of land administration has not happened except for sporadic attempts at allowing Panchayats to administer government or public land [50]. The complete abolition of revenue collection in most states has also significantly weakened land revenue administration [51]. Consequently, a decline in revenue used to cover costs, as well as the emergence of an institutionally complex landscape in terms of government departments with limited coordination, characterizes India's land administration system today [52].

2.4. Structure and Performance of Indian Land Administration

The goal of Indian land administration has long been revenue collection, and this gained heightened legitimacy during colonial times. As such, the Land Revenue Department is a key institutional actor, entrusted with management of land and other allied matters in states. Broadly its role and responsibilities range between policy formulation, policy implementation, and judicial matters, which contradicts the legal doctrine of Separation of Powers [53], an essential principle of democracy that prevents the concentration of power and provides for checks and balances. While not explicitly recognized in the Indian Constitution, the violation of this doctrine is conspicuous around land administration.

The department is concerned with laws related to the state list (Section 2.3). Maintenance of land records, survey for revenue purposes and updating of the Record of Rights (RoR) constitutes one of the major activities of the department. 'Land records' is a generic expression in India and could refer to different kinds of textual documentation related to land, i.e., RoRs (primary land records and not property records), sale deeds (registered transactions for tax/stamp duty collection), rent receipts (annual tax collection documents), and cadastral maps [54]. The frontline revenue apparatus within a district is under the authority of the District Collector and each district is divided into subdivisions headed by Sub-Collectors. The next lower revenue administrative unit is the Tehsil, which functions under the Tehsildar. It is the Tehsildar office where most revenue matters are handled at the lowest level; however, the Revenue Inspector remains the cutting-edge official interacting with farmers, being responsible for a revenue circle, which usually consists of few villages.

The RoR is prepared under the respective State *Survey and Settlement Act and Rules*. After final publication of the RoR, the information is continually updated at the Tehsil level by the process of mutation as per the State's *Mutation Manual*. The updated land records at the Tehsil level become the base data for preparation of RoRs in subsequent settlement/consolidation operations. All transactions of immovable properties under the *Registration Act 1908* takes place in registration offices and are subsequently maintained by these offices. In terms of data maintenance of RoRs, textual data is maintained by the revenue department (which also collects rent) while spatial data (i.e., cadastral maps) is maintained by the survey and land records department; deeds are maintained by the registration department.

These departments tend to work in silos and the lack of coordination (settlement and consolidation vs. registration) entrenches a disconnect between textual and spatial data, resulting in data across departments not being updated properly [54]. Additionally, frontline revenue administration staff is thin and Revenue Officials are overloaded with land and non-land administrative tasks [52]. Following the launch of the Digital India Land Records Modernization Program (http://dilrmp.gov.in/), there has been substantive progress in linking disparate land information databases; nonetheless, challenges remain due to jurisdictional resistance (e.g., the registration department claiming that the state government cannot impose any conditions on the registration procedure) as well practical challenges related to daily operations.

3. FFPLA in India: Three Narratives

The three cases of land formalization initiatives (set out in Table 1) are described here. Although not explicitly framed as FFPLA, all cases have adopted a flexible, technical approach for mapping, echoing FFPLA guidelines: satellite imagery and mobile applications (FRA); unmanned aerial vehicles (UAVs) or 'drones' for rapid and high accuracy orthophoto production as the basis of cadastral mapping (OLRSD and SVAMITVA); participatory boundary demarcation and mapping (all cases); digital data integration and updating (all cases). In the following sections, each initiative is described, presenting an overview of the local land issue, the implementation process and a per-case institutional analysis. It should be noted that, being recently initiated schemes, information on the second and third cases are comparatively limited.

3.1. Scheduled Tribes and Other Traditional Forest Dwellers (Recognition of Forest Rights) Act 2006

3.1.1. Local Land Issue and Community Needs

The rights of India's Indigenous Peoples (Scheduled Tribes) and Other Traditional Forest Dwelling communities over forested areas have long been denied. Colonial and post-independence appropriation of these areas (key sources of culture and livelihood), in conjunction with poor legislative frameworks, resulted in the degradation of forests and the erosion of the rights of Scheduled Tribes and Other Traditional Forest Dwelling communities, leading to marginalization and severe poverty with forest dwelling communities being amongst the most chronically poor in India [55–57].

3.1.2. FRA and Its Decentralized Implementation

The *Scheduled Tribes and Other Traditional Forest Dwellers (Recognition of Forest Rights) Act 2006* (FRA) was enacted as a result of democratic processes driven by demand for recognition of forest rights by forest dwellers. FRA represents a political, demand-based response to reform forest governance, attempting to shift away from a centralized, top-down, locally-insensitive model of land and forest administration that denied the rights of tribal and forest dwellers [56]. It sought to reinstate and recognize the traditional rights of forest dwelling communities over land and resources and establish community-based governance over an estimated 40 million hectares of forests, thereby empowering over 150 million forest dwellers in over 170,000 villages [58]. The main types of formal tenure introduced were: Individual Forest Rights (IFR; the paper focuses only on IFR for comparability across cases, i.e., instances of formalizing individual tenure) and Community Rights to use and access to forest land and resources, and a new category of rights, Community Forest Resource rights, to use, manage and govern forests within the traditional boundaries of villages [59].

An overview of the decentralized land administration structure for FRA is provided in Table 3. FRA mandated a multi-level governance system for adjudicating and verifying claims, as well as dispute resolution, with the Ministry of Tribal Affairs appointed as the responsible department [60]. The Act is particularly significant for seeking to democratize forest governance and empower communities by devolving administrative power to the Gram Sabha (the local self-governance unit and village assembly) who oversees claim-making, vetting and rights recognition [56,59]; the Act also prescribes conditions for inclusion such as gender quotas in the Gram Sabha. This framework for localized administration was intended to contribute to building trust and empowering forest-dependent communities by creating conditions for greater downward accountability, equity and social justice to reduce corruption and ensure inclusion [61].

Table 3. Decentralized land administration structure under the *Recognition of Forest Rights) Act 2006* (FRA).

Governance Level	State Actors	Non-State Actors
Federal	Ministry of Tribal Affairs (nodal agency)	Advocacy Networks
Sub-state	District Level Committee Sub-divisional Committee (representatives from Land Revenue, Forest and tribal departments_	NGOs
Village	Gram Sabha	NGOs

Once accepted by the Gram Sabha, tenure claims are advanced to Sub-Divisional Level Committees who prepare the forest title (RoR). This is then forwarded to the District Level Committee for final approval. Additionally, each state is expected to initiate a state-level monitoring committee for oversight of the process.

3.1.3. Institutional Analysis of Decentralized Land Administration in FRA

The FRA framework reveals the complexity that federalism has imposed on forest land administration, and the significant roles that intra-state politics and non-state actors (especially the activists' coalition) play in pro-poor land institutions [62]. The decentralized environment introduced a new dominant institutional actor in the Ministry of Tribal Affairs and empowered the agency of forest dwelling communities. It simultaneously reduced the role of state forest departments and forest officers within the national Ministry of Environment, Forest and Climate Change, who up till the introduction of FRA, were key actors of a centralized forest governance system.

However, new actors have struggled to establish their legitimacy, evident in a failure to attract resources, e.g., lack of training of officials under Ministry of Tribal Affairs or hiring of additional staff to manage a complex process [63]. There has also been poor oversight and accountability of the various multi-level committees intended to govern and approve forest rights claims, leading to a clumsy and dysfunctional claims process lasting several years [64]. This has led to perceptions of FRA as opaque, with allegations of serious lack of awareness about its provisions among rights holders and duty bearers alike: the District Level Committee and Sub-Divisional Level Committees, who play a critical role in rights recognition, have executive powers that lie only with the chairperson [65]. With a lack of clarity about the weight of other non-governmental members, the interests of forest communities have become subservient to those of bureaucrats and officials [66].

The prioritization of Individual Forest Rights recognition and granting of "land titles" has also been interpreted as a populist measure conveying political benefits. Kumar et al. [58] found that Individual Forest Rights created rent-seeking opportunities for field functionaries and received less resistance from the land-owning forest departments as these lands tend to already be under cultivation. In addition to contests by forest bureaucracy, institutional tensions are also experienced with conservationists, who see democratic decentralization as a threat to the forest conservation paradigm [55,56,58,67].

Consequently, the actual land recognition is far less compared to what has been stated in the Act, with only 13% of this potential area realized as of March 2020 [68]. The recording of land rights has become a contraction of legal access to existing tenure over forest resources and FRA has been reduced to a "beneficiary scheme with patta-giving exercise" (patta is a local term for land title), with many states using it for political benefit with issues around subsequent possession, clear recording and access to entitlements [69].

However, in spite of this contested institutional environment, FRA has had some success, especially in states where pressure has been exerted by civil society organizations and grassroots advocacy groups, and supported by progressive bureaucrats, particularly tribal department officials and district collectors (e.g., in Odisha and Gujarat) or higher-level public servants (e.g., the Governor's office in Maharashtra) [58].

3.2. *Odisha Slum Dweller's Act 2017 and Jaga Mission 2018–2023*

3.2.1. Local land Issue and Community Needs

In Odisha, 'slums' refer to a compact settlement of at least 20 households with kutcha houses, lack of access to drinking water, sanitation and closed drainage (any household will be considered as a "slum-like" household if it satisfies all the four deprivations, i.e., (i) kutcha house (temporary housing), (ii) source of drinking water not available within the premises, (iii) no latrine within premises and (iv) not having closed drainage) [70]. Although Odisha currently has the lowest percentage of slum population in India (comprising almost 4% of total population), most of these are concentrated in larger urban areas like Bhubaneshwar and Cuttack, and 23% of Odisha's urban population live in slums [71].

3.2.2. OLRSD and Its Decentralized Implementation

Land tenure security has been a policy focus in Odisha to address slum challenges since 1980s. In 2011, the government developed the Slum Rehabilitation and Development Policy, "Housing for All", and the Odisha Property Rights to Slum Dwellers and Prevention of New Slums Bill [72,73], aimed at building a slum-free Odisha by 2020 and reducing urban poverty [74]. These policies culminated in the *Odisha Land Rights to Slum Dwellers Act 2017* (OLRSD) and the launch of the "Jaga Mission" in 2018, which aimed to physically upgrade slums into 'livable habitats', but also implements the Act, aiming to grant in-situ land rights (albeit with size limitations) that are mortgageable and inheritable (but not transferrable) to around 0.2 M households living in around 2000 slums across the 109 municipalities and Notified Area Councils in Odisha (http://www.jagamission.org/).

The OLRSD Act is a state-led initiative that draws from its urban and housing policy. It follows a string of influential central schemes such as the Prime Minister's Awas Yojana launched in 2015, which mandated land formalization to enable households' access to housing subsidies. The Act is also a shift from earlier state policies that focused more on people living in identified or recognized slum areas, and less on slum-like households living outside those areas. The Act provides the legal framework to grant Land Rights Certificates to urban areas in the whole of the state of Odisha covering all the 116 Urban Local Bodies (i.e., municipal corporations). A key motivation for slum titling was easier transition to livable habitats: secured individual title and clearly delineated boundaries can ease access to better sanitation, credit, healthcare, education and housing services [75].

An overview of the decentralized land administration structure for OLRSD is provided in Table 4. OLRSD introduced a new consortium of actors and departed from the typical Indian model of land administration led by Revenue Departments. The initiative was spearheaded by Odisha's Department of Housing and Urban Development in collaboration with Tata Trusts (a major funding partner) and more than 27 local civil society organizations, international NGOs (e.g., the Norman Foster Foundation, Omidyar Network and Cadasta Foundation), and technology companies. Whilst OLRSD mandates the creation of an Urban Area Slum Redevelopment and Rehabilitation Committee, led by the District Collector (who sits under the revenue department) for each urban area to govern the settlement process, it goes no further [76]. Instead, as part of implementation, contracted local NGOs were tasked with supporting the creation of local Slum Dwellers Associations, comprising members from each slum household and led by an elected executive to represent local interests [77]. The 'Jaga Mission' also employed Jaga Fellows, individuals attached to various communities to support implementation, but also to build capacity in urban development issues (https://socialservicesindia.com/wp-content/uploads/2019/06/Jaga.pdf).

Table 4. Decentralized land administration structure for OLRSD.

Governance Level	State Actors	Non-State Actors
State	Department of Housing and Urban Affairs (nodal agency) *Land Revenue Department* *	Tata Trusts International NGOs Technology companies
Sub-state	Urban Local Bodies Urban Area Slum Redevelopment and Rehabilitation committees (led by District Collectors under land revenue departments)	Jaga Fellows, NGOs Technology companies Slum Dwellers Associations
Slum	Urban Local Body Wards	

* These actors do not have a formally defined role but are involved in the process.

3.2.3. Institutional Analysis of Decentralized Land Administration in OLRSD

Like many other parts of India, state Revenue Departments have an institutionalized role as the authoritative agency for land administration. In this case, the dominant actor was bypassed, although the state government as an entity remains the implementing agency for OLSRD [76]. Arguably, this case provides a demonstration of significant political will, resources and commitment—the project is estimated to cost USD120M and scheduled to be completed in four years, reflecting efforts by new institutional actors to underscore their appropriateness. Key actors have also successfully publicized the rapid outputs of the scheme (around 50,000 households had received LRCs by 2019) which not only established status, it also led to significant global attention and framed OLSRD as an efficient and effective land administration reform for tackling urban poverty (the scheme was awarded a bronze medal in the 2019 World Habitat Awards (https://world-habitat.org/world-habitat-awards/winners-and-finalists/odisha-liveable-habitat-mission/)). All these maneuvers arguably enhanced and reinforced institutional legitimacy of the scheme and its actors.

Decentralization was operationalized through local NGOs and Slum Dwellers Associations. NGOs were contracted to deliver the types of formalization-related duties that the local Tehsil might typically be involved with NGOs selected based on technical experience in similar work, and training was provided by technical organizations and Tehsildars [77]. Decentralization here is not legally mandated but is instead an implementation policy decision. Slum Dwellers Associations therefore do not have statutory authority, nor is community-based participation prescribed in the rules, but was initiated to ensure inclusion. This has been problematic: a lack of training and clarity about responsibilities means that Slum Dwellers Associations are not performing as expected, leading to NGOs taking over [77]. Poor local performance is also substantiated in a recent study which found that a totalitarian approach led to households not being present at the time of adjudication, leading to errors or exclusion from grant of Land Rights Certificates [78].

Leadership of Urban Local Bodies, involvement of NGOs, formation of Slum Dwellers Associations and the use of Jaga Fellows as agents potentially creates a conducive environment for community participation. Activities like drone flying, community interaction with aerial imagery, house marking, etc., also potentially contribute to trust-building in the project and long-term slum governance. However, in reality, this potential has been diluted by multiple factors during implementation, e.g., the limited capacity of Urban Local Bodies and their upward accountability and lack of autonomy from state government; the lack of capacity and decision-making role of the Slum Dwellers Associations; limited and temporary involvement of NGOs; poor communication between the state and communities; delays or exclusions of certain vulnerable households in terms of Land Rights Certificate allotment; and contraction of households' legally allotted area.

3.3. SVAMITVA Scheme 2020–2024

3.3.1. Local Land Issue and Community Needs

Rural land administration has been an ongoing challenge for India. A historic focus on surveying solely for revenue collection has resulted in the exclusion of 'marginal' areas including abadi (inhabited) village land from formal records, which are often grossly outdated due to mandated rural resurveys not being executed for generations [79]. Consequently, rural India suffers from lack of tenure security, land conflict and landlessness, all of which are linked to high levels of rural poverty [80].

3.3.2. SVAMITVA and Its Decentralized Implementation

In response to this chronic problem of rural land administration, the SVAMITVA (Survey of Villages and Mapping with Improvised Technology in Village Areas) scheme was launched in 2020. Echoing De Soto [81], SVAMITVA sought to stimulate a growth agenda through financialization of 'dead capital', facilitation of land markets, and reducing federal fiscal burden by buttressing the property tax base of Panchayats (Constitutional local governance bodies).

An overview of the decentralized land administration structure for SVAMITVA is provided in Table 5. Drawing from pilots in Maharashtra and Haryana, the scheme is a collaborative project involving the federal Ministry of Panchayati Raj, State Panchayati Raj Departments, State Revenue Departments and the Survey of India. The deliberate move to dissociate from the traditional land revenue department as the nodal agency, as well as incorporating engagement with the Survey of India, drone technology and private sector actors, indicated an intention to enhance the efficiency of land administration through a "business *unusual*" approach. The four-year project aims to map rural land parcels across 0.66M villages in India using GNSS and drone technology to enable the issuing of Sampatti Patrak (Rural Property Cards), available as digital or hard copies. The process will be led by state governments, supported by the federal government with implementation devolved to local Panchayats. Data collected will be used to update the RoRs in the land revenue register (an online registry (every state in India, has its own online land registry portal) supported the Digital India Land Records Modernization Program) and property registers (which are maintained by Panchayats).

Table 5. Decentralized land administration structure for SVAMITVA.

Governance Level	State Actors	Non-State Actors
Federal	Ministry of Panchayati Raj (nodal agency) Survey of India *Ministry of Rural Development (Digital India Land Records Modernization Program)* *	Technology and/or survey companies
State	Panchayati Raj Department *Land Revenue Department* *	Technology and/or survey companies
Village	Gram Panchayat	

* These actors do not have a formally defined role but are involved in the process.

The decentralized data collection process is intended to be democratic and transparent, relying on the communities themselves to draw chunna (white chalk) lines and Panchayats for adjudication and oversight. The maps produced are intended to provide important spatial inputs into the preparation of Gram Panchayat development plans, which are constitutionally mandated and prescribes a participatory process. While orthorectified base maps will be jointly owned by the Survey of India, Ministry of Panchayati Raj and the state government, property data will be owned by the State Revenue Department. Other updated GIS data layers will be shared by the Talathi/Patwari level officer once every year incorporating updates that have been done in the preceding 12 months.

The scheme is currently being piloted in eight states (Maharashtra, Karnataka, Haryana, Uttar Pradesh, Uttarakhand, Madhya Pradesh, Punjab and Rajasthan). In October 2020, 0.1 M property holders from more than 760 villages across six states received the first Rural Property Cards via SMS links.

3.3.3. Institutional Analysis of Decentralized Land Administration in SVAMITVA

SVAMITVA seeks to institutionalize a new rural land administration system and new actors. With leadership by the Ministry of Panchayati Raj and devolution of tasks to local Panchayats, the decentralized model draws immediate institutional legitimacy by realizing constitutional, policy and public consensus that local bodies should be involved with land improvement, implementation of land reforms, land consolidation and soil conservation (in line with the 73rd and 74th Amendments, the Panchayats (Extension to Scheduled Areas) Act (PESA), as well as article 243(G) of Schedule 11 of the Constitution), advancing India's stalled devolution agenda. Further, the introduction of new actors occurs at a time when the role of the land revenue department—a law enforcement agency whose continued role in land administration is a persistent colonial legacy practice—is increasingly delegitimized (as seen in decreasing staff and resources) [79]. However, by leveraging their expertise in survey and settlement, the scheme inadvertently preserves a significant role for state revenue departments.

The use of local Panchayats enrolls an actor with a high level of local acceptance, as well as a demonstrated ability to support land reform programs [82]. Explicit formal (implementation) rules (albeit not laws) clarifies roles and expectations around community engagement in boundary definition and adjudication, ensuring practices are, in theory, standardized and inclusive [83]. There is also a clear incentive for Panchayats to engage: in India, the low level of tax collection by rural local governments has been a challenge in reconciling fiscal federalism, accountability and true devolution, with Panchayats depending on funds from the center/state for 95% of their resources [84]. SVAMITVA could generate improved collection of property tax, which in turn would improve the financial autonomy of Gram Panchayats.

SVAMITVA holds promise to build trust locally by involving a more downward accountable, easily accessible and less corrupt Gram Panchayat in its implementation, while use of drone technology is expected to stimulate participation and ensure transparency. Similarly, the involvement of communities in boundary delineation and the issuing of Rural Property Cards digitally through text messages are potential trust building measures in meeting needs. However, the limited devolution of land administration related powers to Panchayats along with their limited capacity, the ongoing significant role of the land revenue department around Rural Property Card processes, and already reported threats of exclusion and elite capture in survey processes, could undermine community trust during implementation.

4. Decentralization as a Scaling Strategy

Using three cases of FFPLA-like initiatives in India, this paper sought to broadly explore the socio-political aspects of implementing decentralized land administration by critically analyzing the consequences of decentralization on new authority structures, local capacity and autonomy related to new land administration processes, and how this is coordinated and governed across multiple levels.

The three cases demonstrated a range of decentralization strategies at work. Of these, FRA comes closest to an aspired model of devolution; OLRSD appears to behave more like deconcentration, and SVAMITVA more like delegation. An overview of theory vs. practice of decentralization is provided in Table 6, which reveals differing outcomes in authority and democratic dividends at local levels. Despite varying successes, the cases underscore the difficulty of redistributing power; as well, all cases—even FRA—struggle with integrating new processes and new forms of property rights with extant systems, arguably introducing greater institutional fragmentation into Indian land administration.

Table 6. Overview of initiatives and decentralization model/attributes (as defined in [12]).

Initiative and Decentralization Type	Theoretical Attributes	± Authority Outcomes	± Democracy Outcomes
FRA (Most like devolution)	Authority is given to local governments to act, although they can still be held accountable to a central authority	+ Gram Sabha plays major role in making and validating tenure claims + Executive governance at district and sub-district level + Recognition by multi-department committee involving land and forest departments + District support for procedural validation	+ Gram Sabha emerging as powerful actor; state support limited in practice – Gram Sabha needs local NGO support – NGO involved voluntarily; sporadic state invitation/partnership – RoR integration of Individual Forest Right title envisaged but process not clear – New record updating processes unclear
OLRSD (Most like deconcen-tration)	Central responsibilities shifted to local branches	+ Urban Local Bodies lead the process and plays a major role + Department of Housing and Urban Development and state support with policy, partnerships and coordination of involvement of land revenue expert ± Non- state actors (e.g., drone firms and NGOs) are contracted for survey purposes ± District Collector coordinates and issues the Land Rights Certificates (although Land revenue department not usually involved)	+ Slum Dweller Associations assist implementation and helps in local dispute resolution – Role of Slum Dweller Associations not endorsed in act and rule. – RoR integration of Land Rights Certificates – New record updating processes unclear
SVAMITVA (Most like delegation)	Centre facilitates states to act involving local governments	+ Panchayati Raj department plays major role, decides partnerships as well as coordinates with revenue departments; also issues Rural Property Cards + Survey by Survey of India or private firms ± Land revenue department enacts policy, supervises survey and arbitrates – NGO roles not clear	+ Gram Panchayat facilitates implementation and addresses local disputes – Gram Panchayat does not have full devolution of legal power – RoR integration of Rural Property Cards envisaged but process not clear – New record updating processes unclear

4.1. *The need for Institutional Legitimacy for Political Authority*

New legislative frameworks introduced new nodal land administration agencies and established roles, rules and processes around new land administration structures.

A decision to bypass traditional land administration actors potentially signals a bid to rebuild trust in the Indian land administration system; similarly, a policy of decentralization signals a bid to improve democracy. The various institutional analyses (Section 3) however clearly show capacity, autonomy and governance challenges—in some instances, still unresolved—that impact on local benefit and the ability of new processes to meet needs. This underscores the reality that formal 'rules' are only part of the puzzle in producing authority; an ongoing challenge remains in shoring up cultural, cognitive and normative support, i.e., institutional legitimacy, to ensure that new actors at both macro and micro levels have true political authority and technical capacity to act. In FRA, we see the contested role and authority of the Ministry of Tribal Affairs leading to flow-on effects and similar challenges for Gram Sabhas.

A main barrier to building institutional legitimacy for new state actors seems to be the continued explicit or implicit involvement of the Land Revenue Department since it remains the legal custodian of land records and implementer of key land revenue laws. This affects new land administration processes; further, as new tenures are being created based on legacy records and cadastral boundaries, the location and extent of new property rights and documents remains relative to the RoR and cadastral maps. Arguably, this has made the Land Revenue Department more critical in the coordination of new and existing land administration systems, not only for the relative authoritativeness of these documents, also for their intended future use, e.g., for mortgaging and entitlements, as well as subsequent updating. This raises the need to further investigate legal and policy paradoxes in both property rights and administrative structures as it inevitably impacts implementation of FFPLA.

4.2. Local Capacity and Autonomy

Local governing bodies like Gram Sabhas and Panchayats now play a lead role in local and participatory forms of decision making and administration of formalized rights in FRA and SVAMITVA respectively; in OLRSD, Slum Dwellers Associations have been created and Urban Land Bodies are more involved in land administration. Redistribution of power is formalized via a mix of legislative and policy actions, with different resources provided to support new responsibilities.

Although there is limited evidence in SVAMITVA, the experiences in FRA and OLRSD indicate that true empowerment is vital for local autonomy and the ability to meet local needs. In FRA, support from NGOs is identified as a factor in strengthening the performance of Gram Sabhas; in OLRSD, early evidence points to poor training and lack of clarity over the role of Slum Dwellers Associations, which has resulted in poor performance and community outcomes. Although conceived as community-driven development, OLSRD and SVAMITVA appear to be government-initiated and led, with little evidence that thought has been given as to how local institutions might dovetail with new centrally designed institutions (e.g., legal frameworks). If unaddressed, this can create a cycle of not meeting local needs and eroding trust in new decentralized systems.

4.3. The need for Collaborative Governance for Institutional Coordination

Decentralization introduces different coordination and governance challenges that may not have been required previously: indeed, the FRA case shows how poor coordination and multi-level governance has negatively impacted on the new system. The cases demonstrate a reduced role for the state, increased collaboration, and a greater number of non-state actors at multiple levels. This directs a need for a governance method that supports institutional integration and realization of democratic outcomes like integrity and accountability.

Presently, the cases illustrate how existing structures and politics related to decentralized land administration may not be facilitating goals around trust and democracy. In FRA, clear formal rules within the Act for inclusion on multiple fronts are not being complied with or enforced due to limited governance and legitimization of new land administration processes. In fact, a lack of effective enforcement has contributed to the intended model of devolution in FRA becoming dysfunctional.

This seems more apparent in OLRSD and SVAMITVA, where governance frameworks for public entities across multiple institutional levels are set out in policy, and presumably, the use of contracts provides some oversight for NGOs and private sector organizations. However, in FRA, the initial collaborative model did not consider involvement by local NGOs, and despite their positive work, has not been able to evolve to accommodate these actors formally. Given this new landscape of networked actors, the literature suggests a possible solution might be to turn towards a model of collaborative governance, which is a governance arrangement where state and non-state actors interact *formally* in joint and deliberate decision-making, either with regards to policy making or implementation, or management of public programs [85].

4.4. "Fit-for-Purpose" or "Fit-for-People"?

In OLSRD and SVAMITVA, the cases prove that, while they are technically "fit-for-purpose" with their rapid outputs, this may not be the case institutionally. Although decentralization seems to work to kickstart implementation at a significant scale with supporting legislative and policy frameworks, we raise the question as to whether it is necessarily "fit-for-people"? The analysis indicates true authority and autonomy is not being experienced and appropriate governance structures are not in place to coordinate a network of state and non-state actors at multiple levels. Decentralization may be the policy goal but the 'rules of the game' are still very much centrally dictated, which creates a greater need for resourcing and empowerment at the local level, and effective consensus-based governance across levels.

We propose that the notion of "fit-for-people" references both the process and product of implementation: the process should be attendant to local politics and 'rules of the game', i.e., values, norms and dynamics to ensure social equity, build trust and facilitate coordination and collective decision making. The product of implementation should be authority and resources to empower new actors to act to meet local needs in a way that has cultural, cognitive and normative support, which will help build public confidence that the new FFPLA initiatives will address local land issues and needs with integrity.

5. Conclusions

The ideal of FFPLA is becoming normative but there is still a paucity of knowledge on socio-political, i.e., institutional, innovation to scale FFPLA. This paper has contributed to this gap by critically analyzing the use of decentralization as a scaling strategy in India, an approach also endorsed by the FFPLA guide. Decentralization has introduced new actors at both macro and micro levels but the persistent institutional legitimacy of existing land actors, structures and processes, has led to contests over political authority, limited realization of local capacity and democratic dividends, and a lack of effective governance and institutional integration between new and existing land administration systems.

India's experience shows decentralization can work to effectively kickstart the implementation of FFPLA at scale; however, beyond initial policies, there is tremendous work to be done to ensure that implementation is "fit-for-people" to deliver a trustworthy system that redistributes power and delivers critical social justice. Uncovering local realities and the significance of informal 'rules of the game' is therefore an important area for future research; additionally, we acknowledge that the research presented here is limited in evidence base and by context and further research should also aim to extend and enrich this investigation by undertaking empirical research in varying country contexts.

Author Contributions: Conceptualization, S.H., P.R.C., N.H. and R.L.; methodology, S.H.; formal analysis, S.H., P.R.C., N.H. and R.L.; investigation, S.H., P.R.C., N.H. and R.L.; writing—original draft preparation, S.H., P.R.C. and N.H.; writing—review and editing, S.H., P.R.C., N.H. and R.L. All authors have read and agreed to the published version of the manuscript.

Funding: Funding of the publication costs for this article has kindly been provided by the School of Land Administration Studies, University of Twente, in combination with Kadaster International, The Netherlands.

Institutional Review Board Statement: Not applicable.

Informed Consent Statement: Not applicable.

Data Availability Statement: Not applicable.

Acknowledgments: The authors thank the anonymous reviewers for their helpful comments which have resulted in the final version of the paper.

Conflicts of Interest: The authors declare no conflict of interest.

References

1. Enemark, S.; Bell, K.C.; Lemmen, C.; McLaren, R. Fit-for-Purpose Land Administration. In *Joint FIG/World Bank Publication*; FIG Publication: Copenhagen, Denmark, 2014.
2. Enemark, S.; McLaren, R.; Lemmen, C. Fit-for-Purpose Land Administration: Guiding Principles for Country Implementation. 2016. Available online: https://unhabitat.org/fit-for-purpose-land-administration-guiding-principles-for-country-implementation (accessed on 30 December 2020).
3. Dijkstra, P.; Mitchell, D.; Bennett, R.; Prasad, G.P.; Chhatkuli, R.R. Ignition for Fit for Purpose Land Administration in Nepal, Presented at the Surveying the World of Tomorrow—From Digitalisation to Augmented Reality. In Proceedings of the FIG Working Week 2017, Helsinki, Finland, 29 May–2 June 2017. Available online: https://researchbank.swinburne.edu.au/items/2072ed2a-f65a-49a1-aa0b-4f36502aa761/1/ (accessed on 30 December 2020).
4. Musinguzi, M.; Enemark, S. A fit-for-purpose approach to land administration in Africa—Supporting the 2030 global agenda. *Int. J. Technosci. Dev.* **2019**, *4*, 69–89.
5. Bennett, R.M.; Alemie, B.K. Fit-for-purpose land administration: Lessons from urban and rural Ethiopia. *Surv. Rev.* **2016**, *48*, 11–20. [CrossRef]
6. Bennett, R.M.; Yimer, F.A.; Lemmen, C. Toward Fit-For-Purpose Land Consolidation. In *Advances in Responsible Land Administration*; CRC Press: Boca Raton, FL, USA, 2015; pp. 163–182.
7. Koeva, M.; Bennett, R.; Gerke, M.; Crommelinck, S.; Stöcker, C.; Crompvoets, J.; Ho, S.; Schwering, A.; Chipofya, M.; Schultz, C.; et al. Towards innovative geospatial tools for fit-for-purpose land rights mapping. *Int. Arch. Photogramm. Remote Sens. Spat. Inf. Sci. ISPRS Arch.* **2017**, *42*, 37–43. [CrossRef]
8. Rahmatizadeh, S.; Rajabifard, A.; Kalantari, M.; Ho, S. A framework for selecting a fit-for-purpose data collection method in land administration. *Land Use Policy* **2018**, *70*, 162–171. [CrossRef]
9. Panday, U.S.; Joshi, J.R.; Chhatkuli, R.R.; Enemark, S.; Antonio, D.; Deuja, J. Development of Fit-For-Purpose Land Administration Country Strategy: Experience from Nepal. In Proceedings of the FIG Working Week 2019, Hanoi, Vietnam, 22–26 April 2019.
10. Aditya, T.; Maria-Unger, E.; Bennett, R.; Saers, P.; Lukman Syahid, H.; Erwan, D.; Wits, T.; Widjajanti, N.; Budi Santosa, P.; Atunggal, D.; et al. Participatory land administration in Indonesia: Quality and usability assessment. *Land* **2020**, *9*, 79. [CrossRef]
11. Barry, M. Fit-for-purpose land administration—Administration that suits local circumstances or management bumper sticker? *Survey Rev.* **2018**, *50*, 383–385. [CrossRef]
12. Prud'homme, R. The dangers of decentralization. *World Bank Res. Obs.* **1995**, *10*, 201–220. [CrossRef]
13. Stoker, G. Governance as theory: Five propositions. *Int. Soc. Sci. J.* **1998**, *50*, 17–28. [CrossRef]
14. Wunsch, J.S. Institutional analysis and decentralization: Developing an analytical framework for effective third world administrative reform. *Public Adm. Dev.* **1991**, *11*, 431–451. [CrossRef]
15. Kulipossa, F.P. Decentralisation and democracy in developing countries: An overview. *Dev. Pract.* **2004**, *14*, 768–779. [CrossRef]
16. Bergh, S. Democratic decentralisation and local participation: A review of recent research. *Dev. Pract.* **2004**, *14*, 780–790. [CrossRef]
17. Wolman, H. Decentralization: What It is And Why We Should Care. In *Decentralization, Local Governments, and Markets: Towards a Post-Welfare Agend*; Bennett, R.J., Ed.; Clarendon Press: Oxford, UK, 1990; pp. 29–41.
18. FAO. *Good Governance in Land Tenure and Administration*; Food and Agriculture Organization of the United Nations: Rome, Italy, 2007.
19. Hilhorst, T. Decentralization, land tenure reforms and local institutional actors: Building partnerships for equitable and sustainable land governance in Africa. *Land Tenure J.* **2010**, *10*, 35–60.
20. Burns, T.; Grant, C.; Nettle, K.; Brits, A.; Dalrymple, K. Land administration reform: Indicators of success and future challenges. *Agric. Rural Dev. Discuss. Pap.* **2007**, *37*, 1–227.
21. Wily, L.A. *Governance and Land Relations: A Review of Decentralisation of Land Administration and Management in Africa*; IIED: Hertfordshire, UK, 2003. Available online: http://pubs.iied.org/pdfs/9304IIED.pdf. (accessed on 11 June 2020).
22. Wily, L.A. Getting to the Heart of the Matter: Powers over Property. Devolved Land Governance—The Key to Tackling the Land Issue in Kenya? 2008. Available online: https://mokoro.co.uk/land-rights-article/getting-to-the-heart-of-the-matter-powers-over-property-devolved-land-governance-the-key-to-tackling-the-land-issue-in-kenya/ (accessed on 15 December 2020).

23. Ibrahim, A.-S.; Akanbang, B.A.A.; Laube, W. Sustaining decentralized collaborative governance arrangements in Africa: A case study of land management committees in the Upper West Region, Ghana. *GeoJournal* **2020**. [CrossRef]
24. Bevir, M. *Key Concepts in Governance*; SAGE Publications Ltd.: London, UK, 2009.
25. Dick-Sagoe, C. Decentralization for improving the provision of public services in developing countries: A critical review. *Cogent Econ. Financ.* **2020**, *8*, 1804036. [CrossRef]
26. Ho, S.; Rajabifard, A.; Stoter, J.; Kalantari, M. Legal barriers to 3D cadastre implementation: What is the issue? *Land Use Policy* **2013**, *35*, 379–387. [CrossRef]
27. Ho, S.; Zein, T.; Nkerabigwi, P.; Pattyn, V.; Crompvoets, J. Scaling Emerging Geospatial Technologies for Land Administration: Understanding Institutional Innovation Dynamics through a Technological Innovation System Perspective. In Proceedings of the 2019 World Bank Conference on Land and Poverty, Washington, DC, USA, 25–29 March 2019.
28. Khan, M.H. Governance Capabilities and the Property Rights Transition in Developing Countries. In *Monographs and Working Papers*; DFID: London, UK, 2009.
29. Khan, M.H. Power, pacts and political settlements: A reply to Tim Kelsall. *Afr. Aff.* **2018**, *117*, 670–694. [CrossRef]
30. North, D.C. *Institutions, Institutional Change and Economic Performance*; Cambridge University Press: Cambridge, MA, USA, 1990.
31. Scott, W.R. *Institutions and Organizations*, 1st ed.; SAGE Publications: Thousand Oaks, CA, USA, 1995.
32. Hall, P.A.; Taylor, R.C.R. Political science and the three new institutionalisms. *Political Stud.* **1996**, *44*, 936–957. [CrossRef]
33. Deephouse, D.L.; Suchman, M. Legitimacy in Organizational Institutionalism. In *The SAGE Handbook of Organizational Institutionalism*; SAGE Publications Ltd.: London, UK, 2008; pp. 49–77.
34. Sandelowski, M. Real qualitative researchers do not count: The use of numbers in qualitative research. *Res. Nurs. Health* **2001**, *24*, 230–240. [CrossRef]
35. Lincoln, Y.S. Emerging criteria for quality in qualitative and interpretive research. *Qual. Inquity* **1995**, *1*, 275–289. [CrossRef]
36. Morse, J.M.; Barrett, M.; Mayan, M.; Olson, K.; Spiers, J. Verification strategies for establishing reliability and validity in qualitative research. *Int J. Qual. Methods* **2002**, *1*, 13–22. [CrossRef]
37. Zakout, W.; Wehrmann, B.; Törhönen, M.-P. Good Governance in Land Administration Principles and Good Practices. Available online: www.fao.org/3/a-ak375e.pdf (accessed on 11 June 2020).
38. Bruce, J. Decentralization of Land Administration in Sub-Saharan Africa: Recent Experiences and Lessons Learned. In *Agricultural Land Redistribution and Land Administration in Sub-Saharan Africa: Case Studies of Recent Reforms*; The World Bank: Washington, DC, USA, 2014; pp. 55–84.
39. Breslin, S.; Nesadurai, H.E.S. Who governs and how? Non-state actors and transnational governance in Southeast Asia. *J. Contemp. Asia* **2018**, *48*, 187–203. [CrossRef]
40. Kim, A. Decentralization and the Provision of Public Services: Framework and Implementation. In *Working Paper 4503*; World Bank: Washington, DC, USA, 2008.
41. Transparency International. Global Corruption Barometer 2013. Available online: www.transparency.org/gcb2013/report. (accessed on 15 December 2020).
42. Bharti, N.K. *Local Level Land Inequality in India (First Draft)*; School of Economics: Paris, France, 2019.
43. Besley, T.; Leight, J.; Pande, R.; Rao, V. Long-run impacts of land regulation: Evidence from tenancy reform in India. *J. Dev. Econ.* **2016**, *118*, 72–87. [CrossRef]
44. Sharma, K. *The Changing Rural Stratification System: A Comparative Study of Six Villages in Rajasthan, India*; Orient Longman: New Delhi, India, 1974.
45. Government of India. Agriculture Census 2015-16 (Phase-I): Provisional Results', Ministry of Agriculture and Farmers Welfare (Agriculture Census Division). 2018. Available online: www.thehinducentre.com/the-arena/current-issues/article28682480.ece/binary/T1_ac_2015_16.pdf (accessed on 17 December 2020).
46. Kumar, K.; Choudhury, P.R.; Sarangi, S.; Mishra, P.; Behera, S. A Socio-economic and legal study of scheduled tribes land in Orissa. In *Unpubl. Study Comm. World Bank Washington*; The World Bank: Washington, DC, USA, 2005.
47. Tankha, R. *Engendering Rural Livelihoods: Supporting Gender Responsive Implementation of The National Rural Livelihoods Mission*; UN Women Multi-Country Office for India, Bhutan, Maldives & Sri Lanka: New Delhi, India, 2014.
48. Baden-Powell, B.H. *A Short Account of the Land Revenue and Its Administration in British India; With a Sketch of the Land Tenures*; HardPress Publishing: Hawthorne, CA, USA, 2013.
49. Banerjee, A.; Iyer, L. History, institutions, and economic performance: The legacy of colonial land tenure systems in India. *Am. Econ. Rev.* **2005**, *95*, 1190–1213. [CrossRef]
50. Hanstad, T.; Nielsen, R. *West Bengal's Bargadars and Landownership*; Economic and Political Weekly: Mumbai, India, 2004; Volume 39, pp. 853–855.
51. Rothermund, D. The record of rights in British India. *Indian Econ. Soc. Hist. Rev.* **1971**, *8*, 443–461. [CrossRef]
52. Mearns, R.; Sinha, S. *Social Exclusion and land administration in Orissa, India*; Social Science Research Network: Rochester, NY, USA, 1999; Available online: https://papers.ssrn.com/abstract=636207 (accessed on 8 September 2020).
53. Montesquieu, C.L. *The Spirit of Laws*, 1st ed.; Amherst Prometheus: Manhattan, NY, USA, 2002.
54. Mishra, P.; Suhag, R. *Land Records and Titles in India*; PRS Legislative Research: New Delhi, India, 2017.
55. Springate-Baginski, O.; Sarin, M.; Reddy, M.G. Resisting rights: Forest bureaucracy and the tenure transition in India. *Small Scale For.* **2013**, *12*, 107–124. [CrossRef]

56. Kumar, K.; Singh, N.M.; Kerr, J.M. Decentralisation and democratic forest reforms in India: Moving to a rights-based approach. *For. Policy Econ.* **2015**, *51*, 1–8. [CrossRef]
57. Mehta, A.K.; Shah, A. Chronic poverty in India: Incidence, causes and policies. *World Dev.* **2003**, *31*, 491–511. [CrossRef]
58. Kumar, K.; Singh, N.M.; Rao, G. Promise and performance of the forest rights act A ten-year review. *Econ. Political Wkly.* **2017**, *52*, 40–43.
59. Ministry of Tribal Affairs and UNDP. *Forest Rights Act, 2006: Act, Rules and Guidelines*; Ministry of Tribal Affairs, Government of India and United Nations Development Programme: New Delhi, India, 2014; Available online: https://tribal.nic.in/FRA/data/FRARulesBook.pdf (accessed on 12 September 2020).
60. Government of India. *The Scheduled Tribes and Other Traditional Forest Dwellers (Recognition of Forest Rights) Act*. 2006.
61. Deininger, K. *A Strategy for Improving Land Administration in India*; World Bank: Washington, DC, USA, 2008; Volume 33.
62. Bose, I. *How did the Indian Forest Rights Act, 2006, Emerge?* University of Manchester: Manchester, UK, 2010. Available online: https://assets.publishing.service.gov.uk/media/57a08b1440f0b652dd000ab0/dp39.pdf (accessed on 17 December 2020).
63. Government of India. *Report: National Committee on Forest Rights Act. A Joint Committee of Ministry of Environment and Forests and Ministry of Tribal Affairs, Government of India*; Ministry of Environment and Forests and Ministry of Tribal Affairs, Government of India: New Delhi, India, 2010; Available online: https://hrln.org/publication/report-national-committee-on-forest-rights-act-5ec7a9e0bb033 (accessed on 12 October 2020).
64. Sahu, G.; Dash, T.; Dubey, S. Political economy of community forest rights. *Econ. Political Wkly.* **2017**, *52*, 44–47.
65. Bandi, M. Forest rights Act: Is there an underlying pattern in implementation? *Econ. Political Wkly.* **2016**, *51*, 16–17. Available online: www.epw.in/journal/2016/19/commentary/forest-rights-act.html (accessed on 16 December 2020).
66. Kumar, S. Forest Rights Act Enables State Control of Land and Denies Most Adivasis and Forest Dwellers Land Rights. >*Econ. Political Wkly.* **2020**, *55 (online version)*. Available online: www.epw.in/engage/article/forest-rights-act-state-appropriation-land-rights (accessed on 16 December 2020).
67. Das, B.K. Denial of rights continues: How legislation for "democratic decentralisation" of forest governance was subverted in the implementation process of the forest rights act in India. *Eur. J. Dev. Res.* **2019**, *31*, 957–983. [CrossRef]
68. Ministry of Tribal Affairs. *Monthly Update on Status of Implementation of the Scheduled Tribes and Other Traditional Forest Dweller (Recognition of Forest Rights) Act, 2006*; Government of India: New Delhi, India, 2020. Available online: www.fra.org.in/document/FRA%20Status%20Report%20up%20to%2031st%20March%202020-MoTA.pdf (accessed on 23 December 2020).
69. Das, B.K. Are You Talking about "Jungle Patta"?: FRA Implementation in West Bengal. *Econ. Political Wkly* **2019**, *54 (online version)*. Available online: www.epw.in/journal/2019/43/special-articles/are-you-talking-about-%E2%80%98jungle-patta%E2%80%99.html (accessed on 16 December 2020).
70. Government of India. *Report of the Committee on Slum Statistics/Census, Ministry of Housing and Urban Poverty Alleviation*; Ministry of Housing and Urban Poverty Alleviation and National Buildings Organisation: New Delhi, India, 2010. Available online: http://mohua.gov.in/upload/uploadfiles/files/9Slum_Report_NBO(2).pdf (accessed on 11 November 2020).
71. Government of Odisha. *Odisha Economic Survey 2018-19*; Planning and Convergence Department, Government of Odisha: Bhubaneshwar, India, 2020.
72. Government of India. *Draft: Guidelines for Preparation of a Slum Free City Plan of Action Under the Rajiv Awas Yojana*; Ministry of Housing & Urban Poverty Alleviation, Government of India: Cuttack, India, 2011.
73. Government of Odisha Housing and Urban Development Department. *The Odisha Property Rights to Slum Dwellers and Prevention of New Slums Bill*; Government of Odisha: Cuttack, India, 2012.
74. Housing and Urban Development Department, Government of Odisha. *The Orissa Gazette (Extraordinary)*; Odisha Government Press: Cuttack, India, 2011.
75. Leshinsky, R.; Ho, S.; Choudhury, P.R. Social Innovation as a Disruptor of Tenure: Recognising Land Rights of Slum Dwellers in Odisha, India. In *Disruptive Technology, Legal Innovation, and the Future of Real Estate*; Levine-Schnur, R., Ed.; Springer Books: Berlin/Heidelberg, Germany, 2020.
76. Government of Odisha. *The Odisha Land Rights to Slum Dwellers Act, 2017*; Government of Odisha: Cuttack, India, 2017.
77. The Bridgespan Group. Odisha: Land Rights to Slum Dwellers. Observations on detailed work flow process maps'. Report to Government of Odisha. Available online: http://www.jagamission.org/pdf/Compendium%20Land%20Rights.pdf (accessed on 24 January 2020).
78. Killemsetty, N.; Patel, A. Giving Voice to the Slum Dwellers—Understanding the Implication of the Land Rights Act in Odisha state, India. In Proceedings of the Annual World Bank Conference on Land and Poverty, Washington, DC, USA, 16–20 March 2020.
79. The World Bank. *India Land Governance Assessment: National Synthesis Report*; The World Bank Group: Washington, DC, USA, 2015.
80. Hanstad, T.; Haque, T.; Nielsen, R. Improving land access for India's rural poor. *Econ. Political Wkly.* **2008**, *43*, 49–56.
81. de Soto, H. *The Mystery of Capital: Why Capitalism Triumphs in the West and Fails Everywhere Else*; Basic Books: New York, NY, USA, 2000.
82. Abraham, D.T. *Lease Land Farming by Women Collectives: An Enquiry into Earnings of Kudumbashree Groups*; Centre For Women's Development Studies: New Delhi, India, 2019.

83. Ministry of Panchayati Raj. *Framework for Implementation of SVAMITVA Scheme: Survey of Villages and Mapping with Improvised Technology in Village Areas. Guidelines (2020)*; Government of India: New Delhi, India, 2020.
84. Department of Economic Affairs, Ministry of Finance. *Economic Survey 2017-18: Volumes 1 and 2*; Economic Division, Department of Economic Affairs, Ministry of Finance, Government of India: New Delhi, India, 2018.
85. Ansell, C.; Gash, A. Collaborative governance in theory and practice. *J. Public Adm. Res. Theory* **2007**, *18*, 543–571. [CrossRef]

land

MDPI

Article

The Benefits of Fit-for-Purpose Land Administration for Urban Community Resilience in a Time of Climate Change and COVID-19 Pandemic

David Mitchell [1,*], Bernhard Barth [2], Serene Ho [1], M. Siraj Sait [3] and Darryn McEvoy [4]

[1] Geospatial Science, School of Science, RMIT University, Melbourne 3000, Australia; serene.ho2@rmit.edu.au
[2] UN-Habitat, Regional Office for Asia and the Pacific, Fukuoka 810-0001, Japan; bernhard.barth@un.org
[3] Noon Centre for Equality and Diversity, University of East London, London E16 2RD, UK; s.sait@uel.ac.uk
[4] School of Engineering, RMIT University, Melbourne 3000, Australia; darryn.mcevoy@rmit.edu.au
* Correspondence: david.mitchell@rmit.edu.au

Abstract: The major global pressures of rapid urbanization and urban growth are being compounded by climate impacts, resulting in increased vulnerability for urban dwellers, with these vulnerabilities exacerbated during the COVID-19 pandemic. Much of this is concentrated in urban and peri-urban areas where urban development spreads into hazard-prone areas. Often, this development is dominated by poor-quality homes in informal settlements or slums with poor tenure security. Lessons from a resilience-building project in the Pacific shows that a fit-for-purpose (FFP) approach to land administration can provide solutions by increasing the number of households with security of tenure, and consequently, improving resilience outcomes as informal settlements grow. This paper specifically discusses the influence of FFP land administration on reducing vulnerabilities to external shocks, such as climate change and COVID-19. It proposes ways to be better manage urban growth through the responsible governance of land tenure rights and more effective land-use planning to improve resilience to multiple shocks and stresses, hence, delivering improved access to safe land and shelter. Land administration systems can contribute to enhanced resilience to the shocks of climate extremes and pandemics by improving tenure security and enhancing land-use planning controls. It is argued that climate change adaptation and disaster risk reduction need to be better mainstreamed into two major elements of land governance: (i) securing and safeguarding of land rights, and (ii) planning and control of land use.

Keywords: fit-for-purpose land administration; rapid urbanization; climate change; pandemic; urban resilience

Citation: Mitchell, D.; Barth, B.; Ho, S.; Sait, M.S.; McEvoy, D. The Benefits of Fit-for-Purpose Land Administration for Urban Community Resilience in a Time of Climate Change and COVID-19 Pandemic. *Land* **2021**, *10*, 563. https://doi.org/10.3390/land10060563

Academic Editors: Stein T. Holden, Fabrizio Battisti and Stig Enemark

Received: 5 April 2021
Accepted: 18 May 2021
Published: 27 May 2021

1. Introduction

Urbanization is a key driver of both climate vulnerability and tenure insecurity in urban settlements in the Global South [1]. The global pressures of rapid urbanization and urban growth are being compounded by the impact of a changing climate, resulting in the increased vulnerability of urban dwellers. Land-use planning has not controlled the growth of informal settlements, which often occur in highly vulnerable areas [2], with inadequate housing, insecure tenure, and no formal access to water and sanitation, making them particularly sensitive to climate impacts [2].

The recent literature has described the interrelationships between land tenure and climate vulnerability as well as potential land governance responses [3]. Insecure land tenure exacerbates vulnerability to climate-related hazards [3,4], as these households are disconnected from formal governance processes, lack knowledge to inform resilience decisions, and have restricted access to finance for actions to strengthen their adaptive capacity [5]. Those without formal land records are also more likely to be excluded from post-disaster reconstruction programs and grants [6,7]. Climate impacts lead to human mobility (migration, displacement, and resettlement) with impacts to tenure security [3].

The importance of 'responsible' land governance to secure access to land for shelter and livelihoods and reduce disaster vulnerability is recognized in the Committee on World Food Security (CFS)-endorsed 'Voluntary Guidelines on the Responsible Governance of Tenure of Land, Fisheries and Forests in the Context of National Food Security' (VGGTs). The VGGTs call on states to ensure that legitimate tenure rights are respected and protected for effective land-use planning [8]. To improve both resilience and tenure security at scale, climate adaptation and disaster risk management need to be better mainstreamed into two major elements of responsible land governance: (i) securing and safeguarding of all formal and informal land tenure rights, and (ii) effective planning and control of land use to prevent housing in the most hazard-prone locations [3,6]. Taking a responsible land governance approach informed by the VGGTs provides a pathway to address tenure security in the process of strengthening urban resilience, especially for marginalized communities.

Urban housing needs to be more climate resilient to meet the Global 2030 Agenda; providing secure land tenure for all should be a fundamental aim. With the ongoing impact of the global COVID-19 pandemic, what is less well understood are the interlinkages between land tenure, climate vulnerability, and pandemics. As the 'UN Special Rapporteur on the right to adequate housing' noted, "By ensuring access to secure housing with adequate sanitation, States will not only protect the lives of those who are homeless or living in informal settlements but will help protect the entire world's population by flattening the curve of CV19" [9]. While the recent literature has focused on understanding the interrelationships between land tenure and pandemics [10,11], there has been little to no research yet on the interlinkages between land tenure and multiple shocks, such as the impact of climate extremes and pandemics.

The emphasis on land tenure in the context of climate and pandemic shocks draws attention to the potential application of fit-for-purpose land administration (FFP LAS) in the context of both climate change and pandemics, where there is significant literature on how the FFP LAS approach can improve tenure security at scale [12,13] by (i) using methods that fit the context, (ii) being flexible in terms of accuracy requirements of land tenure information and adopting a continuum of land rights approach, and (iii) starting with low-cost approaches and allowing for incremental improvement. The approach involves developing the core FFP LAS components of the spatial, legal, and institutional frameworks. This paper focuses on the overall aim of tenure security at scale through responsible land governance, and on using high-resolution imagery rather than field surveys to develop the spatial framework component of FFP LAS [12].

A significant post-earthquake project in Nepal illustrated the benefits of applying the FFP LAS approach to document existing people—land relationships to support improved tenure security and disaster and climate resilience efforts [4,7]. The application of FFP LAS and other tools, such as participatory enumeration and the Social Tenure Domain Model (both discussed later), provided the tools to record and recognize all existing land tenure rights to inform both land administration and disaster reconstruction activities. However, to the authors' knowledge, there is no existing literature on the benefits of FFP LAS to support responses to both climate and pandemic impacts. This paper addresses this gap by using the context of an ongoing research project in Honiara, Solomon Islands, to consider how FFP LAS can support climate resilience building to address vulnerability to both climate and pandemic impacts.

The aim of this paper is to describe how improving tenure security at scale, using the FFP LAS approach, can enhance climate resilience to both climate and pandemic impacts. This contributes to the literature through introducing new principles and methods for applying FFP LAS to urban resilience initiatives.

Several of the authors have a long-standing engagement through a UN-Habitat-led climate adaptation planning process, culminating in the Honiara Urban Resilience and Climate Action Plan (HURCAP) [14]. The research methods include an extensive literature review of the interrelationships between land tenure, climate change and pandemics. The empirical data are drawn from two sources. Firstly, data are derived from participatory action research that commenced in 2012 under UN-Habitat 'Cities and Climate Change Initiative' and is continuing through the Climate Resilient Honiara (CRH) project support by the UNFCCC Adaptation Fund, which commenced in 2018 [15]. The lessons from this engagement in Honiara include experiences prior to and during the COVID-19 pandemic.

Secondly, data are also derived from a recent rapid assessment of COVID-19 carried out by UN-Habitat, which provides insights into the socioeconomic impacts on residents in five communities across Honiara [16]. This rapid assessment involved surveys with 100 households across four wards in Honiara during the period 17–31 August 2020. Respondents were randomly selected from informal settlements participating in the ongoing CRH project. The study examined six key areas: (1) livelihood security and household income, (ii) food security, (iii) access to health care, (iv) knowledge, attitudes and practices related to COVID-19, (v) climate related hazards and COVID-19, and (vi) tenure security [16].

2. Pandemics and Land Tenure Rights

Global changes in land-use patterns and an accelerating rate of land conversion are recognized as contributing factors in increased pandemic risk and the emergence of new infectious diseases, due to diminishing natural habitats and ecological disruption [17,18]. Health emergencies, such as the 2003 severe acute respiratory syndrome (SARS), 2009 H1N1 influenza (or "swine flu"), the 2014 West African Ebola crisis, and more recently, the 2016 Zika outbreak in the Americas [19], demonstrate not merely similarities between climate change and health risks, but their intertwined trajectories (see [20–22]). COVID-19 underscores the need to rethink land-use change and the preparedness of health systems by closing critical knowledge gaps and fostering society-wide engagement in pandemic risk reduction in the new 'pandemic era' [23]. A case study of slum dwellers in Liberia shows how multiple vulnerabilities arise out of their location—exposure to climate risks, the impacts of past epidemics, such as Ebola, and socioeconomic profiles given prolonged civil war and displacement [11]. Instead of a linear approach that creates a dichotomy between health responses and other vital societal adaptation, the common socioeconomic and ecological determinants that disproportionately affect certain categories by gender, age, ethnicity and landlessness, and other vulnerabilities need to be addressed concurrently.

Recent research attest to how unsustainable urbanization—with its knock-on effects on human health and wellbeing—is a critical part of reducing the risk of future pandemics [24–27]. Human development choices directly impact the natural world, and biodiversity and natural habitats are conditioned on sustainable urbanization and the responsible human consumption of animal products to prevent new, communicable zoonotic diseases. While pandemics are often viewed as health crises, the socioeconomic implications are under-researched, often overlooking urban resilience and sustainability. Enforced border shutdowns, travel restrictions and quarantines have highlighted the impact of the virus on the global economy, affecting well-being, employment opportunities, and food security [28]. Thus, responsible land administration approaches in the future will have to respond to concerns across various levels.

The nature of pandemic risks, in relation to land tenure rights, intersects the loss of livelihoods, threats of eviction, and changes to human mobility patterns. Pandemics exacerbate unaffordability of adequate housing and existing characteristics in many informal settlements—poor sanitation, high density housing, insecure tenure rights, and mobile populations—and therefore, contribute to the complexity in addressing health risks [29,30]. Loss of livelihoods, affecting the ability to pay rent and mortgages, eviction, and human mobility, while elements of the stressors of urbanization and climate-related impacts became more critical issues during the COVID-19 pandemic. Impacts on access to drinking water and sanitation for vulnerable households, as well as overcrowding in affected households, are also risk factors [31].

The restrictions imposed during the pandemic interrupted urban services, such as access to drinking water and sanitation, and the capacity of local actors to intervene to redress these risks due to social distancing, lockdowns, and diminishing resources [32,33].

An effective way of tackling complex land, housing, environmental and health challenges is to understand the linkages and pooling of resources through local perspectives and community-led action.

As with earlier epidemics, such as Ebola and HIV, COVID-19 exposed how the urban poor (including migrants and slum communities) struggle with household size, housing costs, livelihoods and tenure security, while encountering unfavorable structural, economic, and political conditions [19]. Strategies, such as hand washing, self-isolation, or self-quarantine after exposure to the virus, physical distancing, and 'work from home' advice are based on elitist assumptions. Those living in urban informal settlements need additional support, as well as local knowledge to create equitable systems for the most vulnerable populations [34]. The 'new normal' for future cities and communities requires a shift toward a 'new social contract' that fosters rights-based, well-planned, inclusive and climate-resilient cities [31]. Building back better in the face of climate change and the COVID-19 pandemic will need be centered upon future innovative land-use decisions and sustainable urban development practices.

The drivers and pandemic vulnerabilities discussed above are summarized in Table 1 below.

Table 1. Major issues and drivers impacting pandemic vulnerability.

Issue/Driver	Pandemic Vulnerabilities
Urbanization leading to unplanned urban growth in slums and informal settlements. Increased density of development	Increased vulnerability to disease. Limited ventilation between buildings exacerbates disease.
Slums and informal settlement with limited access to formal water supply and sanitation	Poor water supply and sanitation impacts health and spread of disease.
Informal settlement occupants may not be included in DRR, CCA, resilience or disaster reconstruction programs	May result in some households not receiving government pandemic grants and support.
Poor quality house construction and materials	Densely populated settlements impact pandemic responses and spread of disease.
Human mobility as an adaptive response	Pandemic restrictions limit adaptive human mobility opportunities.
Livelihood options and food security	Existing livelihood options restricted affecting household income and household food security.

3. Climate Change, COVID-19 and Land Administration in Pacific Islands Countries

3.1. The Socio-Economic Impact of Urbanization, Climate Extremes and the COVID-19 Pandemic on Urban Systems in the Pacific Island Countries

Pacific Island Countries (PICs) are highly exposed to natural hazards, such as earthquakes, cyclones, and tsunamis. They are affected by the El Niño-Southern Oscillation (ENSO), which creates climate variability and impacts each PIC differently. These 'natural' regional climate cycles influence extreme events, such as drought, flooding, and tropical cyclones [35]. During April 2020, the PICs faced the widespread destruction caused by Tropic Cyclone Harold in the Solomon Islands, Vanuatu, Fiji, and Tonga, which compounded the impacts of COVID-19 and presented additional challenges through damages to crops, homes, buildings, and roads [16].

Rapid border closures and swiftly imposed lockdowns curtailed the impact of COVID-19 in the region and at the end of 2020, only four of the 14 PICs had confirmed cases. Nonetheless, the experience of Papua New Guinea this year underscores the fragility of this stability and how quickly PICs can be overwhelmed due to existing and entrenched development challenges. Indeed, the use of widespread lockdowns and states of emergency, while undoubtedly saving lives, have resulted in extensive externalities in terms of macroeconomic pressures for governments as well as myriad socioeconomic impacts for PIC communities, particularly urban ones.

Most people in PICs live in cities and towns. Urban growth rates continue to exceed annual population growth rates in nearly all Pacific economies [16]. This growth is especially significant in Melanesia, where the urban growth rate of some countries, such as the Solomon Islands, exceeds 5% and its capital, Honiara, has a population density of nearly 6000 persons per square kilometer [36]. The total population of Pacific countries is forecast to grow by more than 60% by 2050 (to almost 20 million), propelled by growth in just four countries: Vanuatu, Kiribati, the Solomon Islands, and Papua New Guinea [37]. This presents challenges for every development sector.

Pacific urbanization is characterized by social, cultural, linguistic, political, economic, and environmental diversity across Melanesia, Micronesia, and Polynesia. Climate impacts, informal settlement, urban poverty, and infrastructure deficiencies undermine urban resilience. Within informal settlements, the challenges include evictions and discrimination, with the more vulnerable and marginalized being most affected. The degree of resilience of households is a major factor in how they transition from an informal settlement to a formally recorded settlement [35].

Currently, around one in four PIC residents live below national poverty lines; for seven out of 11 PICs, this is more likely to be the experience for urban populations rather than rural ones [38]. Livelihoods are heavily reliant on informal and subsistence economies: limited labor statistics for PICs show that informal employment rates can range between around 30% (e.g., Cook Islands) to 80% (e.g., Tonga). Hence, the informal economy is significant for PICs and although accounting of the impact is difficult, a case study in Fiji shows that the informal economy contributes around 15% toward GDP [39,40]. During COVID-19, restricted movements and lockdowns meant that these economies ground to a halt and many were forced to turn to subsistence rather than cash economies. Although subsistence economies are already dominant in PICs such as Solomon Islands and Papua New Guinea, especially in rural areas (ILO Office for Pacific Island Countries), for urban communities, this undoubtedly placed more pressure on already scarce (and often contested) land resources. However, during COVID-19, many PIC governments mandated a return to rural islands under state-of-emergency powers, which reversed longstanding patterns of rural–urban and inter-island migration.

Demographic trends in the region mean that half of the population are below 24 years of age, especially in the Melanesian countries of Solomon Islands, Vanuatu, and Papua New Guinea. The youth bulge tends to be concentrated in urban areas, and youth in these areas are more likely to face increased poverty, chronic health issues, poor educational outcomes, unemployment, and higher risks of political and socioeconomic grievance [41]. Additionally, geography is a significant factor impacting development, especially in countries such as the Federated States of Micronesia, which comprises 607 islands spread across a large area, imposing unique challenges for coordination and distribution of services. A recent study demonstrates that 50% of Pacific people live within one km of the coast and 90% live within five km of the coast (excluding Papua New Guinea, as its exponentially greater population numbers skews the analysis) [42]. In recent years, a growing trend in rural-to-urban migration and inter-island migration has led to an increase in informal settlements as a dominant urban form throughout PICs [39]. These coastal urban areas are exposed to an array of climate-related impacts, including sea level rise, storm surges and cyclones.

These environmental risks are exacerbated by limited adaptive capacities. Some of these are consequences of physical isolation and the physical sizes of the countries themselves but PICs also experience chronic structural issues that produce weak governance systems, infrastructure deficits, and a lack of diversity in internal and external trade markets, all of which contribute to poor socioeconomic development outcomes for urban residents [43]. Consequently, the 2020 World Risk Index identifies three PICs as being in the top five most-at-risk countries in the world: Vanuatu (first), Tonga (second), and Solomon Islands (fifth).

The introduction of a pandemic into such contexts severely tests the limits and capacities of urban systems to cope. Furthermore, institutional ambiguity over the governance of urban informal settlements often creates conflict over who is responsible for land governance and basic service provision, resulting in many settlements living without access to basic sanitation and water facilities. Urban informal settlements are, therefore, particularly vulnerable to the health and socioeconomic impacts of COVID-19: crowded housing, the existing prevalence of vector-borne diseases, coupled with limited access to clean water and sanitation, and higher rates of chronic health conditions. All serve to create ideal conditions for a virus to spread.

3.2. Land Tenure Issues and Vulnerability to Multiple Shocks and Stresses in Honiara

3.2.1. Tenure Security Issues in Honiara

Honiara, the capital city of the Solomon Islands, is situated on a narrow coastal strip, spreading out into a series of rugged hills and valleys to the south (see Figure 1). It faces a wide range of severe climate hazards exacerbated by shortcomings in urban development and infrastructure. Flood events, extreme heat, drought, sea level rise, and landslides are projected to increase in intensity and frequency due to a changing climate [44]. Informal settlements are often located in exposed coastal areas, river floodplains, or steep hilly terrain. As such, they are highly exposed to climate-related hazards, adding to the existing underlying vulnerabilities [5,45]. The high levels of exposure and sensitivity to extreme events and limited adaptive capacity mean that the current-day vulnerabilities of informal settlements are considered priorities over future events, with local actions needed to increase urban resilience to a combination of urbanization and climate-related drivers [5].

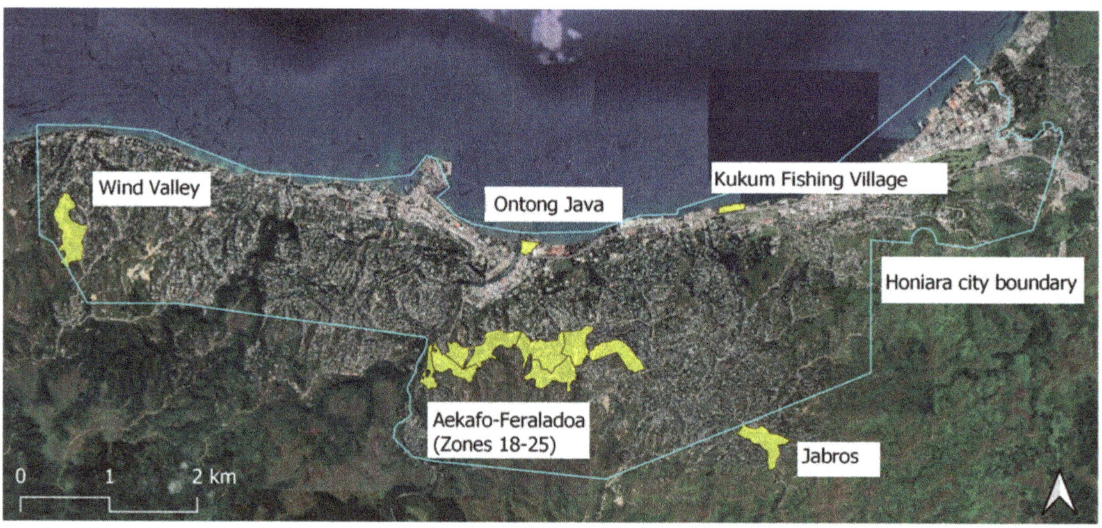

Figure 1. Honiara map of informal settlements (author-generated using data from Google Earth and Ministry of Lands, Housing and Survey).

Within the municipal boundary, formal land tenure is granted through 50-year period Fixed-Term Estate leases (FTEs) of government-held land, or 3-year Temporary Occupation Licenses (TOLs). TOLs were originally introduced in the 1970s to manage unplanned urban migration; however, most have lapsed. Government efforts to convert TOL areas to FTE through surveying, subdivision and valuation are underway but the conversion is not keeping up with the informal urban growth [5]. In addition, disputes remain, with some customary landowners challenging the location of the boundaries of the city established in 1978 [46].

Informal settlements cover about 15% of the city's total land area, housing approximately 35–40% of the population with population densities typically much higher than the rest of the city [16]. The Ontong Java settlement, one of the CRH project's vulnerability 'hotspots', has a density of 21,800 residents per square kilometer. Households in informal settlements also have inadequate access to water and sanitation infrastructures and limited access to other essential services [46].

More recently, the UNFCCC Adaptation Fund supported the CRH project to scope and implement actions identified by the HURCAP. CRH has four work packages at the community level: (i) producing comprehensive community profiles, (ii) developing climate action plans in hotspot settlements, (iii) designing engineering solutions based on community needs, and (iv) awareness raising and capacity development [5]. The community profiling provides important socioeconomic information to inform climate and land responses. The process was based on the vulnerability framework outlined in the HURCAP, and the household survey included 54 main questions about the three factors that influence vulnerability (exposure, sensitivity, and adaptive capacity) as well as six themes: household, livelihoods, housing conditions, utilities, land tenure, and climate change and disaster experiences.

An important element of both the HURCAP and CRH projects has been regular community workshops (see Figure 2) to better understand community needs, validate the vulnerability assessment and action plans, and understand the land tenure implications.

Figure 2. Community workshop in Ontong Java settlement (Photo credit: McEvoy).

In Honiara, the Ministry of Land Housing and Surveys' (MLHS) decision to take no action against informal settlers and those with lapsed TOL was an important first step in improving tenure security. This was evident in the recent COVID-19 survey with almost all (99 percent) respondents perceiving that they were safe from eviction, and 95 percent responding they had not been threatened with eviction due to COVID-19 in the 30 days prior to the survey [45].

The HURCAP and CRH projects build on ongoing initiatives, such as the Participatory Slum Upgrading Programme (PSUP), supported by UN-Habitat since 2008 and the Solomon Islands Government's efforts to 'formalize' housing through a more intensive process of subdivision and converting crown land to leased land, though with mixed success to date [5,40]. The process of land-use planning, PSUP and design of new subdivisions has benefited from the existing LiDAR data and high-resolution imagery from drones, creating a strong spatial framework. The drivers and climate vulnerabilities discussed above are summarized in Table 2 below, with the interlinkages with land issues further summarized in Table 3.

Table 2. Major issues and drivers impacting climate vulnerability.

Issue/Driver	Climate Vulnerabilities
Urbanization leading to unplanned urban growth in slums and informal settlements. Increased density of development	More people exposed to climate impacts.
Slums and informal settlement with limited access to formal water supply and sanitation	Poor water supply and sanitation increases climate vulnerability.
Informal settlement occupants may not be included in DRR, CCA, resilience or disaster reconstruction programs.	May result in disaster affected households not receiving disaster recovery and reconstruction assistance.
Poor quality house construction and materials	Increased sensitivity to disasters and climate impacts.
Human mobility as an adaptive response	Displacement due to disasters or climate impacts.
Livelihood options and food security	Disasters lead to loss of some existing livelihood options and impacting household income and food security.
	DRR and CCA may lead to decision to resettle households impacting livelihood options.

Table 3. Key interlinkages between land tenure and climate and pandemic stressors.

Issue/Driver	Climate and Pandemic Vulnerabilities	Land Issues in the Context of Climate Change and Pandemics
Urbanization leading to unplanned urban growth in slums and informal settlements. Increased density of development	Climate—more people exposed to climate impacts.	Ineffective urban planning and lack of enforcement of building codes allows unplanned high-density development. Increased potential for land disputes
	Pandemic—increased vulnerability to disease. Limited ventilation between buildings exacerbates disease.	
Slums and informal settlement with limited access to formal water supply and sanitation	Climate—poor water supply and sanitation increases climate vulnerability.	Impact on tenure security, potential increased threat of eviction and land disputes.
	Pandemic—poor water supply and sanitation impacts health and spread of disease.	
Informal settlement occupants may not be included in DRR, CCA, resilience or disaster reconstruction programs	Climate—may result in disaster affected households not receiving disaster recovery and reconstruction assistance.	Tenure insecurity can lead to exclusion from government resilience and other programs.
	Pandemic—may result in some households not receiving government pandemic grants and support.	
Poor quality house construction and materials	Climate—increased sensitivity to disasters and climate impacts.	Ineffective urban planning and enforcement of building codes.
	Pandemic—densely populated settlements impact pandemic responses and spread of disease.	

Table 3. *Cont.*

Issue/Driver	Climate and Pandemic Vulnerabilities	Land Issues in the Context of Climate Change and Pandemics
Human mobility as an adaptive response	Climate—displacement due to disasters or climate impacts.	Human mobility can lead to tenure insecurity and landlessness
	Pandemic—pandemic restrictions limit adaptive human mobility opportunities.	
Livelihood options and food security	Climate—disasters lead to loss of some existing livelihood options, impacting household income and food security.	Reduced ability to pay rent or mortgage payments may lead to eviction, migration, or landlessness. Potential for increased tenure insecurity.
	Climate—DRR and CCA may lead to decisions to resettle households impacting livelihood options.	
	Pandemic—existing livelihood options restricted affecting household income and household food security.	

3.2.2. COVID-19 Issues in Honiara

The 2020 UN-Habitat survey found that the average household income dropped due to the pandemic across all the surveyed settlements (see Figure 3). More than half the respondents were concerned about food running out at home, and 65% of those who received financial support used the money to buy food [45].

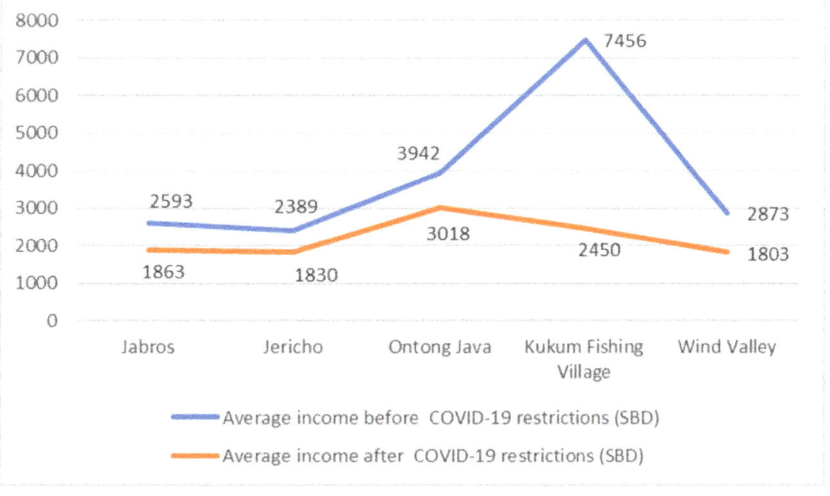

Figure 3. Average income before and after COVID-19 restrictions by settlement (Data from [16]).

One response has been to seek alternative livelihood options, such as farming and fishing for food (see Figure 4). The emphasis on self-sufficiency is reinforced by the findings from recent workshops in Honiara on gender and food security, which highlighted access to urban gardens as a 'survival' mechanism. However, this expansion of land use for fishing, agriculture or home gardens further increases the risk of land disputes (many bush gardens are outside the municipal boundary and on customary land).

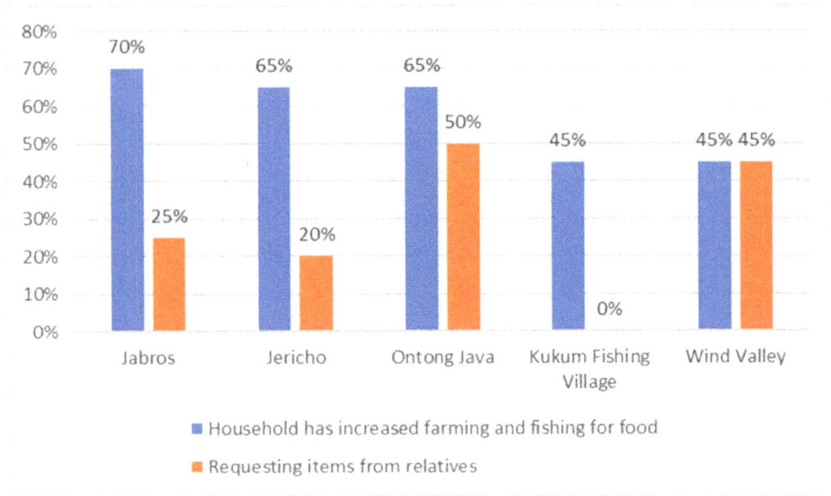

Figure 4. Traditional forms of social protection by settlement (Data from [16]).

The pandemic has highlighted many underlying socioeconomic problems, including impacts to household income and food security, exposing the fragility of informal settlements and slums. Households rely on limited livelihood options for daily subsistence and are frequently unable to afford soap, food, or medical treatment during movement restrictions or lockdowns [45]. The outbreak, and associated restrictions, have the potential to push many more into poverty and increase their climate vulnerability.

3.3. Key Interlinkages between Urban Land Tenure and Climate and Pandemic Stressors

As summarized in Tables 1 and 2 above, the urban drivers and interrelationships between underlying vulnerabilities to pandemics and climate change are complex. These include poverty, hazard-prone dwellings, lack of access to water and sanitation and basic services, overcrowded informal settlements, and limited open space. These vulnerabilities are compounded when a disaster event and a pandemic occur at the same time.

Urbanization and a lack of affordable and safe land means that informal settlements are often located in hotspots of natural hazards and in areas prone to flooding due to poor drainage. Houses built from poor-quality materials create new vulnerabilities to current and future climate impacts and pandemics. Densely populated housing, lack of formal connection to water and sanitation services, and poverty are underlying vulnerabilities for both COVID-19 and climate extremes and can contribute to the spread of infectious diseases. The urban poor and vulnerable groups within informal settlements can be more isolated from social networks and government services, and the informal settlement of land makes settlers more vulnerable to eviction. Larger household sizes create additional pressures during a pandemic lockdown, as houses also become the workplace for more family members, as well as a place for schoolwork and for health care. The result for households has been social, structural, economic, and political impacts with exaggerated vulnerabilities to multiple crises.

While voluntary human mobility can be an important adaptation response, pandemic lockdowns can make this unavailable, and climate impacts can lead to involuntary displacement or resettlement, potentially leading to tenure insecurity or landlessness.

Climate extremes and pandemics can lead to loss of livelihoods and impact on food security. Urbanization and climate drivers became more critical issues during the COVID-19 pandemic. Loss of livelihoods affected the ability of households to pay mortgages or rent, sometimes forcing human mobility decisions that impacted tenure security and the quality of shelter. Reduced cash income during the pandemic reduced the purchasing capacities of the urban poor for basic food items, creating a decline in food security. Alternative livelihood strategies, including expanding agriculture or home gardens, increases the potential for land disputes. Indeed, land disputes and conflict can lead to increased landlessness as well as undermining climate adaptation efforts.

The key drivers and vulnerabilities discussed above and summarized in Tables 1 and 2 are combined in Table 3 below, followed by a discussion of each column.

The first column of Table 1 includes the major drivers impacting vulnerability to climate change and pandemics as well as tenure security as discussed in previous sections. These can be summarized as the following:

1. Urbanization manifests as unplanned urban growth in slums and informal settlements, often on hazard-prone land. An implication is the increased density of development, making more people exposed to natural disasters.
2. Slums and informal settlements often have limited access to formal water supply and sanitation.
3. Informal settlers may not be included in DRR, CCA, resilience or disaster reconstruction programs. This is at odds with the aim of leaving no one behind in resilience efforts.
4. Poor quality house construction and materials makes houses more sensitive to climate impacts.
5. Human mobility is often an important adaptive response. However, disasters can lead to involuntary displacement or resettlement, and pandemics can restrict human mobility.
6. Livelihood options and food security.

The second column summarizes the climate and pandemic vulnerabilities associated with each of these major drivers, as discussed in earlier sections. What is also evident from Table 3 is that there are many land issues common to climate and pandemic vulnerabilities. The remainder of this discussion will focus on addressing these common land issues and how FFP LAS can help address these in the context of climate change and pandemics.

4. FFP LAS for Urban Resilience

Urban resilience is a key objective in managing a combination of urbanization, climate, pandemic, and land issues. It promotes an integrated approach to addressing multiple shocks and stresses, such as climate extremes and pandemics, that impact urban systems both now and into the future [35]. Urban resilience is enhanced by explicitly considering insecure tenure and vulnerability to multiple stressors and, as informal settlements are a dominant form of new housing in cities of the Global South, any process to improve tenure security at scale must include all existing forms of land tenure as reflected in the 'continuum of land rights' concept (see Figure 5). This reinforces the fact that typically, a range of informal and formal tenure systems exist, varying in tenure security [47].

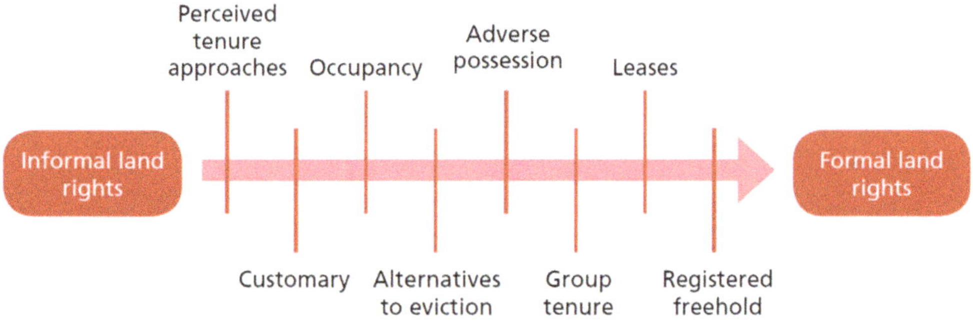

Figure 5. The continuum of land rights (Data from [47]).

In line with the aim of the paper, this section describes how improving tenure security at scale using the FFP LAS approach can enhance climate resilience to both climate and pandemic impacts. The FFP LAS approach facilitates improving tenure security at scale, while recognizing a continuum of land rights, including all the existing people-to-land relationships. The FFP LAS approach to developing a spatial framework typically involves large-scale and high-resolution imagery that supports the adoption of visible (physical) boundaries [12].

Table 4 below builds on the land issues identified in previous tables and recommends appropriate land governance responses, and how FFP LAS can support improved land governance, based on project experience.

Table 4. Land governance and FFPLAS responses to land issues in the context of climate change and pandemics.

Land Issues in the Context of Climate Change and Pandemics	Land Governance Response, How FFP LAS Can Help, and Related Land Tools
Ineffective urban planning and lack of enforcement of building codes allows unplanned high-density development. Increased potential for land disputes	**Effective land-use planning and control**; FFP LAS: participatory enumeration, visible boundaries defined on high resolution imagery; tenure-responsive land-use planning.
Impact on tenure security, potential increased threat of eviction and land disputes.	**Securing and safeguarding land tenure rights, Effective land use planning and control**; FFP LAS: improving tenure security at scale using methods that fit the context—continuum of land rights, participatory enumerations, visible boundaries defined on high resolution imagery, tenure-responsive land-use planning.
Tenure insecurity can lead to exclusion from government resilience and other programs.	**Securing and safeguarding land tenure rights, Effective land use planning and control**; FFP LAS: improving tenure security at scale using methods that fit the context -continuum of land rights, participatory enumeration, visible boundaries defined on high resolution imagery.
Ineffective urban planning and enforcement of building codes.	**Effective land use planning and control**; FFP LAS: continuum of land rights, participatory enumeration, visible boundaries defined on high resolution imagery; tenure-responsive land-use planning.
Human mobility can lead to tenure insecurity and landlessness	**Securing and safeguarding land tenure rights**; FFP LAS: improving tenure security at scale using methods that fit the context—continuum of land rights, participatory enumeration, visible boundaries defined on high resolution imagery.
Reduced ability to pay rent or mortgage payments may lead to eviction, migration, or landlessness. Potential for increased tenure insecurity.	**Securing and safeguarding land tenure rights**; FFP LAS: improving tenure security at scale using methods that fit the context—continuum of land rights, participatory enumeration, visible boundaries defined on high resolution imagery.

The first column lists the major land issues to be addressed related to each driver and vulnerability. These can be consolidated into the following major land issues to be addressed:

i. Ineffective urban planning and enforcement of building codes.

ii. Tenure insecurity, loss of access to land, landlessness, potential increased threat of eviction and land disputes.
iii. A lack of formal land records can lead to exclusion from government resilience and other programs.
iv. Human mobility can lead to tenure insecurity or landlessness due to migration, involuntary resettlement, or displacement.

The following section discussed how these FFP LAS responses can support improved resilience. The right-hand column summarizes the appropriate land governance response, how FFP LAS can help address the land issues, and related land tools that support the FFP LAS approach. As discussed earlier, responsible land governance can be considered to comprise two major elements:

- **Effective land-use planning and control**: In Table 4, the common impacts due to poor land-use planning and control affecting both climate and pandemic vulnerability are informal settlement in hazard-prone areas, high housing density and poor housing quality. Lack of connection to formal water supply and sanitation are also common in informal settlements. Lack of formal records due to informal settlements can mean that households do not receive financial support during natural disasters and pandemics. FFP LAS can support effective land-use planning and control through recognizing, mapping, and recording all existing land tenure rights, using participatory enumeration, with visible boundaries defined on high resolution imagery. This information on land tenure systems informs a tenure-responsive approach to land-use planning.
- **Securing and safeguarding land tenure rights:** In Table 4, the common impacts due to poor tenure security affecting both climate and pandemic vulnerability include eviction, displacement, and involuntary resettlement, as well as loss of livelihood options. FFP LAS can support approaches to improve tenure security at scale, using methods that fit the context. This includes adopting the continuum of land rights to support the aim of tenure security for all, participatory enumeration to recognize and record existing land tenure rights, and mapping visible boundaries defined on high resolution imagery.

Building from project experience in Honiara, there are some clear lessons for the role of FFP LAS in resilience actions at the city level to improve tenure security at scale to support city-wide improvement to resilience to multiple stressors. This can be broadly considered to comprise three main stages:

1. Assessing the climate, pandemic, and land vulnerabilities and risk factors.
2. Resilience action planning.
3. Enhancing resilience through responsible land governance.

How FFP LAS potentially supports each of these stages is described in the following sections.

4.1. Assessing the Climate, Pandemic, and Land Vulnerabilities and Risk Factors to Support Recognizing and Recording Land Tenure Rights as well as Climate Action Planning

The UN-Habitat Cities and Climate Change Initiative (CCCI) has supported city-wide climate change Vulnerability and Risk Assessments (VRA), identifying 'hotspots', which often correspond with informal settlements. Adopting a system's response to risk and vulnerability, the VRA supports the development of Climate Action Plans to build the resilience of communities [45]. Based on internationally recognized methodologies, the VRA analyzes climate hazard characteristics, exposure, sensitivity, and adaptive capacities (see Figure 6).

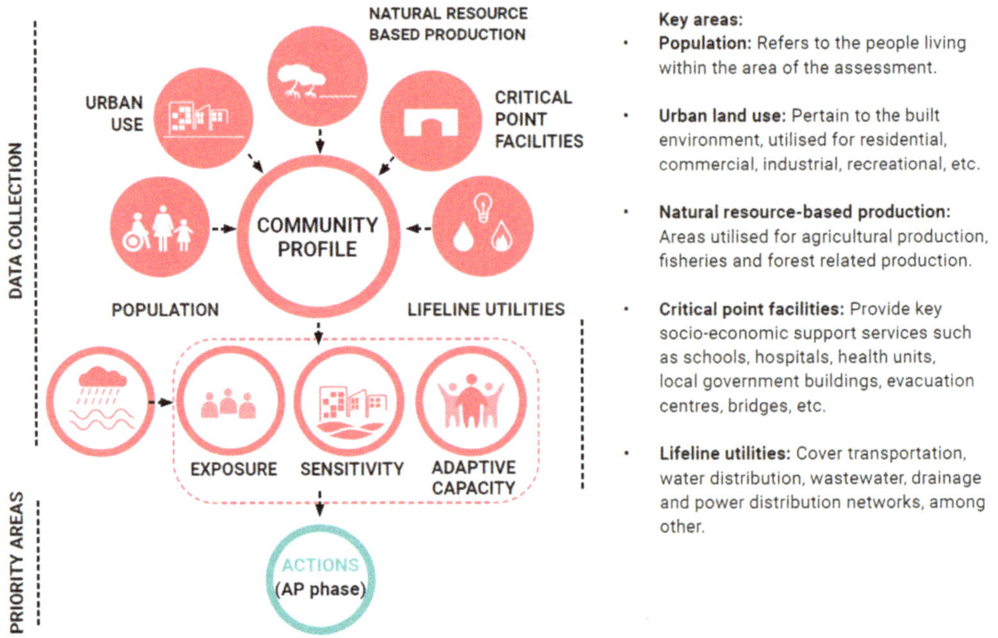

Figure 6. Vulnerability and risk assessment and action planning overview (Data from [45]).

The development of climate- and pandemic-resilient action plans require a deep understanding of the existing risk factors and vulnerabilities faced by households and other stakeholders. Resilience planning and implementation must apply to all existing households, not just those with formal land records. This household-level and settlement-level understanding supports decisions on housing and tenure security, explicitly informed by climate impacts and a vulnerability assessment of affected urban informal settlements.

4.1.1. Community Profiling Supported by Participatory Enumeration to Recognize and Record Land Tenure and Vulnerability to Multiple Stressors

Community profiling based on a VRA also helps to design resilience action plans so that both climate and pandemic responses are integrated into a broader resilience and urban development process (UN-Habitat, 2020). Given the similarity of vulnerabilities caused by pandemics and climate impacts, this approach also fits a multi-stressor approach. These inform actions, including climate change adaptation, resilience building, urban planning, and infrastructure provision. A multi-stressor VRA supports a consultative, bottom-up, and more inclusive approach to the development of action plans. As the CRH project demonstrates, community profiling provides valuable understanding of the diversity of communities and their existing vulnerabilities. The profiling is based on participatory enumeration, using household surveys at the settlement level, supported by spatial information that provides mapping of the houses, public buildings, and other infrastructure in the settlement. Questions related to existing tenure arrangements and perceptions of tenure security allows assessment of the existing land tenure rights as the first stage in a FFP LAS approach to improving tenure security at scale.

4.1.2. FFP LAS Approach to Building the Spatial Framework

A key component of FFP LAS is the large-scale spatial framework comprising mapping that identifies the buildings and land parcels across the continuum of land rights to support decisions on securing land-tenure rights and land-use control [12]. In the Honiara case study, large-scale LiDAR imagery supported by small scale imagery from drones provided an important spatial framework to support land-use planning, slum upgrading and the design of new subdivisions. This spatial framework also supported the community profiling and VRA. A low-cost approach using high-resolution imagery and adopting the principle of visible boundaries allows expansion of the land administration system at scale, using this imagery to support registration processes [6]. Once the spatial framework is developed, the imagery data are also very useful for supporting community profiling and vulnerability assessment processes, assessing tenure security, as well as supporting community consultation discussions on action plans. Mapping of tenure security based on the concept of the continuum of land rights is possible in the community profiling process. It provides the necessary baseline data to inform tenure-responsive land-use planning, using FFP LAS approaches [5].

4.2. Action Planning

The findings from the VRA lay the foundation for the resilience action plans (AP). The aim of the AP is to support decision making on community-based interventions to strengthen resilience to climate change, as well as to support local development [45]. Community profiling and a FFP LAS approach to developing a spatial framework provide the basis for the design of action plans and community and stakeholder consultation to validate these. This detailed understanding and mapping is also very useful for supporting the resolution of disputes over land. This detailed baseline information allows harmonization between resilience action planning and actions to improve tenure security, such as slum upgrading programs or land readjustment projects. The mapping and community profiling can also support systematic and sporadic formal land recording. The resilience action planning stage is where land issues can be mainstreamed into climate and pandemic action plans and, conversely, hazard-risk information can be mainstreamed into land administration and land-use planning.

4.3. Enhancing Resilience through Responsible Land Governance

4.3.1. Securing and Safeguarding Land Tenure Rights at Scale

Safeguarding all land tenure rights involves understanding, recording, and recognizing the complex, long-established and accepted social tenure relationships. Slum upgrading is a typical approach to improve the security of tenure and upgrade infrastructure and facilities in settlements, as well as reduce hazard risk. UN-Habitat's Participatory Slum Upgrading Programme (PSUP) seeks to address vulnerability and marginalization during informal settlements upgrading [45].

Safeguarding all land tenure rights requires all existing people-to-land relationships to be recorded and recognized. A FFP LAS approach is necessary to do this complex task at scale, based on the participatory enumeration that informs community profiles as discussed above. FFP LAS supports the data collection and recording of informal land rights as well as formal land rights. Land tools for recording the people-to-land relationship, including social tenures, can be used to support the recognizing and recording of land tenure rights, especially for poor and informal settlers. One example is the Social Tenure Domain Model (STDM), which can record complex land–people relationships based on the Land Administration Domain Model and can later be upgraded for inclusion in the formal register of land records.

The FFP LAS approach was a piloted project in Nepal involving the post-earthquake data collection and recordation of customary and informal land rights, using STDM. The documented land information was certified and used to inform decisions on the allocation or reconstruction of grants, and support the land tools used to improve tenure security, as well as support the processes of relocation and reconstruction in four settlements in the Dolakha district in Nepal [7].

4.3.2. Tenure-Responsive Land-Use Planning in the Context of Climate Change and Pandemics

Community profiling, VRA and action plans help to mainstream climate and pandemic considerations in land-use planning. Effective land-use planning is necessary to restrict housing built in hazard-prone locations, and to support slum upgrading where informal settlement is well established. Improvement in community resilience to climate and pandemic stressors requires improving housing quality, connection to formal water supply and sanitation, and tenure insecurity in informal settlements. The community profiling based on VRA and an assessment of tenure security, supported by the FFP LAS spatial framework and recording of the people–land relationships is a strong basis for effective land-use planning and control. This means that land-use planning can be both tenure-responsive and cognizant of vulnerabilities and risks.

5. Conclusions

Based on an extensive literature review, and participatory action research based in Honiara, this paper discussed how improving tenure security at scale using the FFP LAS approach can enhance community resilience. Vulnerability to climate change and pandemics is a widespread challenge in slums and informal settlements. As tenure insecurity is an important factor in vulnerability, resilience efforts must address both vulnerability and tenure insecurity at scale. FFP LAS is necessary for improved tenure security at scale, with an approach that includes emphasizing adopting visible boundaries and the use of large-scale imagery to support the spatial, legal, and institutional components of land administration. Lessons from a current resilience project shows that the FFP LAS, informed by vulnerability assessments and community profiling, can support interventions aiming to scale up the number of households with security of tenure and improved resilience outcomes as informal settlements grow.

This paper describes how improving tenure security at scale using the FFP LAS approach can enhance climate resilience to both climate and pandemic impacts. This can be achieved at the city and settlement levels by including tenure in vulnerability and risk assessments (VRA) and the development of resilience action plans, as well as using VRA to inform efforts to improve tenure security, reduce land disputes and make land-use planning more effective.

We described how FFP LAS can support the two components of responsible land governance: (i) effective land-use planning and control, and (ii) securing and safeguarding land tenure rights. This includes developing a spatial framework based on imagery to support both VRA and action planning, as well as land-use planning, slum upgrading and subdivision design. Adopting the continuum of land rights, the other key aspect of the FFP LAS approach is using low-cost participatory enumeration methods and tools, such as the STDM to record all existing people-to-land relationships to allow all de facto and de jure tenures to be included in efforts to improve resilience and tenure security.

We contend that the FFP land administration approach, informed by participatory enumeration of the complexities of urban land tenure, can support scaling up efforts to improve tenure security and deliver more effective and equitable climate resilience actions for vulnerable urban communities in the Global South. Given the similarities in the vulnerabilities due to climate change and pandemics, the FFP LAS approach has broad applications in development efforts to reduce risk and improve resilience.

Author Contributions: Conceptualization, D.M. (David Mitchell) and D.M. (Darryn McEvoy); methodology, D.M. (David Mitchell), B.B., M.S.S., S.H. and D.M. (Darryn McEvoy); formal analysis, D.M. (David Mitchell), B.B., M.S.S., S.H. and D.M. (Darryn McEvoy); investigation, D.M. (David Mitchell), B.B., M.S.S., S.H. and D.M. (Darryn McEvoy).; writing—original draft preparation, D.M. (David Mitchell), B.B., M.S.S., S.H. and (Darryn McEvoy); writing—review and editing, D.M. (David Mitchell), B.B., M.S.S., S.H. and D.M. (Darryn McEvoy). All authors have read and agreed to the published version of the manuscript.

Funding: The Climate Resilient Honiara is funded by the UNFCCC Adaptation Fund and administered by UN-Habitat. Funding of the publication costs for this article has kindly been provided by the School of Land Administration Studies, University of Twente, in combination with Kadaster International. the Netherlands.

Institutional Review Board Statement: Not applicable.

Informed Consent Statement: Not applicable.

Data Availability Statement: Not applicable.

Acknowledgments: Climate Resilient Honiara is funded by the UNFCCC Adaptation Fund and administered by UN-Habitat.

Conflicts of Interest: The authors declare no conflict of interest.

References

1. Yohe, G.W.; Lasco, Q.K.; Ahmad, N.W.; Arnell, S.J.; Cohen, C.; Janetos, A.C.; Perez, R.T. Perspectives on climate change and sustainability. In *Climate Change 2007: Impacts, Adaptation and Vulnerability. Contribution of Working Group II to the Fourth Assessment Report of the Intergovernmental Panel on Climate*; Parry, M.L., Canziani, O.F., Palutikof, J.P., van der Linden, P.J., Hanson, C.E., Eds.; Cambridge University Press: Cambridge, UK, 2007; pp. 811–841.
2. Dodman, D.; Brown, D.; Francis, K.; Hardoy, J.; Johnson, C.; Satterthwaite, D. *Understanding the Nature and Scale of Urban Risk in Low- and Middle-income Countries and Its Implications for Humanitarian Preparedness, Planning and Response*; Human Settlements Discussion Paper Series—Climate Change and Cities 4; International Institute for Environment and Development: London, UK, 2013.
3. Mitchell, D.; McEvoy, D. *Land Tenure and Climate Vulnerability*; UN-Habitat: Nairobi, Kenya, 2019.
4. Unger, E.-M.; Zevenbergen, J.; Bennett, R. On the need for pro-poor land administration in disaster risk management. *Surv. Rev.* **2016**, *49*, 437–448. [CrossRef]
5. McEvoy, D.; Mitchell, D.; Trundle, A. Land tenure and urban climate resilience in the South Pacific. *Clim. Dev.* **2020**, *12*, 1–11. [CrossRef]
6. Mitchell, D.; Enemark, S.; van der Molen, P. Climate resilient urban development: Why responsible land governance is important. *Land Use Policy* **2015**, *48*, 190–198. [CrossRef]
7. Unger, E.; Chhatkuli, R.R.; Antonio, D.; Lemmen, C.; Zevenbergen, J.; Bennett, R.; Dijkstra, P. Creating Resilience to Natural Disasters Through Fit-For-Purpose Land Administration in Nepal. In Proceedings of the 2019 World Bank Conference on Land and Poverty, Washington, DC, USA, 25–29 March 2019.
8. CFS; FAO. *Voluntary Guidelines on the Responsible Governance of Tenure of Land, Fisheries and Forests in the Context of National Food Security*; Committee on World Food Security ; UN FAO: Rome, Italy, 2012.
9. Farha, L. Housing, the Front-Line Defence against the COVID-19 Outbreak. Says UN Expert, OHCHR. 2021. Available online: https://www.ohchr.org/EN/NewsEvents/Pages/DisplayNews.aspx?NewsID=25727 (accessed on 12 February 2021).
10. USAID. Impact of COVID-19 on Women's Customary Land Rights and Livelihoods in Southern Africa, ABA, ARICA. 2020. Available online: https://www.usaid.gov/sites/default/files/documents/ABA-ARISA_Impact_of_COVID_19_on_Womens_Customary_Land_Rights_and_Livelihoods_Dec_2020.pdf (accessed on 22 April 2021).
11. Sait, M.S. Should Monrovian Communities Agree to Voluntary Slum Relocations: Land, Gender and Urban Governance. In *Land Issues for Urban Governance in Sub-Saharan Africa*; Home, R., Ed.; Springer: Cham, Switzerland, 2021; pp. 339–354.
12. Enemark, S.; McLaren, R.; Lemmen, C. *Fit-for-Purpose Land Administration: Guiding Principles for Country Level Implementation*; UN-Habitat; GLTN: Nairobi, Kenya, 2016.
13. Lengoiboni, M.; Richter, C.; Asperen, P.; Zevenbergen, J. Initial Insights on Land Adjudication in a Fit-for-Purpose Land Administration. *Land* **2021**, *10*, 414. [CrossRef]
14. Trundle, A.; McEvoy, D. *Honiara Urban Resilience and Climate Action Plan*; UN-Habitat: Fukuoka, Japan, 2017.
15. McEvoy, D.; Barth, B.; Trundle, A.; Mitchell, D. Reflecting on a journey from climate change vulnerability assessments to the implementation of climate resilience actions: Honiara, Solomon Islands. In *Urban Resilience in the Asia Pacific: Challenges and Opportunities for Addressing Disasters and Climate Change in the Built Environment*; Routledge: Oxford, UK, 2020.
16. UN-Habitat. *Rapid Assessment of Covid-19 in Informal Settlements in Solomon Islands Insights into Socio-Economic Impacts on Residents in 5 Communities across Honiara, October 2020*; UN-Habitat: Suva, Fiji, 2020.
17. Patz, J.A.; Daszak, P.; Tabor, G.M.; Aguirre, A.A.; Pearl, M.; Epstein, J.; Wolfe, N.D.; Kilpatrick, A.M.; Foufopoulos, J.; Molyneux, D.; et al. Unhealthy Landscapes: Policy Recommendations on Land Use Change and Infectious Disease Emergence. *Environ. Health Perspect.* **2004**, *112*, 1092–1098. [CrossRef]
18. Murray, K.A.; Daszak, P. Human ecology in pathogenic landscapes: Two hypotheses on how land use change drives viral emergence. *Curr. Opin. Virol.* **2013**, *3*, 79–83. [CrossRef]
19. Acuto, M. COVID-19: Lessons for an Urban(izing) World. *One Earth* **2020**, *2*, 317–319. [CrossRef]

20. Moraci, F.; Errigo, M.F.; Fazia, C.; Campisi, T.; Castelli, F. Cities under pressure: Strategies and tools to face climate change and pandemic. *Sustainability* **2020**, *12*, 7743. [CrossRef]
21. Cooper, D.H.; Nagel, J. Lessons from the pandemic: Climate change and COVID-19. *Int. J. Sociol. Soc. Policy* **2021**. [CrossRef]
22. El Hamichi, S.; Gold, A.; Murray, T.G.; Graversen, V.K. Pandemics, climate change, and the eye. *Graefe's Arch. Clin. Exp. Ophthalmol.* **2020**, *258*, 2597–2601. [CrossRef]
23. The Lancet Planetary Health. A Pandemic Era. *Lancet Planet Health* **2021**, *5*, e1. [CrossRef]
24. Di Marco, M.; Baker, M.L.; Daszak, P.; De Barro, P.; Eskew, E.A.; Godde, C.M.; Harwood, T.D.; Herrero, M.; Hoskins, A.J.; Johnson, E.; et al. Opinion: Sustainable development must account for pandemic risk. *Proc. Natl. Acad. Sci. USA* **2020**, *117*, 3888–3892. [CrossRef] [PubMed]
25. Terraube, J.; Fernández-Llamazares, Á. Strengthening Protected Areas to Halt Biodiversity Loss and Mitigate Pandemic Risks. *Curr. Opin. Environ. Sustain.* **2020**, *46*, 35–38. [CrossRef] [PubMed]
26. Chin, A.; Cui, X.; Gillson, L.; Nelson, D.; Taylor, M.P.; Vanacker, V.; Wang, E. Anthropocene in an age of pandemics. *Anthropocene* **2020**. [CrossRef]
27. Plowright, R.K.; Reaser, J.K.; Locke, H.; Woodley, S.J.; Patz, J.A.; Becker, D.J.; Oppler, G.; Hudson, P.J.; Tabor, G.M. Land use-induced spillover: A call to action to safeguard environmental, animal, and human health. *Lancet Planet. Health* **2021**, *5*, e237–e245. [CrossRef]
28. Nicola, M.; Alsafi, Z.; Sohrabi, C.; Kerwan, A.; Al-Jabir, A.; Iosifidis, C.; Agha, M.; Agha, R. The socio-economic implications of the coronavirus and COVID-19 pandemic: A review. *Int. J. Surg.* **2020**, *78*, 185–193. [CrossRef]
29. Huchzermeyer, M.; Karam, A. The continuing challenge of informal settlements: An introduction. In *Informal Settlements: A Perpetual Challenge*; Huchzermeyer, M., Karam, A., Eds.; Juta and Company Ltd.: Johannesburg, South Africa, 2006; Volume 1, pp. 1–16.
30. Penrose, K.; De Castro, M.C.; Werema, J.; Ryan, E.T. Informal Urban Settlements and Cholera Risk in Dar es Salaam, Tanzania. *PLoS Negl. Trop. Dis.* **2010**, *4*, e631. [CrossRef]
31. UN Habitat. *Cities and Pandemics: Towards a More Just, Green and Healthy Future*; UN Habitat: Nairobi, Kenya, 2021.
32. Wilkinson, A.; Ali, H.; Bedford, J.; Boonyabancha, S.; Connolly, C.; Conteh, A.; Dean, L.; Decorte, F.; Dercon, B.; Dias, S.; et al. Local response in health emergencies: Key considerations for addressing the COVID-19 pandemic in informal urban settlements. *Environ. Urban.* **2020**, *32*, 503–522. [CrossRef]
33. Chigbu, U.E.; Onyebueke, V.U. The COVID-19 pandemic in informal settlements: (re)considering urban planning interventions. *Town Plan. Rev.* **2021**, *92*, 115–121. [CrossRef]
34. Corburn, J.; Vlahov, D.; Mberu, B.; Riley, L.; Caiaffa, W.T.; Rashid, S.F.; Ko, A.; Patel, S.; Jukur, S.; Martínez-Herrera, E.; et al. Slum Health: Arresting COVID-19 and Improving Well-Being in Urban Informal Settlements. *J. Hered.* **2020**, *97*, 348–357. [CrossRef]
35. Kiddle, G.L.; McEvoy, D.; Mitchell, D.; Jones, P.; Mecartney, S. Unpacking the Pacific Urban Agenda: Resilience Challenges and Opportunities. *Sustainability* **2017**, *9*, 1878. [CrossRef]
36. National Statistics Office. Provisional Count: 2019 National Population and Housing Census. Ministry of Finance and Treasury, Government of Solomon Islands. 2020. Available online: https://www.solomonchamber.com.sb/media/1997/provisional_count-2019_census_result.pdf (accessed on 12 March 2021).
37. Firth, S. Instability in the Pacific Islands: A Status Report. Lowy Institute. 2018. Available online: https://www.lowyinstitute.org/publications/instability-pacific-islands-status-report (accessed on 12 March 2021).
38. Oliver-Smith, A.; and de Sherbinin, A. Resettlement in the Twenty-First Century, Crisis, Forced Migration Review 45. 2014. Available online: https://www.fmreview.org/crisis/oliversmith-desherbinin (accessed on 16 March 2021).
39. Jones, P. Pacific Urbanisation and the Rise of Informal Settlements: Trends and Implications from Port Moresby. *Urban Policy Res.* **2012**, *30*, 145–160. [CrossRef]
40. ADB. Pacific Economic Monitor: July 2014—Midyear Review (Cook Islands, Fiji, Kiribati, Marshall Islands, Micronesia, Federated States of, Nauru, Palau, Papua New Guinea, Samoa, Solomon Islands, Timor-Leste, Tonga, Tuvalu, Vanuatu; Issue July 2014). Asian Development Bank. 2014. Available online: https://www.adb.org/publications/pacific-economic-monitor-july-2014 (accessed on 14 February 2021).
41. Wilson, C. Demanding the Future: Navigating the Pacific's Youth Bulge. Lowy Institute. 2020. Available online: https://www.lowyinstitute.org/publications/demanding-future-navigating-pacific-youth-bulge (accessed on 22 March 2021).
42. Andrew, N.L.; Bright, P.; De La Rua, L.; Teoh, S.J.; Vickers, M. Coastal Proximity of Populations in 22 Pacific Island Countries and Territories. *PLoS ONE* **2019**, *14*, e0223249. [CrossRef]
43. SPREP; APAN. Report on Adaptation Challenges in Pacific Island Countries. SPREP. 2013. Available online: https://www.sprep.org/attachments/Publications/Adaptation_challenge_PICs_13_hr.pdf (accessed on 22 March 2021).
44. PACCSAP. Climate Variability, Extremes and Change in the Western Tropical Pacific: New Science and Updated Country Reports 2014, Solomon Islands. 2014. Available online: https://www.pacificclimatechangescience.org/publications/reports/climate-variability-extremesand-change-in-the-western-tropical-pacific-2014/ (accessed on 24 March 2021).

45. UN-Habitat. *Climate Change Vulnerability and Risk—A Guide for Community Assessments, Action Planning and Implementation, Rise Up: Resilient Settlements for the Poor*; UN-Habitat: Nairobi, Kenya, 2020.
46. Foukona, J.D. Urban Land in Honiara: Strategies and Rights to the City. *J. Pac. Hist.* **2015**, *50*, 504–518. [CrossRef]
47. UN-Habitat; IIRR; GLTN. *Handling Land: Innovative Tools for Land Governance and Secure Tenure*; UN-Habitat: Nairobi, Kenya, 2012; Available online: https://gltn.net/download/handling-land-innovative-tools-for-land-governance-and-secure-tenure-eng-2012/ (accessed on 11 February 2021).

land

MDPI

Article

Applying the FFP Approach to Wider Land Management Functions

Kathrine Kelm [1], Sarah Antos [1] and Robin McLaren [2,*]

[1] Urban, Disaster Risk Management, Resilience & Land, World Bank, Washington, DC 20433, USA; kkelm@worldbank.org (K.K.); santos1@worldbankgroup.org (S.A.)
[2] Know Edge Ltd., Edinburgh EH14 1AY, UK
* Correspondence: robin.mclaren@KnowEdge.com; Tel.: +44-78-03163137

Abstract: The initial focus of implementing the Fit-for-Purpose Land Administration (FFPLA) methodology was to address the significant, global security of tenure divide. We argue that this land tenure methodology is proving successful in scaling up the provision of security of tenure for developing countries. The increasing adoption of the FFPLA methodology has also opened opportunities and provided flexibility for the innovative use of emerging technologies to accelerate the global roll out of security of tenure, such as the use of autonomous drones and machine learning techniques applied to image analysis. Despite wider adoption of participatory approaches to the recording of land tenure, similar FFP solutions for the other components of land administration services (land value, land use and land development) and land management functions are still evolving. This article therefore explores how the FFP approach can be applied to this wider set of land administration services and land management functions. A case study methodology, using three case studies, is used to determine if the case study approaches meet the FFP criteria. The focus is on the urban environment, drawing mostly from experiences and case studies in the Urban, Disaster Risk Management, Resilience & Land Global Practice of the World Bank. These opportunities for the wider application of the FFP approach and associated principles are being triggered by the innovative use of emerging new data capture technology developments. The paper examines the innovative use of these emerging technologies to identify a common set of data capture techniques and geospatial data that can be shared across a range of urban land administration and management activities. Finally, the paper discusses how individual land projects could be integrated into a more holistic land administration and management program approach and deliver a significant set of socio-economic benefits more quickly. It is found that the FFP approach can be more widely adopted across land administration and land management and in many cases can share a common set of geospatial data. The authors argue that the wider adoption and integration of these new, innovative FFP urban management approaches will require a significant cultural, professional, and institutional change from all stakeholders. Future work will explore more deeply these institutional weaknesses, which will provide a basis for guidance to the World Bank and similar institutions.

Keywords: fit-for-purpose land management; innovative technology; aerial and street level imagery; machine learning; integrated land programs

Citation: Kelm, K.; Antos, S.; McLaren, R. Applying the FFP Approach to Wider Land Management Functions. *Land* **2021**, *10*, 723. https://doi.org/10.3390/land10070723

Academic Editor: Rohan Bennett

Received: 25 May 2021
Accepted: 30 June 2021
Published: 9 July 2021

1. Introduction

1.1. *Significance of Land Administration to the Sustainable Development Goals (SDGs)*

The World Bank provides an excellent example of the significant role land administration has in supporting the Sustainable Development Goals (SDGs). In 2013 the World Bank adopted the Twin Goals to: End Extreme Poverty—Reduce the percentage of people living on less than USD 1.90 a day to 3 percent by 2030; and Promote Shared Prosperity—Foster income growth of bottom 40 percent of the population in every country. Achieving the Goals in a Sustainable Manner- Securing the long-term future of the planet and its resources, ensuring social inclusion, and limiting the economic burdens on future generations underpin efforts to achieve the two goals. The goals are aligned with the 2030 Sustainable Development Agenda, and the principles and targets embodied in the SDGs. Land, as a natural resource and strategic asset, features prominently in the SDGs and land administration, especially in securing land rights, and thus performs a fundamental role in reducing poverty and promoting shared prosperity. The World Bank is the custodian agency for key SDG indicators on land tenure security:

- SDG 1 Poverty, Target: 1.4: By 2030, ensure that all men and women, in particular the poor and the vulnerable, have equal rights to economic resources, as well as access to basic services, ownership and control over land and other forms of property, inheritance, natural resources, appropriate new technology and financial services, including microfinance; and
- Indicator: C.2: Percentage of people with secure tenure rights to land (out of total adult population), with legally recognized documentation and who perceive their rights to land as secure, by sex and by type of tenure.

World Bank programs and projects provide financing, capacity building and technical support for land administration and land management investments.

1.2. The Emergence of FFPLA

The delivery of security of tenure to citizens across the globe, especially in developing countries, through land administration services has often been inefficient and its outreach remains limited; it is estimated that around 75 percent of the world's population do not have access to formal systems [1] to register and safeguard their land rights—the majority of these are the poor and the most vulnerable in society [2]. The underlying reasons for this limited success include: the continued influence of the colonial legacy on land administration institutions, inequalities in land distribution, and legal/regulatory frameworks; reluctance of land professionals to adopt a new information management role and allow 'barefoot surveyors' to use new, participatory technologies and lower specifications for the spatial accuracy of the capture and maintenance of land rights information; endemic corruption and the protection of the elite; significant lack of capacity at all levels; aversion to integrate informal land rights into the formal land administration system; and lack of sustained political support for reforms.

However, the new global agenda and associated targets around the SDGs have triggered a strong, positive response from global land professionals since land governance is on the critical path of delivering 10 of the SDG targets [3]. The land administration sector is experiencing a significant transition in the adoption of emerging technologies and new flexible, faster and cheaper approaches [4]. The previous straitjacket of high quality and strict conventions is being relaxed and the flexibility of a continuum of approaches is increasingly being observed in literature and practice [5–8]. These new approaches are being based and designed around Fit-For-Purpose Land Administration (FFPLA) solutions that reflect the characteristics of being flexible, inclusive, participatory, affordable, reliable, attainable, and upgradeable. Cost effective and sustainable solutions can then be achieved in a relatively short timeframe [2]. This approach was defined originally by the International Federation of Surveyors (FIG) and the World Bank [2] and subsequently by a more detailed country implementation guide supported by the Global Land Tool Network [9]. Nationwide programs have included: the rural land certification programs of Ethiopia [7], Rwanda's Land Tenure Regularization (LTR) program [10], Kadaster International's fast and effective land administration program supporting the implementation of the Colombian Peace Agreements [11], a program to accelerate land rights redefinition and reconstruction in Nepal after the 2015 disastrous earthquakes [12], and the World Bank funded program to support the registration of all land parcels in Indonesia by 2025 [13]. These exemplary, mostly nationwide programs clearly reflect what FFPLA can achieve when there is top level government support.

1.3. New Innovative Technology Supporting FFPLA

One of the key drivers to support FFPLA, and a key element of participatory solutions, has been the inventive use of emerging, innovative technology [14]. A good example is the use of mobile phones by non-land professionals to capture land parcel boundaries and associated legal information to define, record and adjudicate land rights [15]. For example, USAID have embraced this technology to underpin their Mobile Application to Secure Tenure (MAST) project in Tanzania to improve land governance and lower the cost of land certification programs [16].

Unmanned Aerial Vehicles (UAVs) or drones are another game changing technology to significantly impact and accelerate the adoption of FFPLA approaches. Over 20 years ago, drones that started in military applications were adopted as hobby gadgets. However, they have quickly become applicable and very effective in commercial and scientific contexts. For example, as soon as regulations permit, Amazon plans to have a Prime Air Service that uses drones to deliver parcels up to five pounds (2.3 kg) in 30 min [17]. The delivery of life-saving medical supplies in adverse environments is increasingly feasible by drones [18]. Drones are also having a major impact on how geospatial information is being captured. These platforms are increasingly autonomous, able to work out of sight, and for long durations. They are also being supported by an increasingly rich portfolio of digital photogrammetry tools to flight plan and process in real-time or post-process. They can carry a wide range of sensors and are being increasingly used by both the public and private organizations and individuals. Drones can generate high-resolution imagery (often better than 10 cm resolution). Quality geospatial information can be derived, including orthoimages, digital elevation models and 3D point clouds. Drones have therefore created a complimentary niche between traditional ground surveying, aerial surveys and satellite imagery [19].

The costs of producing conventional aerial photographs, digital orthophotos and cadastral base maps often amount to a significant part of a project budget and there is often a considerable delay between data capture and delivery of the processed products. Drones offer a new approach, working at a smaller geographic coverage but producing high-resolution products as and when needed, thus ensuring that the information is current. As drone technology has become a tool adopted by local mapping companies and private surveyors, the time and cost to produce spatial data for land administration projects have dropped. The World Bank has concluded that embedding mapping capacity, including the use of drones, at the local level facilitates a FFP approach to respond more dynamically and economically to land tenure activities [20]. The World Bank was an early adopter and advocate of the use of drones to capture land parcel boundaries in a participatory approach within their land tenure programs. Drones were originally piloted in Albania (2014) then used in Kosovo (2016), Vietnam (2018) and the Philippines (2021).

Despite drones being relatively widely adopted in a range of applications [21], there are still some key issues that are currently inhibiting the greater use of this platform, including regulations [22] and safety concerns with the aviation sector, privacy, licensing, professional capacity, and pushback from land professionals over the perceived threat to their livelihoods. The use of drone and street level imagery has been further enhanced through the use of Machine Learning (ML) to automatically or semi-automatically extract objects from the imagery. The World Bank has developed ML solutions to extract information from georeferenced images [23]. So far, these solutions are being used for assessing building structures, including building materials, number of floors and windows, which can then be used for vulnerability assessment and disaster risk analysis, valuation, and other applications [23]. Research is maturing in the automatic extraction of land parcel boundaries from imagery to accelerate the mapping and registration of large numbers of unrecorded land rights globally [24]. Research has applied and compared the ability of rule-based systems within Object-Based Image Analysis (OBIA), as opposed to human analysis, to extract visible cadastral boundaries from very high-resolution World View-2 images, in both rural and urban settings [25].

1.4. Context of FFPLA in Land Governance

Land governance [26] is about sustainably and transparently managing land, property and natural resources. This is achieved through three fundamental components: Land Policies and strategies, Land Administration Services and Land Management Functions, and Land Information Infrastructures (Figure 1).

Comprehensive and robust land governance [27] depends upon a harmonized legal and regulatory framework, institutional structures with clear roles and responsibilities, and capacity to implement land related policies and strategies consistently across a country. Land administration services and associated land information infrastructure provide a country with the means to implement land policies and supporting land management strategies aligned with the SDGs. These land administration services include land tenure (securing and transferring rights in land and natural resources), land value (valuation and taxation of land and properties), land use (planning and control of the use of land and natural resources), and land development (implementing utilities, infrastructure, construction works, and urban and rural developments) [26]. These land administration services are fundamentally facilitated by access to a comprehensive land information infrastructure; one that consists of information on the built and natural (including marine) environment and is an integral part of a National Spatial Data Infrastructure (NSDI).

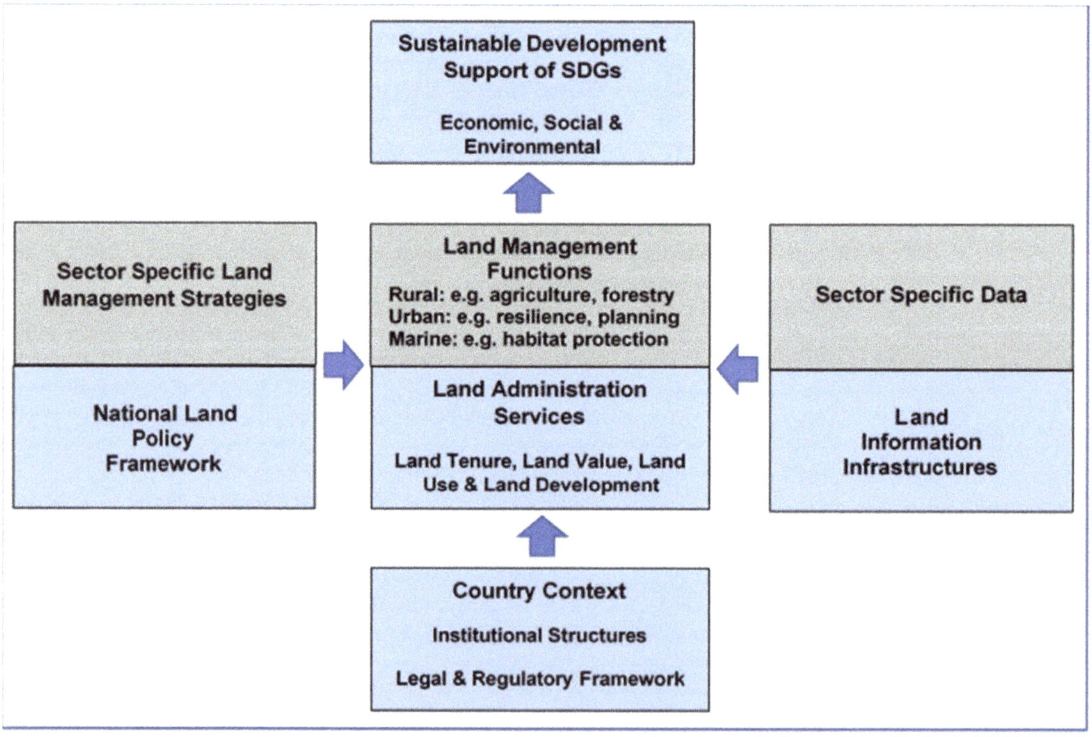

Figure 1. Land Governance Ecosystem (Adapted from The Land Management Paradigm in blue [28] (p. 171)).

Land management [28] (pp. 116–117) is primarily guided by an overall National Land Policy [29] that supports the formulation of sector specific land management strategies, such as forestry and agriculture in the rural context, city resilience and disaster risk management in the urban context, habitat protection and pollution control in the marine context, and climate change mitigation in an over-arching context. Sound land management functions ensures that decisions on land (rural, urban, and marine) are truly evidence based, using the underpinning land administration services and associated land information infrastructures in conjunction with land sector specific data. This overall land governance ecosystem (Figure 1) ensures that sustainable development can be achieved.

Currently, the focus of applying the FFP paradigm [2] is land tenure. However, the use of innovative technologies is creating opportunities for the paradigm to be applied across a much broader set of land sector applications. Through case studies in the urban environment, this paper investigates how the FFP methodology can be applied to a wider set of land administration services and land management functions. The paper also evaluates if a common set of data capture techniques and geospatial data can be shared across a range of urban land administration and management activities. Finally, the paper discusses how individual land projects could be integrated into a more holistic land administration and management program approach and thus deliver a significant set of socio-economic benefits more quickly.

2. Approach and Methodology

The concept of the FFP approach emerged because the traditional approaches to land tenure projects were judged to be too expensive, took a long time to implement, the process was too rigid, and there were insufficient professional resources to implement and maintain at scale. Therefore, the FFP approach adopted the model of a Minimum Viable Product (MVP) where an entry level solution is developed that meets the basic needs of customers and one that can then be incrementally improved over time, where there is demand. In the case of FFP land tenure this involved reducing the accuracy specifications, using uncomplicated, participatory tools for data capture and being inclusive of all types of tenure.

The original definition of the FFP approach, outlined by FIG and the World Bank in 2014 [2], identified seven elements (Table 1) to be used in assessing the compliance of a proposed solution as being FFP.

Table 1. Characteristics of the FFP Approach [2].

Flexible	in the spatial data capture approaches to provide for varying use and occupation.
Inclusive	in scope to cover all tenure and all land.
Participatory	in approach to data capture and use to ensure community support.
Affordable	or the government to establish and operate, and for society to use.
Reliable	in terms of information that is authoritative and up-to-date.
Attainable	to establish the system within a short timeframe and within available resources.
Upgradeable	with regard to incremental improvement over time in response to social and legal needs and emerging economic opportunities.

These seven elements, detailed in Table 1, are used in the adopted methodology as a set of criteria to evaluate three urban case studies of new, innovative land administration services and land management functions to determine if they can be classified as FFP and used as cheaper and faster entry level solutions to manage the sustainable development of urban environments more effectively.

Wherever possible, the approaches used to support these new FFP solutions should mirror the tools and data being used for FFP land tenure solutions. This commonality will then allow more integrated and cost-effective FFP solutions to be created. Therefore, as a second step in the methodology, we evaluate the data capture technology and corresponding geospatial data requirements needed to support the FFP approach to land tenure against the three land administration services and land management functions case studies. This then enables us to determine if there is a common set of geospatial data that can be captured once and shared across many other land services and functions in the urban environment. The starting assumption for the elements included in this common set of geospatial data are listed in Table 2:

Table 2. Elements of Common Set of Geospatial Data.

Drone imagery with a pixel resolution <10 cm from RGB digital camera and spatially corrected through Ground Control Points or Real-Time Kinematic (RTK).
Street level imagery with a resolution of 30 megapixels stitched together into georeferenced street panoramas and spatially corrected through Ground Control Points or RTK.
Digital Elevation Model/Digital Terrain Model (DEM/DTM).
3D photogrammetric models to support data capture from the drone and street level imagery through digital photogrammetry.
3D point cloud (either from the Structure from Motion (SfM) approach or using LiDAR).
Digital orthophoto.

To effectively allow these data to be shared and support interoperability then all the data components must be compatible with international or national geospatial data standards, such as those published by the Open Geospatial Consortium [30].

This then leads to a discussion around the opportunity to have a more holistic approach to integrating the implementation of FFP land administration services and land management functions within an integrated land program rather than the current practice of implementing completely independent land projects.

The FFPLA concept is composed of three frameworks: spatial, legal and institutional. This study has primarily focused on the spatial framework and identified how emerging technologies are facilitating the wider use of the FFP approach. However, the successful implementation of these new FFP land administration services and FFP land management functions will require supportive changes to the legal and institutional frameworks. Although these non-technical transformations are not a core part of this article, they are important and are briefly included as part of the discussion of the article.

3. Emerging Innovative Technologies Supporting FFP Land Tenure

Disruptive and innovative technologies are being applied to the FFP land tenure approach to semi-automate elements of data capture, reduce the costs of obtaining imagery, lessen the need for highly skilled resources and provide high fidelity 3-D models. These technologies are reviewed here to determine the characteristics of their data outputs and therefore evaluate their applicability in other FFP land administration services and land management functions.

3.1. Drones

Drones were not initially designed for professional geomatics and land administration applications. However, new technologies, including a wide range of miniaturized sensors (Table 3) have been developed for civilian application, opening new use opportunities. These enhancements have improved the flexibility (fewer restrictions in terms of a sensor's installation), performance (more duration, better aerodynamics profile, better navigation system), and planning tools (new tools have been developed for planning and control of the UAV operations) [31]. Fidelity has been revolutionized through the use of real-time RTK services and the emergence of 'RTK from the sky' [32]. The use of SfM applications that allow a 3D model to be derived from a sequence of images captured from different points of view [33] has further enhanced their use.

Table 3. Sensors Onboard Drones: Auxiliary and Specific [21]. Reprinted with permission from the American Society for Photogrammetry & Remote Sensing, Bethesda, Maryland, www.asprs.org/, accessed on 6 January 2021.

Auxiliary	Specific	
• Ultraviolet spectrometer.	• Thermal cameras.	• Ultraviolet spectrometer.
• Multi-gas detector.	• Infrared cameras.	• Multi-gas detector.
• Sonar.	• FLIR.	• Sonar.
• Smartphone.	• LIDAR (Laser scanner).	• Smartphone.
• Aerosol sampling.	• Irradiance.	• Photometer, aethalometer.
• Pyranometer.	• Radar/SAR.	• Aerosol sampling.
• Particle counters (optical, condensation).	• Multi-Hyperspectral (HyperUAS).	• Probes (temperature, humidity, pressure).
• Photometer, aethalometer.	• Radiometer (multi-frequency).	• Cloud droplet spectrometer.
• Probes (temperature, humidity, pressure).	• Infrared spectroscopy.	• Pyranometer.
• Cloud droplet spectrometer.	• Video cameras (visible spectrum): EOS, stereoscopic, omnidirectional, fisheye lens.	• Particle counters (optical, condensation). • Electrostatic collector.
	• VCSEL.	• Magnetic sensor.
	• WMS.	• Ultraviolet flame detector.
	• Electronic nose.	• Gas/smoke detector.
		• Radiation gauge.

There are several types of drones (Table 4) with various sizes and payloads. The majority of drones used for FFP land tenure have been in the nano and micro categories.

Table 4. Drone Classification [31].

Category	Range (km)	Flight Height (m)	Duration (h)	MTOW [1] (kg)
Nano	<1	<100	<1	<0.025
Micro	<10	250	1	<5
Mini	<10	150–300	<2	150
Close range	10–30	3000	2–4	150
Short range	30–70	3000	3–6	200

[1] Maximum Take-off Weight.

Various studies highlight the applicability of drones as an efficient tool for: (1) mapping customary land rights in Namibia [34]; (2) capturing cadastral boundary data in Indonesia [35]; and (3) Vietnam where UAVs produced survey grade orthophotos with positional accuracy compliant with existing surveying regulations [36]. In some cases, the legal and regulatory framework has been modified to accommodate a reduction in the surveying regulations and thus support the FFP methodology.

The typical outputs from the use of drones used in FFP land tenure applications are reflected in Table 2 and include:

- Imagery with a pixel resolution <10 cms from RGB digital camera;
- Spatially correct data through Ground Control Points/RTK post processing/Realtime RTK/RTK from Sky;
- DEM;
- 3D photogrammetric models to support data capture through digital photogrammetry;

- 3D point cloud (the SfM approach creates a low-density point cloud compared to the use of LiDAR);
- Digital orthophoto.

The key product derived from the drone to support the FFP land tenure methodology is the orthophoto. This is used to engage with citizens and communities to identify and record visual parcel boundaries. The orthophoto can be produced in paper form or used directly in digital form using mobile phones or tablets technologies. Machine Learning (ML) has been successfully used to extract parcel boundary objects [37,38] in order to better support the citizen engagement process.

Although drones can produce high-resolution imagery and derived 2D and 3D data very efficiently and cheaply with less skilled resources, there remain some constraints. For example, the duration of flights is limited by battery and payload capacity, and aeronautical regulations can constrain flying areas. Despite these restrictions, drones have become a game changer in efficiently capturing geospatial data over large areas.

3.2. LiDAR

LiDAR is a remote sensing technology, which emits lasers to collect measurements that can later be used to create 3D models and maps of objects and environments. LiDAR shoots pulses of light (up to a million pulses per second), which bounce off a surface and return to the sensor. The sensor then calculates the time it took for the ray of light to return and which direction it came from. This process ultimately creates a point cloud map of the scanned surroundings—reality capture [38]. LiDAR is increasingly being adopted by autonomous vehicles and this application is accelerating the research and capabilities of the technology. The key characteristics and value add of LiDAR are the inherent accuracy capability, the generation of a high degree of information completeness of reality and the ability to integrate a variety of additional sensor data that together provides a point cloud with high-fidelity visualization—highly accurate, high-definition and increased-density point cloud data with equidistant point patterns [39]. This allows Machine Learning and geospatial analytics to derive and extract spatial objects.

The typical outputs from the use of LiDAR to support FFP land tenure applications include:

- Spatially correct 3D point cloud with a high resolution (180 points/m^2) [40];
- DEM/DTM.

The key product derived from LiDAR to support the FFP land tenure methodology is the 3D point cloud that is used to derive the DEM, DTM and orthophoto, helps 3D visualization, and aids ML to extract objects, such as parcel boundaries [25]. LiDAR produces large volumes of accurate, consistent data quickly, but the technology is complex in nature and requires a deep understanding of the sensor and associated technical know-how.

3.3. Street Level Imagery

In addition to imagery being captured by airborne sensors, imagery from the street level is increasingly being used, for example, to support navigation and autonomous vehicles. Geo-located images taken by citizens and obtained at scale from 360-degree cameras, usually attached to cars and motorcycles, are stitched together to form interactive street panoramas. Google 'Street View', launched in 2007, has the most comprehensive coverage and by 2017 had captured more than 16 million kilometres of Google Street View imagery across 83 countries [41]. The other source of Google Street View imagery is from users. Google introduced this feature in 2017 to allow contributors to add their own images to the Google Street View database for possible inclusion into the 'map' [42]. The analysis of this imagery is highly restricted through licensing. Mapillary [43] and CartaView [44] (shared under a Creative Commons license) are other services sharing crowdsourced, geotagged street level images.

An example of this approach is a World Bank Global Program for Resilient Housing project in Padang, West Sumatra, Indonesia that captured drone and street-view imagery to create a building inventory for a study that supported a home improvement subsidy program, known as Bantuan Stimulan Perumahan Swadaya (BSPS). Normally, drones are perceived to be capable of just gathering information in small areas of about 2–5 km^2, depending on the spatial resolution, flight pattern and drone type. However, the project area of 90 km^2 was covered by drone and street-level imagery at a rate of 3 km^2 per day [23]. Street view imagery captured using a Trimble MX7 360 camera mounted on top of an SUV that has six, 5 megapixel cameras, the Trimble Applanix GNSS and inertial geo-referencing modules. This generated 30-megapixel geo-referenced images. Using the georeferenced, overlapping images it was possible to get robust object positioning, perform measurements within the images and create 3D models. The georeferenced street panorama can be feed into platforms like Mapillary for visualization and navigation, and also used to generate a housing stock inventory that includes ML derived characteristics such as building material, use, construction type, and vintage. This inventory in turn can be used to inform home improvement programs (see Figure 2), property valuation estimations, and disaster risk management exposure models, for example [23]. The typical outputs from the use of street level imagery include:

- Imagery stitched together into georeferenced street panoramas;
- Spatially correct data through RTK post processing/Realtime RTK;
- 3D photogrammetric models to support data capture through digital photogrammetry;
- Digital orthophoto.

Figure 2. Results of building quality assessment, Penang, Indonesia [source World Bank].

The key product derived from street level imagery to support the FFP land tenure methodology is the georeferenced street panorama Although FFP land tenure can be completed using just drone imagery, the georeferenced street panorama derived from street view imagery can add further support to the FFP land tenure process. It can be used to perform measurements, help interpretation of the urban landscape and support allocation of street addresses. However, the street view imagery supports ML to extract building characteristics from imagery [23] and consequently facilitates new FFP land administration services and FFP land management functions.

3.4. Object Extraction from Imagery Using Machine Learning (ML)

Artificial intelligence as defined by Moore 'is the science and engineering of making computers behave in ways that, until recently, we thought required human intelligence' [45]. However, ML is a branch of artificial intelligence, and as defined by Computer Scientist and machine learning pioneer Mitchell: 'Machine learning is the study of computer algorithms that allow computer programs to automatically improve through experience' [46]. There are three types of ML: supervised, unsupervised and reinforced. Supervised learning algorithms 'try to model relationship and dependencies between the target prediction output and the input features, such that we can predict the output values for new data based on those relationships, which it has learned from previous datasets' [47]. This type of ML is the most commonly used to support the extraction of features from imagery datasets to support land administration services and land management functions. For example, windows, doors and materials of buildings can be identified from street-level imagery and recorded to support the valuation of buildings once the algorithms have been trained through examples [23]. ML is playing an increasingly important role in the automatic capture of spatial data—particularly in the context of citizen science projects. For example, Varas-Munoz et al. [48] demonstrate the benefit of using convoluted neural networks (CNN) to support what is currently a human centered approach to creating content for OpenStreetMap. They review CNN techniques that have been used to automatically extract roads from imagery, and to identify geometric and semantic errors in the data entry process. The ambition is to incorporate ML techniques into various editing suits in order to increase both the speed and quality of data captured.

Cadastral intelligence [49] is the ability to apply spatial intelligence [50] to identify visible cadastral parcel boundaries. This is a key principle supporting approaches to FFP land tenure where citizens are provided with orthophotomaps to identify their properties and delineate the parcel boundaries. This is most successful where the parcel boundaries align or overlap with visual or topographic objects. However, to accelerate this process, ML is now being applied to extract cadastral parcel boundaries automatically from imagery and point clouds, and automated cadastral intelligence [25] is now being delivered. Significant research is taking place within the geospatial information and land administration domains to achieve this holy grail. Although tests have achieved around 50 percent success in rural areas, with poorer results in urban areas [49], the use of point clouds in the analysis could significantly improve this success rate [37]. However, even these initial, relatively low rates of automatic parcel boundary extraction will support and accelerate the current process of humans visually identifying the parcel boundaries.

The Technology & Innovation (TI) Lab of the World Bank have investigated the feasibility of using ML to automate the extraction of parcel features for parcel mapping procedures in Punjab, Pakistan. Three methods were prototyped: segmented and classified image using Support Vector Machine (supervised machine learning methods used for classification, regression analysis and outlier detection); edge detection; and segmented and classified image, using Deep Learning (a subset of ML that is data-driven modelling, which leverages the use of Neural Networks). The most promising results were obtained through the Deep Learning method [24] where there is significant research interest in this state of the art approach.

4. FFP Assessment of Wider Land Administration and Management Opportunities

The review of emerging innovative technology has identified that the technology is supporting and accelerating the adoption of FFP land tenure through data capture automation and acquisition of low-cost imagery, for example. However, can these tools and the associated data be directly applied to a wider set of land administration services and land management functions where the key FFP principles (Table 1) can be applied? Three case studies from across the land sector, sourced from the World Bank, are investigated to review the opportunities of a new and wider set of FFP solutions that can share common geospatial data.

Although these case study projects have been directly managed, financed and often resourced by the World Bank, there is a clear objective of the World Bank to allow countries, especially low-income countries, to adopt these approaches without the direct support of the World Bank. This will allow countries to effectively manage private sector companies to capture the drone and the street level imagery and to feel comfortable to hire and oversee Machine Learning experts to analyze the imagery. This greater accessibility will then allow more countries to take advantage of these new approaches and create higher degrees of sustainability of their projects. This is already happening in countries like Colombia and Indonesia.

4.1. Case Study 1: Valuation and Property Tax Revenues

One of the main sources of revenue for Local Government is through property taxation. There are many ways to categorize and value land and property for taxation, but many countries adopt a market value and comparable sales approach [23]. However, this assumes that there is a vibrant land market operational and there are real estate experts and property taxation experts. However, in many developing countries this is not the case and Local Governments miss out on this crucial source of income. A FFP valuation approach being tested by the World Bank [23] involves extracting building characteristics from drone and street-level imagery using Machine Learning (ML) techniques to support a basic model (Minimum Viable Product) for land and property valuation. The information extracted from the aerial and street-level images includes land coverage and use, and building location, size, height, number of stories, building materials of roof and walls, windows and doors. Billboards and other signs can be used to derive the use of properties. This limited set of measurable tax value parameters can be efficiently maintained, is transparent and can be easily understood by all the stakeholders.

One of the cities used to test this approach by the World Bank's Global Program for Resilient Housing was a 7 km^2 section of Bogota in Colombia. Aerial images were collected using an 'ebee' drone with Post-Processed Kinematic (PPK) and base stations. Street view images were collected approximately every 2 m using a high-resolution (30 megapixel) car-mounted 360-degree or street view camera. Each image (JPG) file was delivered so that it included the "captured at" timestamp and corresponding WGS84 latitude and longitude coordinates. Metadata included the camera facing direction as actually recorded or systematically derived by an interpolation method and were capable of being stored in the image EXIF standard. The survey team used a Trimble MX7 with a Ladybug camera nestled inside a Trimble mobile mapping system. This portable system took 6 pictures every few seconds for a panoramic photo of 30 megapixels. This apparatus also allowed for an increase in geospatial accuracy. The Trimble MX7 system used tightly coupled GNSS and an inertial referencing system. In addition, it was also equipped with GNSS Azimuth Measurement Subsystem (GAMS) to continuously calibrate the inertial measurement unit (IMU) and ensure that azimuth did not drift—maintaining heading accuracy. The following lessons learned were derived from the project:

Advantages:

- A horizontal accuracy of 5–10 cms and a vertical accuracy of 10–20 cms were achieved for the dataset over this urban area.
- Many different types of building objects and characteristics were extracted and registered to support the valuation process.
- The minimal set of tax value parameters is easy to explain and creates a just and transparent taxation process.
- The ML process can be less expensive (particularly at scale), more flexible and faster than the traditional field surveying approach. Due to the low data acquisition costs,

the approach allows for more frequent data collection/maintenance cycles to be established, thus helping to build trust in the building valuations with the citizens.

- The tax value parameters are easy to maintain and can be augmented to support a more sophistication of the valuation model over time when there is demand.
- Standard Geographic Information System (GIS) tools can be used to process, categorize and visualize the data (Figure 2).
- Usually, there is a considerable lag in time between the completion of land registration and cadastral projects and corresponding property valuation projects. In the above scenario, it is feasible to combine the projects and use the same data capture techniques and share the data. This will allow property tax revenues to be brought on-line more quickly.

Disadvantages:

- The project area included properties with rooftops that were very tightly packed together, and the resulting ML polygons sometimes required manual disaggregation.
- The details of the owners/occupiers still require visits to the properties to carry out a survey. However, this can be an integral part of community meetings where the building objects extracted by ML can be verified.
- The capture of the drone and street level imagery requires good ground control or RTK with good access to the GNSS network. This can often be interrupted by foliage and buildings.
- The approach is markedly different from conventional approaches to valuation and professionals will need to be convinced of its adoption.

4.1.1. Check 1: FFP Approach Compliance

The FFP criteria (Table 1) was used to assess whether this case study meets the FFP kitemark and the following checks were carried out (Table 5):

Table 5. FFP Approach Compliance Check.

Flexible	The types of land and building features extracted from the imagery can be modified through ML to support regional variation in building characteristics and a variety of simple valuation models.
Inclusive	The approach involves obtaining drone and street level imagery for all land and buildings in a city, including informal settlements where motorcycles are used in place of cars to obtain street level imagery. In addition, the valuation model is simple and easy to understand by all citizens.
Participatory	The capture process is not participatory, but all citizens and businesses are able to view and understand the land and building features used to calculate their valuation and tax.
Affordable	The costs of capturing and processing the drone and street level imagery are low in comparison to the formal aerial photogrammetry, ground surveying and manual creation of a building cadastre.
Reliable	The land and building features extracted from the imagery are obtained through ML and the algorithms ensure consistency and reliability. In addition, the data is derived quickly after the capture of the imagery, ensuring the data are up to date.
Attainable	The World Bank experience in building local capacity to manage these types of projects and use the associated toolkits indicates that it is not onerous and can be sustained with local resources tapping into remote technical support.
Upgradeable	When the land market matures then the simple approach to valuation modeling can be replaced with a market value or comparable sales approach, for example.

This valuation, land administrative service complies with the FFP criteria and is clearly a cost effective and sustainable solution that can be achieved in a relatively short timeframe and merits the FFP kitemark.

4.1.2. Check 2: Common Supporting Data Compliance

The data captured to support the valuation case study exactly matches the data elements outlined in Table 2. In this case, the point cloud was derived from SfM rather than using a LiDAR sensor. This FFPLA valuation service therefore uses a common set of geospatial data that can be captured once and shared.

4.2. Case Study 2: Resilient Housing

In the context of rapid global urbanization, it is estimated that over the next 15 years the number of people living in substandard housing will double to 3 billion. Disasters, such as hurricanes, floods and earthquakes, are increasing and significantly threaten these homes. Access to formal, affordable, and safe housing is not keeping pace with this extraordinary growth in urbanization. Sustainable communities can be built through making housing safe and resilient to disasters and help to protect lives and livelihoods. The World Bank has a Global Program for Resilient Housing that has developed a methodology 'to predict which houses are at risk of getting damaged by natural hazards, to identify which can be made safe before it is too late, and to connect them with government subsidies and private capital'. The goal of the program is 'to help communities 'BuildBetterBefore' a disaster strikes and to save lives and protect livelihoods after one occurs' [51].

Urbanization across all countries around the world is estimated to increase in the following decades, but at varied rates. It is projected that 68 percent of the world's population will live in urban areas by 2050; an increase from 54 percent in 2016 [52]. India's small to medium sized towns are growing at a rate ranging between 48 to 185 percent [53]. This growth is outpacing the rate at which cities can respond to housing needs and in many countries this urban expansion manifests itself in slums that grow out of control. It is estimated that 900 million people currently live in slums and one in every four people will live in a slum by 2030 [54]. For many cities, just knowing where slums exist, and their condition is a difficult enough problem to solve before cities design interventions to mitigate the problem. However, when drone and street-level images are captured for neighborhoods, Machine Learning (ML) algorithms can be trained to extract specific characteristics of each property (such as façade and rooftop size, condition, material, density and regularity) allowing slums to be recognized. This fundamental information allows neighborhoods to be categorized within cities and interventions prioritized, especially in areas of risk.

An example of this approach is a World Bank resilient housing project in Colombia [55] that addresses the Colombian government's priority of creating resilient and inclusive housing; approximately 23 percent of all Colombian households currently live in substandard and inadequate housing. Quantitative improvements in housing directly reduce poverty and improve living standards. The COVID-19 crisis has highlighted even more the fundamental value of quality housing. Improved housing also mitigates disaster risks and encourages climate sustainability. Colombia ranks 10th globally in terms of economic risk posed by three or more hazards and has the highest recurrence of extreme events in South America [56]. To address this qualitative housing deficit, the government is introducing two housing subsidy programs, including one to incrementally improve housing. 'Experience has shown that under certain conditions it is more cost effective and life-saving to strengthen (retrofit) the existing housing stock and to construct robust buildings than it is to repair damaged buildings after a disaster: to Build Better Before. Studies have shown that every $1 in disaster mitigation saves from $4 to $10 in post-disaster reconstruction costs' [23]. The World Bank has piloted the use of drone and street level images (Figure 3) in combination with ML to identify buildings with specific construction material, size, and use characteristics to create information and monitoring systems for the optimization of housing subsidy allocation.

Figure 3. Street view panoramic from Colombia [source World Bank].

Advantages:

- Specific construction material, size, and use characteristics of properties were extracted to estimate the resilience of the property and to support the housing subsidy allocation process.
- The minimal set of resilience parameters was used to estimate the resilience of buildings. This made it easy to explain create a fair subsidy program.
- The ML process can be less expensive and faster than conducting a traditional survey approach, particularly at scale. The low acquisition costs allows data maintenance cycles to be established and ensures that the housing resilience model is current.
- If there is a need to change the housing resilience model then further property characteristics can be extracted from the imagery using ML.
- Standard Geographic Information System (GIS) tools can be used to process, categorize and visualize the data (Figure 2).
- This approach directly supports the 'BuildBetterBefore' approach being adopted as a priority for many governments and accelerates its implementation.

Disadvantages:

- The results of ML extraction of the building characteristics need to be verified through ground truthing.
- The capture of the drone and street level imagery requires good ground control or RTK with strong access to the GNSS network. This can often be interrupted by foliage and buildings.

4.2.1. Check 1: FFP Approach Compliance

The FFP criteria (Table 1) was used to assess whether this case study meets the FFP kitemark and the following checks have been carried out (Table 6):

Table 6. FFP Approach Compliance Check.

Flexible	The types of building features extracted from the imagery can be modified through ML to analyze the resilience of buildings to different types of disasters, e.g., flood, earthquake, hurricanes.
Inclusive	The approach involves obtaining drone and street level imagery for all buildings in a city, including informal settlements. This supports the formulation of strategies for entire cities rather than just regions of the city.
Participatory	The capture process is not participatory, but the citizens and businesses in most need of improving the resilience of their buildings will be targeted for housing subsidy.

Table 6. *Cont.*

Affordable:	The costs of capturing and processing the drone and street level imagery are low in comparison to manually creating an inventory of all buildings in a city.
Reliable	The building features extracted from the imagery are obtained through ML and the algorithms ensure consistency and reliability. The building feature data are derived quickly after the capture of the imagery, ensuring the data are up to date.
Attainable	The World Bank experience in building local capacity to manage these types of projects and use the toolkits indicates that it can be sustained with local resources tapping into remote technical support.
Upgradeable	The model and the building features used to evaluate building resilience can be upgraded over time and extracted with ML.

This housing resilience, land management function complies with the FFP criteria and is an affordable and easily maintained solution that can be achieved quickly and merits the FFP kitemark. A further example of this effective land management function is described below.

4.2.2. Check 2: Common Supporting Data Compliance

The data captured to support the housing resilience case study exactly matches the data elements outlined in Table 2. Although in this case, the point cloud was derived from SfM rather than using a LiDAR sensor and thus has a lower density. This FFP housing resilience service to support the optimization of housing subsidy allocation therefore uses a common set of geospatial data that can be captured once and shared.

4.3. *Case Study 3: Solid Waste Management*

It is estimated that the annual global damage of plastic pollution to marine ecosystems exceeds USD 13 billion. In the Asia-Pacific Economic Cooperation (APEC) alone, this causes over USD 1.3 billion per year to be lost to the tourism, fishing, and shipping industries [57]. A World Bank project in Cambodia identified that plastic waste pollution forms a particularly crucial part of solid waste mismanagement. The Mekong River is amongst the most polluted rivers worldwide and in Phnom Penh alone, about 10 million plastic bags are used every day [58]. However, dependable statistics on the varieties and quantities of plastic in the Cambodian rivers and sea are not available. These statistics are essential to establish robust policies and investment strategies to reduce plastic waste and prevent damage to the environment and the economy.

The World Bank's Cambodian project [59] piloted an innovative approach to monitoring and gathering these essential statistics. An RGB camera was mounted on a drone and images of polluted rivers, beaches, and urban canals were collected at different flying heights. The images were then analyzed using Machine Learning (ML) (Convolutional Neural Networks), and the automated image analysis enabled a wide range of the plastic pollution characteristics to be derived, including the detection of pollution hotspots, the size of areas covered with plastic, volume estimation, and importantly, classification of more than 10 different types of plastics. This remote sensing approach was combined with scientific field surveys to ground truth and improve the process of classification. A second phase of piloting is planned, and this will involve the use of a multispectral sensor to improve the overall data quality in terms of detection and classification of plastic waste.

The introduction of solid waste management policies requires significant investment and a key, sustainable source of this financing is normally achieved by municipalities through collecting waste fees from households. This requires the registration of properties and the creation of street/property gazetteers to underpin the fee collection system. This geospatial and property information can then be re-used by municipalities to support a range of their core activities, including the collection of property taxes, spatial/urban planning as well as environmental resilience planning [58]. In this Cambodian example, the introduction of solid waste management is the policy driver that is generating the base data identified in this study to support a wide range of other FFP land administration and management applications.

Advantages:

- A wide range of the plastic pollution characteristics was derived to help formulate a more effective set of solid waste management policies.
- These inputs to policy formulation through the ML process are much cheaper and faster than conducting a traditional survey approach.
- The ability to identify and classify more than 10 specific, different types of plastics allows a much more targeted and effective anti-pollution policy to be formulated and implemented.
- The on-going monitoring of the plastic pollution allows the impact and effectiveness of the policies to be measured.
- The level of resolution of the drone imagery required for the ML process is high to allow the identification of specific plastic pollutants. This is fully compatible with the resolution required for FFP land tenure applications.
- To attempt to achieve the monitoring and classification of plastics in rivers by manual means would be problematic, resource intensive and expensive, normally beyond the means of the local environmental agencies involved. This FFP approach provides a very low-cost solution and delivers significant benefits through more effective policies to reduce the level of plastic waste.

Disadvantages:

- The results of ML extraction of the plastic characteristics need to be verified through ground truthing, although the future use of a multispectral sensor will improve the overall data quality.
- This application does not need to have a high degree of positional accuracy when the drone imagery is georeferenced. To impose this requirement on capturing the drone imagery with good ground control or RTK would impose overheads on the project.
- Properties need to be registered and a street/property gazetteer created to underpin the fee collection system. However, the gazetteer can have many other uses within urban management, including property tax revenues.

4.3.1. Check 1: FFP Approach Compliance

The FFP criteria (Table 1) was used to assess whether this case study meets the FFP kitemark and the following checks have been carried out (Table 7):

Table 7. FFP Approach Compliance Check.

Flexible	The types of waste material extracted and classified from the imagery can be modified through ML to be attuned with the types of pollution occurring in the area of interest.
Inclusive	The monitoring of pollution can be performed across an entire city or just known hotspots.
Participatory	Communities could be involved in ground truthing the results of ML.
Affordable	The costs of capturing and processing the drone are low in comparison to manually creating an inventory of pollutants.
Reliable	The waste material features extracted from the imagery are obtained through ML and the algorithms ensure consistency and reliability.
Attainable	The World Bank experience in building local capacity to manage these types of projects and use the toolkits indicates that it can be sustained with local resources tapping into remote technical support.
Upgradeable	The pollutant types used to formulate solid waste management policies can be upgraded over time and extracted with ML.

This solid waste management land management function complies with the FFP criteria and is affordable, provides ongoing monitoring of the impact of policies and can be achieved quickly and merits the FFP kitemark.

4.3.2. Check 2: Common Supporting Data Compliance

The data captured to support the solid waste management case study does not fully match the data elements outlined in Table 2. The georeferencing of the drone imagery is less rigorous than that required for FFP land tenure and there is less of a requirement to create sufficient overlaps of the imagery to support 3D photogrammetric models. To apply these more stringent requirements on the data capture process will increase the complexity and costs of the project. However, these costs could be shared across a number of other projects using the data. This FFP solid waste management service to support the more targeted solid waste management policies can use a common set of geospatial data. However, this will impose higher costs on the project and required more skilled personnel.

4.4. *Other FFP Land Management Function Opportunities*

The generation of drone and street level imagery along with the derived data outlined in the FFP land administration services and FFP land management functions above, provide opportunities to support a wide variety of derivative, innovative urban land management functions. Examples of these are provided below.

4.4.1. Master Planning

The UN estimates that India currently accounts for 11 percent of the world's population and this will increase to around 15 percent by 2030. The Census of India has identified that the urban population has more than double from 14 percent at the time of Independence to around 32 percent in 2011 and the number of cities/towns has increased from 5161 to 7933 over this period. Urbanization will increase to more than 50 percent by 2051. India has been finding it increasingly difficult to plan the development of its cities. The majority of urban settlements in India can be considered unplanned with uncontrolled growth leading to loss of agricultural land in peri-urban areas, large numbers of informal settlements, significant environmental damage, and a poor quality of life for its citizens [53].

The Master Plan/Development Plan is a fundamental element of urban land management and facilitates sustainable development through land use allocation. These Master/Development plans are designed to manage development over a 20-year period with periodic reviews and revisions every five years. These plans require base mapping to be prepared to support the collection of existing land use surveys and socio-economic data. However, less than a third of the over 8000 towns and cities have master/development plans. The main reasons for this backlog are the lack of quality geospatial data along with lack of capacity and funding in small to medium sized towns that make up 93 percent of the towns and cities. The master planning of these small and medium sized towns is a priority for the Indian government as they are growing at a rate ranging between 48 and 185 percent. The development of these urban areas would support a reduction in the considerable migration to large cities and help to sustain growth in the surrounding rural areas [53].

To accelerate the master planning process across small and medium sized towns, the Ministry of Housing and Urban Affairs (MoHUA) has decided to promote the use of drones to generate the base maps required for the master planning process. This cheaper alternative to manned aerial photography will unlock this capability. The MoHUA has produced a draft set of process designs and standards to use drones in the generation of 1:1000 standardized mapping. This was required to support geospatial data of high accuracy and better spectral and spatial resolution at low cost as well as the 3D models derived from the sensor data provided by drones. In addition, the new standards supported more effective interoperability and seamless vertical integration from Local to Regional Plans These specifications will be tested in ten towns during 2021 before being scaled up across India [53]. This drone-based initiative in India, primarily focused on supporting master planning, opens up significant opportunities to re-use the drone data to support a wide range of land administration services and land management functions highlighted in this paper. For example, Machine Learning could be used to support change detection crucial in modeling city growth patterns.

4.4.2. Urban Digital Twins

Digital twins are a digital, virtual representation of reality and over the past decade have been used to create 3D, often real-time, representations of cities. Singapore, Glasgow, Boston, Jeonju and Jaipur were the early adopters. It is projected that by 2025, the number of urban digital twins globally will be over 500 and the market will have grown from USD 3.8 billion in 2019 to USD 35.8 billion per year [59]. These integrated and comprehensive geospatial information systems are given the accolade of providing the continuous innovation needed to construct new smart cities.

Urban digital twins have a wide set of applications; utilities can plan infrastructure coverage; environment managers can conduct scenario analyses through the simulation of the potential impact of natural disasters like flooding; telecommunications companies can more effectively plan their dense networks of 5G antennae in 3D; urban planners can perform better land use analysis and engage with citizens through immersive visualization experiences with AR/VR applications; the car industry is designing navigation solutions for autonomous vehicles; green energy providers can optimize energy savings and solar capacity; emergency responders can conduct disaster readiness simulations to build city resilience; and the entertainment industry can build seamless computer-generated imagery integration [60]. The result is much more resilient and efficient cities—smart cities.

These 3D urban digital twins are created from multiple sources, including optical and satellite imagery, LiDAR, IoT and crowdsourcing from the citizens. The data created by the use of the emerging, innovative technology being applied to FFP land tenure can be directly used to build these urban digital twins and accelerate the creation of smart cities.

4.4.3. Visualizations to Support Gaming

The point clouds and 3D models generated by this innovative technology are also being adopted by the gaming sector to create more realistic, virtual worlds for enhancing the gaming experience. Ordnance Survey GB have been involved in an initiative to use the OS OpenData products to create a Minecraft world representing over 224,000 km^2 of Great Britain [61]. HERE Technologies has released high-fidelity, 3D models of 75 city centers from around the world to give software developers the geospatial data required: data from over 100,000 sources and with 80 billion API calls per month. This provides the entertainment industry with seamless computer-generated imagery integration to build real-world visualizations of cities [60].

4.4.4. Support of Financial Risk Prediction

The base datasets being derived from FFP land tenure also have the potential to be used commercially. This could generate revenue to help finance the implementation and maintenance of land administration and land management applications. For example, street level imagery is finding increasingly innovative uses. Researchers [62] have used images of residences from Google Street View to predict the likelihood that a policyholder would make a car accident claim. Residences are classified using the type (detached house, terraced house, block of flats, etc.), age, and condition extracted from the imagery and the characteristics used to model and predict the likelihood of a car insurance claim. The correlation was found to be surprisingly strong and when these residence factors were included in the insurer's risk model, the predictive power was increased by 2 percent. This use of residences could equally be applied to improve predicting risk for insurance companies, financial services, and health-care organizations [62]. This use of street level imagery raises important issues around data privacy laws and whether clients' consent given to a company to store their addresses also provides consent to store information about the appearance of their houses. Many countries, including Germany and Austria, have raised privacy concerns over Google's Street View product and coverage has stopped in India [63]. Users of street level imagery need to be aware of the potential limitations imposed through licensing restrictions and privacy laws.

4.4.5. Disaster Risk Management

There were 28 million new displacements associated with conflict (10.8 M) and disasters (17.2 M), including hurricanes, droughts and floods, across 148 countries and territories in 2018 [64], and there is now growing evidence that the impact of natural disasters and associated vulnerability are intensifying. This influence is seriously challenging development and the goal of achieving the SDGs [65]. Countries must implement disaster risk reduction and resilience building measures, rather than just respond to one-off disaster events [65] and provide communities with greater support to adapt to these demanding environmental conditions.

Disaster Risk Reduction is 'the concept and practice of reducing disaster risks through systematic efforts to analyze and manage the causal factors of disasters, including through reduced exposure to hazards, lessened vulnerability of people and property, wise management of land and the environment, and improved preparedness for adverse events' [66]. The management of disaster risk involves a whole range of phases involved in the lifecycles of disasters, including disaster prediction (simulation and visualization), prevention, preparedness and mitigation, emergency response, evacuation planning, search and rescue, shelter operations, and post-disaster restoration and monitoring [67]. Fundamental to managing all of these phases is access to geospatial data and information contained within land administration services. For example, prediction through flood modelling requires DEM and DTM information, the restitution of properties following a disaster requires land registration and cadastral information to re-establish security of tenure for citizens, and financial compensation schemes need valuation information about the impacted land and properties. Without this fundamental information, the ability to manage disaster risk is seriously reduced.

The baseline geospatial data, a key part of the Minimum Viable Product identified in this study, can be captured efficiently and cheaply to derive the information need for land administration and management that can then be used to manage resilience and disaster risks. The recovery from the effects of a hazard can subsequently be achieved more quickly and efficiently, allowing essential basic structures and functions to be restored. The drone and streel level imagery alone are a valuable information asset to support many phases of the disaster management process.

These geospatial and land information services records and data need to be safeguarded to avoid any physical destruction and the information needs to be digitalized and critically, good practice data management practices applied.

5. Discussion

These case studies reveal that the FFP paradigm that has predominantly been applied to the land tenure aspects of land administration has been successfully adopted and implemented in wider land administration services and land management functions. A key element of the FFP approach is the use of a well-defined Minimum Viable Product as an entry level solution that meets the requirements of stakeholders, is easily understood, affordable and fast to implement. The Valuation and Property Tax Revenues case study is an excellent example of how to apply a Minimum Viable Product [23]. The adopted Minimum Viable Product valuation model is simple, easily understood, efficiently captured and maintain through Machine Learning (ML), and can be upgraded when required. Solutions should not be driven indiscriminately by technology, but should be primarily guided by user needs. However, in the case of these emerging FFP solutions, the innovative technologies described in this study have formed a symbiotic relationship with the FFP paradigm and are creating opportunities for implementing new FFP approaches that would not otherwise be feasible.

The use of these innovative technologies requires specialized skills and this capacity to effectively use and maintain the technology is not normally found within the governments most in need of adopting the technology. However, the World Bank's approach is to provide management skills and guidance to governments to allow the effective management of partners in the private and academic sectors to provide these sustainable services. This is providing opportunities for private sector companies to build this specific capacity.

The analysis of the emerging technologies and data used to support the FFP case study solutions has revealed that there is a common set of geospatial datasets that can be captured once and shared across many other land administration services and land management functions in an urban environment. FFP land tenure solutions capture data with the most demanding specifications and when these data are augmented with street level imagery then the combined data portfolio supports a surprisingly wide range of uses. This study has identified support for a range of land management functions, including urban digital twins, master planning, disaster risk management and financial risk prediction. So, although there are financial and resource overheads associated with data capture in the initial FFP solution implemented in an urban environment, the return on investment is significant since many more FFP solutions can be implemented using this shared portfolio of datasets.

Most land tenure programs simply focus on securing land rights for citizens, but this is just the start of their journey to leverage the economic potential of their land and property. For example, smallholder farmers (there are an estimated 500 million smallholder farmer families worldwide [68]) also require access to technical and financial services to improve their livelihoods. The incremental and fragmented delivery of these services makes their sustainable prosperity much more challenging. Therefore, an integrated package of technical and financial services, including security of tenure, would achieve much higher benefits from a portfolio of interventions. However, donors and Non-Governmental Organizations (NGOs) find this more holistic approach difficult to adopt due to their organizational divisions, silos of professional skills, difficulty in managing multi-faceted programs and complexity of interfacing with a range of recipient government ministries and departments. None of these current perceived restrictions are showstoppers, but adopting new approaches will require significant cultural and institutional change.

The FFP paradigm and the emerging, enabling technologies provide compelling opportunities to rethink how land administration and land management programs are designed, integrated and implemented. This research has identified that a common set of base geospatial data, collected by and derived from drone and street level imagery, can support a wide set of FFP land administration services and management functions, ranging from land registration and cadastre to valuation and urban resilience. Therefore, it is feasible that land intervention programs, initially involved in a single land administration or land management activity within a region, could consider capturing this identified base geospatial data to a threshold of quality that would then also support a wide variety of complementary land administration services and land management functions. This more holistic approach would provide the opportunity for single land interventions projects to be integrated into a wider program of land administration services and land management functions delivering a more significant set of socio-economic benefits. For example, a land and property valuation project could be implemented in parallel with other projects to help finance this wider set of projects through property-based tax revenues. Although FFP land tenure projects deliver a fundamental framework of land rights information that normally supports subsequent land administration and land management projects, it is still feasible to reverse this order and build FFP land tenure projects on the back of other land projects.

This integrated approach to program implementation has significant implications for the future design of FFPLA projects where there will be opportunities to deliver an integrated portfolio of land administration services and land management functions more quickly and efficiently compared to the traditional approach of delivering independent, single projects.

Too often, land administration projects are provided with the capital budget to implement the project, but lack the revenue budgets, resources and processes for on-going maintenance of data underlying the services. An example of this situation occurred in Rwanda where, after the very successful implementation of a national FFPLA project in just five years, subsequent poor maintenance of the land rights jeopardized the original investment [69]. Since the FFP approach involves defining a Minimum Viable Product [9] as a starting point, the corresponding data models to support the FFP projects are much simpler with the range and complexity of the data limited to meet initial needs. This makes the maintenance of the data less onerous and more achievable with fewer resources. A further advantage is achieved with the proposed integrated FFP approach where a common set of baseline data supports many FFP land administration services and land management functions. In addition, many of the data are semi-automatically extracted from the imagery using ML, leading to even more efficient data capture for maintenance.

The FFPLA concept is composed of three frameworks: spatial, legal and institutional. This study has primarily focused on the spatial framework and identified how emerging technologies are facilitating the wider use of the FFP approach in land management functions. However, this wider adoption is dependent upon similarly supportive legal and institutional frameworks. For example, how can institutions be encouraged or mandated to share both geospatial datasets and the costs of their collection and maintenance? Does the legal and regulatory framework within a country need to be adapted to accommodate these new FFP approaches, including valuation models, data quality specifications and the move from national mapping to more autonomous city mapping? How can the current project siloes be dismantled to be replaced by integrated programs sharing data, costs and resources? Would the results of a socio-economic impact assessment of implementing a FFP approach deliver a positive return on investment in comparison to conventional approaches and thus convince politicians to invest? Further research is required on how legal and institutional frameworks should be adapted to be more compliant with this new vision.

6. Conclusions

FFPLA, in particular FFP land tenure, is gaining acceptance around the world as a fundamental methodology to scale up the provision of security of tenure [5–8] and hopefully coalesce the currently excluded 70 percent of the world's population into formal land administration systems within a generation. There are many aspects of the FFP methodology that support scalability, including participatory, flexible, inclusive and affordable. However, emerging innovative technologies [14] are enablers and will continue to significantly accelerate the implementation of the FFP methodology. Drone and street level imagery are affordable, require less professional support to operate, can adequately cover small and medium sized towns, provide the required resolution and accuracy to support cadastral applications, and point clouds can be derived or directly captured with LiDAR. Orthophotos can then be directly used in the field in a participatory process with the citizens to identify and validate visible parcel boundaries. Machine Learning can increasingly be used to automatically extract parcel boundaries from the imagery, with support from the point clouds, to aid in the parcel boundary validation process with citizens.

The base data captured for FFP land tenure services can be augmented with street level imagery and then be directly re-used to support a much wider set of FFP land administration services and land management functions. For example, building features and characteristics can be automatically extracted from the drone and street level imagery using Machine Learning to investigate aspects of city resilience and to better target funds to mitigate risks and improve resilience. The data can also be used to create urban digital twins that support a range of opportunities for the sustainable management of urban environments.

The base data also provide opportunities to generate revenue to support and accelerate the implementation of land administration services and land management functions. For example, Machine Learning derived building characteristics can support land and property valuation and the downstream generation of property tax revenues, and images of properties could be sold to insurance companies, financial services, and health-care organizations to improve risk prediction.

The ability to directly re-use the identified base data across a wide range of FPP land administration services and land management functions will allow international funding institutions and development agencies involved in financing land initiatives to reconsider how their land related programs are sequenced, integrated and financed. Traditionally, land tenure projects are prioritized followed sequentially by further land administration services and land management functions. However, the sharing of the base data across a number of FFP solutions can support the parallel implementation of these FFP solutions. This will accelerate the implementation, reduce the associated costs and support more effective maintenance of the data. A more holistic approach to administering and managing land through the delivery of an integrated portfolio of land interventions will leverage the economic potential of land and property more effectively and quickly. This new method will impact the approach to land governance and the institutional arrangements that deliver sustainably managed land, property and natural resources. These alterations should be reflected in analytical tools, such as the World Bank's Land Governance Assessment Framework (LGAF) [70], to encourage these transitions. However, this will require a significant cultural, professional and institutional change from all stakeholders involved. The administration and management of land is normally highly fragmented across a wide number of land professions in specific niches. However, the wider adoption of this FFP approach provides an opportunity for the land professionals to reconsider their professional domains and re-align their professional roles in a more effective, integrated network of services. Although challenging, it is a great opportunity that must be embraced to ensure that land is sustainably managed and delivers significant value to societies and their economies.

Author Contributions: K.K.: visualization; conceptualization; validation; investigation; writing—review and editing. S.A.: visualization; conceptualization; validation; methodology; investigation; writing—review and editing. R.M.: visualization; conceptualization; methodology; investigation; formal analysis; writing—original draft preparation. All authors have read and agreed to the published version of the manuscript.

Funding: This research received no external funding.

Institutional Review Board Statement: Not applicable. All the material used in the study has already been published in World Bank Group reports and no new studies were undertaken.

Data Availability Statement: All the material is referenced in the manuscript and is available on the web.

Acknowledgments: The authors would like to thank Stig Enemark for encouraging the publication of this paper, Simon Wills for providing help with the references, and William Mackaness for providing excellent guidance.

Conflicts of Interest: The authors declare no conflict of interest.

References

1. McLaren, R. How Big Is Global Insecurity of Tenure? GIM International. Available online: http://member.gim-international.com/Geomares/magazine/gim/magazine.jsp (accessed on 2 February 2021).
2. Enemark, S.; Bell, K.C.; Lemmen, C.; McLaren, R. *Fit-for-Purpose Land Administration, In Joint FIG/World Bank Publication, FIG Publication 60*; International Federation of Surveyors (FIG): Copenhagen, Denmark, 2014; p. 44.
3. Wehrmann, D. *Land Governance: A Review and Analysis of Key International Frameworks*; United Nations Human Settlements Programme (UN-Habitat): Nairobi, Kenya, 2017; p. 70.
4. McLaren, R.; Bennett, R.M.; Lemmen, C.H.J. World Bank conference on land and poverty 2013. *GIM Int.* **2013**, *27*, 38–39.
5. UN-HABITAT. *Secure Land Rights for All*; United Nations Human Settlements Programme (UN-HABITAT): Nairobi, Kenya, 2008; p. 47.
6. UN-HABITAT. *Designing a Land Records System for the Poor*; United Nations Human Settlements Programme (UN-HABITAT): Nairobi, Kenya, 2012; p. 39.
7. Bennett, R.M.; Alemie, B.K. Fit-for-purpose land administration: Lessons from urban and rural Ethiopia. *Surv. Rev.* **2016**, *48*, 11–20. [CrossRef]
8. Zevenbergen, J.; Augustinus, C.; Antonio, D.; Bennett, R. Pro-poor land administration: Principles for recording the land rights of the underrepresented. *Land Use Policy* **2013**, *31*, 595–604. [CrossRef]
9. Enemark, S.; McLaren, R.; Lemmen, C. *Fit-For-Purpose Land Administration: Guiding Principles for Country Implementation*; United Nations Human Settlements Programme (UN-Habitat): Nairobi, Kenya, 2016; p. 132.
10. Gillingham, P.; Buckle, F. *Rwanda Land Tenure Regularisation Case Study*; Evidence on Demand: London, UK, 2014; p. 40.
11. TIERRA EN PAZ. Land in Peace. Available online: https://tierraenpaz.com/home (accessed on 6 January 2021).
12. UN-HABITAT; GLTN. *Fit-For-Purpose Land Administration—A Country Level Implementation Strategy for Nepal*; United Nations Human Settlements Programme (UN-HABITAT): Nairobi, Kenya, 2018; p. 62.
13. World Bank. Nearly 4.3 Million to Benefit from Indonesia's Sustainable Land Management. Available online: https://www.worldbank.org/en/news/press-release/2018/07/20/indonesia-sustainable-land-management (accessed on 6 January 2021).
14. McLaren, R.; Fairlie, K.; D'Souza, G.; Kelm, K. New Technology and Emerging Trends: The State of Play for Land Administration. In Proceedings of the World Bank Conference on Land and Poverty, Washington, DC, USA, 19–23 March 2018.
15. McLaren, R. *Crowdsourcing Support of Land Administration—A New, Collaborative Partnership between Citizens and Land Professionals*; Royal Institute of Chartered Surveyors: London, UK, 2011; p. 12.
16. Persha, L.; Zodrow, G. *Cost and Time Effectiveness Study—Final Report, Mobile Application to Secure Tenure Pilot in Tanzania*; United States Agency for International Development (USAID): Washington, DC, USA, 2017; p. 90.
17. Amazon. Prime Air. Available online: https://www.amazon.com/Amazon-Prime-Air/b?ie=UTF8&node=8037720011 (accessed on 11 January 2021).
18. Mesar, T.; Lessig, A.; King, D.R. Use of Drone Technology for Delivery of Medical Supplies during Prolonged Field Care. *J. Spec. Oper. Med.* **2018**, *18*, 34–35. [PubMed]
19. Stöcker, C.; Ho, S.; Nkerabigwi, P.; Schmidt, C.; Koeva, M.; Bennett, R.; Zevenbergen, J. Unmanned Aerial System Imagery, Land Data and User Needs: A Socio-Technical Assessment in Rwanda. *Remote Sens.* **2019**, *11*, 1035. [CrossRef]
20. Kelm, K.; Tonchovska, R.; Volkman, W. Drones for Peace: Part II Fast and Inexpensive Spatial Data Capture for Multi-Purpose. In Proceedings of the World Bank Conference on Land and Poverty, Washington, DC, USA, 24–27 March 2014.
21. Pajares, G. Overview and Current Status of Remote Sensing Applications Based on Unmanned Aerial Vehicles (UAVs). *Photogramm. Eng. Remote Sens.* **2015**, *81*, 281–330. [CrossRef]
22. Stöcker, C.; Bennett, R.; Nex, F.; Gerke, M.; Zevenbergen, J. Review of the Current State of UAV Regulations. *Remote Sens.* **2017**, *9*, 459. [CrossRef]
23. Triveno, L.; Antos, S.A.; Koers, J.; Endo, V. Coupling Imagery from Drones and Street-View with Proper Incentives To Promote Sustainable Urban Cadasters in Developing Countries. In Proceedings of the World Bank Conference on Land and Poverty, Washington, DC, USA, 25–29 March 2019.
24. World Bank. *Use Case Viability Report: AI for Parcel Mapping. Technology and Innovations Lab*; Version 0.1; 14 December 2020; World Bank Group: Washington, DC, USA, 2020.
25. Nyandwi, E.; Koeva, M.; Kohli, D.; Bennett, R. Comparing Human versus Machine-Driven Cadastral Boundary Feature Extraction. *Remote Sens.* **2019**, *11*, 1662. [CrossRef]
26. UN-FAO; UN-Habitat. *Towards Improved Land Governance*; FAO: Rome, Italy, 2009; pp. 1–55.
27. UN-FAO. *Voluntary Guidelines on the Responsible Governance of Tenure of Land, Fisheries and Forests in the Context of National Food Security (VGGTs)*; FAO: Rome, Italy, 2012; pp. 3–9.
28. Williamson, I.; Enemark, S.; Rajabifard, A. *Land Administration for Sustainable Development*; ESRI Press: Dublin, Ireland, 2010; p. 512.
29. Deininger, K. *Land Policies for Growth and Poverty Reduction*; World Bank Group: Washington, DC, USA, 2003; Executive Summary; pp. ix–xxii. Available online: http://documents.worldbank.org/curated/en/485171468309336484/Land-policies-for-growth-and-poverty-reduction (accessed on 4 May 2021).
30. Open Geospatial Consortium. OGC Standards. Available online: https://www.ogc.org/docs/is (accessed on 15 June 2021).

31. Giordan, D.; Adams, M.S.; Aicardi, I.; Alicandro, M.; Allasia, P.; Baldo, M.; De Berardinis, P.; Dominici, D.; Godone, D.; Hobbs, P.; et al. The use of unmanned aerial vehicles (UAVs) for engineering geology applications. *Bull. Eng. Geol. Environ.* **2020**, *79*, 3437–3481. [CrossRef]
32. Hexagon. Hexagon's 'RTK from the Sky' Brings Instant GNSS Accuracy Worldwide. Available online: https://eijournal.com/news/products-2/hexagons-rtk-from-the-sky-brings-instant-gnss-accuracy-worldwide (accessed on 29 November 2020).
33. Westoby, M.J.; Brasington, J.; Glasser, N.F.; Hambrey, M.J.; Reynolds, J.M. 'Structure-from-Motion' photogrammetry: A low-cost, effective tool for geoscience applications. *Geomorphology* **2012**, *179*, 300–314. [CrossRef]
34. Mumbone, M.; Bennett, R.M.; Gerke, M.; Volkmann, W. Innovations in boundary mapping: Namibia, customary lands and UAVs. In Proceedings of the 16th Annual World Bank Conference on Land and Poverty, Washington, DC, USA, 23–27 March 2015.
35. Ramadhani, S.A.; Bennett, R.M.; Nex, F.C. Exploring UAV in Indonesian cadastral boundary data acquisition. *Earth Sci. Inform.* **2018**, *11*, 129–146. [CrossRef]
36. World Bank. *UAV Pilot in Ho Chi Minh City. Final Report under Vietnam Land Governance ASA*; World Bank Group: Washington, DC, USA, 2019.
37. Luo, X.; Bennett, R.; Koeva, M.; Quadros, N. Cadastral Boundaries from Point Clouds? Towards Semi-Automated Cadastral Boundary Extraction from ALS Data. Available online: https://www.gim-international.com/content/article/cadastral-boundaries-from-point-clouds (accessed on 26 November 2020).
38. Ingedata. How to Teach Vision to Computers: 3D Point Cloud Annotation and LiDAR. Available online: https://www.ingedata.net/blog/LiDAR-3d-point-cloud (accessed on 26 November 2020).
39. Van Rens, J. The Future of Lidar is Critical to the Future of Our World. Available online: https://www.gim-international.com/content/article/the-future-of-lidar-is-critical-to-the-future-of-our-world?utm_source=newsletter&utm_medium=email&utm_campaign=GIM+International (accessed on 26 November 2020).
40. Salach, A.; Bakuła, K.; Pilarska, M.; Ostrowski, W.; Górski, K.; Kurczyński, Z. Accuracy Assessment of Point Clouds from LiDAR and Dense Image Matching Acquired Using the UAV Platform for DTM Creation. *ISPRS Int. J. Geo Inf.* **2018**, *7*, 342. [CrossRef]
41. Google. Cheers to Street View's 10th birthday! Available online: https://www.blog.google/products/maps/cheers-street-views-10th-birthday/ (accessed on 26 November 2020).
42. Wikipedia. Google Street View. Available online: https://en.wikipedia.org/wiki/Google_Street_View (accessed on 26 November 2020).
43. Mapillary. Train Recognition Models for Street Scenes. Available online: https://www.mapillary.com/datasets (accessed on 15 June 2021).
44. KartaView. Explore the World. Available online: https://kartaview.org/map/@-6.407518221863476,107.04770438649952,8z (accessed on 15 June 2021).
45. Moore, A.; Artificial Intelligence at Carnegie Mellon University. Machine Learning Department at Carnegie Mellon University, USA. 2018. Available online: https://www.youtube.com/watch?v=HH-FPH0vpVE (accessed on 26 November 2020).
46. Mitchell, T.M. *The Discipline of Machine Learning*; School of Computer Science, Carnegie Mellon University: Pittsburg, PA, USA, 2006; p. 9.
47. Iriondo, R. Machine Learning (ML) vs. Artificial Intelligence (AI)—Crucial Differences. Available online: https://pub.towardsai.net/differences-between-ai-and-machine-learning-and-why-it-matters-1255b182fc6 (accessed on 26 November 2020).
48. Vargas-Munoz, J.E.; Srivastava, S.; Tuia, D.; Falcão, A.X. OpenStreetMap: Challenges and Opportunities in Machine Learning and Remote Sensing. *IEEE Geosci. Remote Sens. Mag.* **2021**, *9*, 184–199. [CrossRef]
49. Bennett, R. Cadastral Intelligence, Mandated Mobs, and the Rise of the Cadastrobots. In Proceedings of the FIG Working Week 2016, Recovery from Disaster, Christchurch, New Zealand, 2–6 May 2016; p. 3.
50. Campbell, L.C.; Campbell, B.; Dickinson, D. *Teaching and Learning Through Multiple Intelligences*, 3rd ed.; Allyn and Bacon: Boston, MA, USA, 2004; p. 368.
51. World Bank. Global Program for Resilient Housing. Available online: https://www.worldbank.org/en/topic/disasterriskmanagement/brief/global-program-for-resilient-housing (accessed on 19 February 2020).
52. Ritchie, H.; Roser, M. Urbanization. *Our World Data*. 2018. Available online: https://ourworldindata.org/urbanization (accessed on 1 July 2021).
53. Government of India. *Formulation of GIS Based Master Plans for Small and Medium Towns—Design and Standards for Application of Drone/UAV Technology*; Town & Country Planning Organization, Ministry of Housing and Urban Affairs: New Delhi, India, 2020; p. 153.
54. Habitat for Humanity. What Is a Slum? Available online: https://www.habitatforhumanity.org.uk/what-we-do/slum-rehabilitation/what-is-a-slum/ (accessed on 20 January 2021).
55. World Bank. *Colombia: Resilient and Inclusive Housing Project (P172535)*; World Bank Group: Washington, DC, USA, 2020; p. 11.
56. World Bank. Climate Change Knowledge Portal, Columbia, Vulnerability. Available online: https://climateknowledgeportal.worldbank.org/country/colombia/vulnerability (accessed on 15 June 2021).
57. UNEP. *Valuing Plastics: The Business Case for Measuring, Managing and Disclosing Plastic Use in the Consumer Goods Industry*; United Nations Environment Programme (UNEP): Geneva, Switzerland, 2014; p. 12.
58. World Bank. *Cambodia—Improving Financial and Environmental Sustainability in Solid Waste Management and Plastics*; World Bank Group: Washington, DC, USA, 2020.

59. Spatial Source. Adaptation to Covid-19 to Drive Digital Twin Uptake: Report. Available online: https://www.spatialsource.com.au/company-industry/adaptation-to-covid-19-to-drive-digital-twin-uptake-report?utm_medium=email&utm_campaign=SS%20Newsletter%20140121&utm_content=SS%20Newsletter%20140121+CID_c17edd3b3aa80b4e716cd82d7936deb1&utm_source=Campaign%20Monitor&utm_term=READ%20MORE (accessed on 19 January 2021).
60. GIM International. HERE Unveils 3D City Models for Industries to Build Reality-Based Applications. Available online: https://www.gim-international.com/content/news/here-unveils-3d-city-models-for-industries-to-build-reality-based-applications (accessed on 19 January 2021).
61. Ordnance Survey. Minecraft: Creating a Map of Great Britain. Available online: https://www.ordnancesurvey.co.uk/newsroom/blog/minecraft-creating-a-map-of-great-britain (accessed on 20 January 2021).
62. Emerging Technology from the arXiv. How a Google Street View Image of Your House Predicts Your Risk of a Car Accident. MIT Technology Review. Available online: https://www.technologyreview.com/2019/04/30/135556/how-a-google-street-view-image-of-your-house-predicts-your-risk-of-a-car-accident/ (accessed on 10 December 2020).
63. EPIC. Investigations of Google Street View. Available online: https://epic.org/privacy/streetview/ (accessed on 2 February 2021).
64. Internal Displacement Monitoring Centre. *Global Report on Internal Displacement. Summary May 2019*; Internal Displacement Monitoring Centre: Geneva, Switzerland, 2019; p. 16.
65. UNISDR; WMO. *Disaster Risk and Resilience*; United Nations International Strategy for Disaster Reduction (UNISDR): Geneva, Switzerland, 2012; p. 13.
66. UNISDR. *2009 UNISDR Terminology on Disaster Risk Reduction*; United Nations International Strategy for Disaster Reduction (UNISDR): Geneva, Switzerland, 2009; p. 13.
67. World Bank. *Improving Resilience and Resilience Impact of National Land and Geospatial Systems*; Centre for Spatial Data Infrastructures and Land Administration, the University of Melbourne: Parkville, Australia, 2019; p. 24.
68. World Bank. A Year in the Lives of Smallholder Farmers. Available online: https://www.worldbank.org/en/news/feature/2016/02/25/a-year-in-the-lives-of-smallholder-farming-families (accessed on 15 June 2021).
69. English, C.; Locke, A.; Quan, J.; Feyertag, J. *Securing Land Rights at Scale. Lessons and Guiding Principles from DFID Land Tenure Regularisation and Land Sector Support Programmes*; Land: Enhancing Governance for Economic Development (LEGEND): Department for International Development: London, UK, 2019; p. 46.
70. World Bank. Land Governance Assessment Framework. Available online: https://www.worldbank.org/en/programs/land-governance-assessment-framework (accessed on 16 June 2021).

land

MDPI

Article

Fit-for-Purpose, Private-Sector Led Land Regularization and Financing of Informal Settlements in Brazil

Malcolm Childress [1,*], Selina Carter [1] and Edgard Barki [2]

[1] Global Land Alliance, Washington, DC 20005, USA; scarter@landallianceinc.org
[2] Fundação Getulio Vargas, São Paulo 01332-000, Brazil; Edgard.barki@fgv.br
* Correspondence: mchildress1@landallianceinc.org

Abstract: This paper aims to analyze the financial and operational approach to land regularization and financing used in Brazil by an innovative private social enterprise in order to demonstrate that the approach widens the concept fit-for-purpose land regularization to include fit-for-purpose land financing, with relevance for wider efforts in informal settlement regularization and upgrading. In this approach, the enterprise acts as a coordinator and broker to organize the residents of informal settlements to regularize their settlements by negotiating buyouts of the underlying private owners at discounted values, handling titling and registration of the occupants, and coordinating with municipal governments to provide infrastructure. The analysis of parcel-level repayment and price data provides evidence of the sustainability of the business model and increase of property values of the regularized parcels. The results presented from the enterprise's own repayment data demonstrate that under (non-pandemic) historical conditions residents are largely able to pay an affordable monthly payment over 7–10 years to the enterprise for the service to purchase the plots and maintain the enterprise. In operation since 2001, the enterprise has regularized over 20,000 parcels in more than 30 settlements, primarily in the cities of Sao Paolo and Curitiba in Brazil. The approach suggests that it could be widely replicable and add to the set of options for regularizing informal settlements, especially when purchase of private land is required.

Keywords: fit-for-purpose land administration; land tenure security; informal settlements; urban development; Brazil

Citation: Childress, M.; Carter, S.; Barki, E. Fit-for-Purpose, Private-Sector Led Land Regularization and Financing of Informal Settlements in Brazil. *Land* **2021**, *10*, 797. https://doi.org/10.3390/land10080797

Academic Editors: Stig Enemark, Robin McLaren and Christiaan Lemmen

Received: 13 May 2021
Accepted: 19 July 2021
Published: 29 July 2021

1. Introduction

This paper aims to analyze the financial and operational approach to land regularization and financing used in Brazil by an innovative private social enterprise called Terra Nova, in order to demonstrate that the approach widens the concept fit-for-purpose land regularization to include fit-for-purpose land financing. The feasibility of the approach has relevance for wider efforts in informal settlement regularization and upgrading.

First, the paper reviews the literature on informal settlement expansion and regularization. This review identifies the need for a broader set of approaches to informal settlement regularization given the continued proliferation of this type of settlement globally. The review also identifies financing for land and housing as a barrier which prevents regularization of informal settlements on a wider scale and impedes the linkage between informal and formal systems for urban services. Multiple observers have hypothesized that improved land financing is a key to resolving the challenge of informal settlement regularization. However, land financing for poor populations is challenging because of low and inconsistent incomes, and requires specific fit-for-purpose arrangements which traditional financing institutions have been reluctant to provide.

The paper then introduces the case of the private social enterprise Terra Nova and presents the central features of its bundled approach to land regularization and land financing, with emphasis on the ways in which the approach conforms to the principles of fit-for-purpose land regularization and land financing. In this approach, the enterprise acts as a coordinator and broker to organize the residents of informal settlements to regularize their settlements by negotiating buyouts of the underlying private owners at discounted values, handling titling and registration of the occupants, and coordinating with municipal governments to provide infrastructure.

The paper then presents empirical results about the performance of Terra Nova's approach. The analysis of parcel-level repayment and price data provides evidence of the sustainability of the business model and increase of property values of the regularized parcels. The results presented from the enterprise's own repayment data demonstrate that under (non-pandemic) historical conditions residents are largely able to pay an affordable monthly payment over 7–10 years to the enterprise for the service to purchase the plots and maintain the enterprise. In operation since 2001, the enterprise has regularized over 20,000 parcels in more than 30 settlements, primarily in the cities of Sao Paolo and Curitiba in Brazil.

The paper concludes that approach expands the toolkit of fit-for-purpose land administration to include the bundling together of fit-for-purpose land administration and land financing. The conclusion suggests that this bundled approach of fit-for-purpose land administration and land financing holds the potential to be replicable in many contexts and add to the set of options for regularizing informal settlements, especially when purchase of private land is required.

2. Literature Review

2.1. Urban Informal Settlements Pose a Global Challenge for Development

Urban population growth is arguably the biggest demographic trend of the last hundred years. Around the year 2000 more than half of the world population lived in urban areas. As of 2015, nearly 4 billion people—or 54 per cent of the world's population—was urban, and the number was expected to reach 5 billion by 2030 [1].

Much of the growing urban population is involved in the informal economy. The informal economy includes actors and activities "unregulated by the institutions of society in a legal and social environment in which similar activities are regulated" [2]. The informal economy is a major provider of housing and livelihoods for large populations of urban workers in low-income countries [3]. Approximately a billion people live in informal settlements or approximately 1 in 10 people globally, live in informal settlements [4]. Their numbers are expected to continue to grow to about 2 billion people by 2030 [1]. The prevalence of informal urban settlement is global. Fifty-nine percent of the urban population in sub-Saharan Africa lives in informal settlements, as does 28 percent of the urban population in Asia, and 21 percent of the urban population in Latin America and the Caribbean [5].

The continuing prevalence of informal settlements as a predominant form of housing and urban services in low and middle-income countries is a major development challenge for urban growth. Informal settlement puts high social and economic burdens on residents and incurs large public costs. These burdens and costs include insecurity of tenure, lack of public services, stigmatization and discrimination by others including law enforcement, environmental and health hazards, inequitable civil rights, direct costs for local governments in upgrading programs and foregone revenues, and indirect costs when coping with other impacts of informality, such as public health (e.g., COVID-19), criminal violence, and related social problems [6].

2.2. Regularization of Informal Settlements Has Been a Global Development Policy Priority for Decades, but Regularization Has Fallen behind the Growth of New Informal Settlements

Informal settlements are perpetuated due to high land costs in comparison to incomes, exclusion from access to credit for low-income groups, and insufficient affordable housing options in the formal sector (i.e., housing in comformity with planning requirements and property rights recognized by land administration authorities) [3]. Regularization of informal settlements has been characterized as a "perpetual problem" without a no sufficient policy or market response [7–9]. As global population growth and urbanization has increased during the second half of the 20th century and first decades of the 21st century, the issue of informal settlement regularization has been a continual focus of international development policy concern and targeting for decades, featuring in the Millenium Development Goals, the, the Sustainable Development Goals (SDG) goals (principally goal 11), and New Urban Agenda.

Over the years, policy and programmatic responses to informal settlement regularization by national and municipal governments, international development partners, and non-governmental and community-based organizations have resulted in some relative improvements in conditions but not in absolute numbers. The proportion of the urban population living in slums (i.e., informal settlements with the worst living conditions) worldwide fell from 28 percent in 2000 to 23 percent in 2014. Despite this improvement in the relative proportion of the informal settlement population, however, the absolute number of residents of informal settlements has continued to grow in the face of accelerating urbanization, population growth, and lack of appropriate land and housing alternatives. In 2014, an estimated 880 million urban residents lived in slums, compared to 792 million in 2000 [10].

Commenting on the lack of achievement of the goals of informal settlement in the Millenium Development Goals, Durand-Lesserve [11] observes policy failure at the level of governments and the lack of success in integrating informal markets for land and housing with formal markets in the following way:

> The global rise of urban poverty and insecure occupancy status is taking place in a context [. . .] of massive government disengagement from the urban and housing sector. Attempts to integrate informal markets—including land and housing markets—within the sphere of the formal market economy, especially through large-scale land ownership registration and titling programs, along with the lack of, or inefficiency of, safety net programs and poverty alleviation policies, have resulted in increased inequalities in the distribution of wealth and resources at all levels. In most countries, the public sector no longer contributes to the provision of serviced land or housing for low-income groups. Furthermore, the private sector targets its land and housing development activities at high-income and middle-income groups with regular employment and access to formal credit [11].

In 2015 the Millennium Development Goals were replaced by the Sustainable Development Goals (SDGs). SDG 11 aims to ensure access for all to adequate, safe and affordable housing and basic services and upgrade slums (Target 11.1, Indicator 11.1.1) [10]. By making informal settlement regularization and SDG target, the UN member states sought to elevate the issue to the top priority of development policy action. As of 2020, progress on SDG 11.1 was deteriorating, with the COVID-19 pandemic contributing to an increase in the absolute number of people in informal settlements. The increase was most acute in Northern Africa, sub-Saharan Africa, and in Western Asia [12].

Echoing the Millennium Development Goals and the SDGs, The New Urban Agenda, adopted by the UN General Assembly in 2016, aims a specific agenda item at informal settlements, calling for strategies for their prevention, regularization, upgrading, and monitoring [13].

In spite of the high priority accorded to informal settlements in these long-standing global development agendas and similar national development agendas, analysts like Millington and Cleland [14] have observed that despite 50 years of efforts, the growth of slums and informal settlements is increasing, particularly in developing countries, and the total number of slum dwellers has increased, undermining the ability of cities in developing countries to economically grow, prosper, and generate wealth for all inhabitants.

2.3. Models for the Prevention or Regularization of Informal Settlements That Can Keep up with Demand Have Been Inadequate

Given the longstanding and globally-prioritized problem of informal settlement proliferation and inadequate regularization, it is relevant to ask why the problem has been so intractable by conventional policies. Measures to address the problems of informal settlements in developing countries have had several different areas of emphasis over the last 20 years. Some measures have emphasized physical upgrading of settlements, others provision of housing subsidy [7], clearing of informal settlements and relocating of the residents to public housing, as well as participatory regularization and attempts at addressing market imperfections [15]. Observers have noted low levels of awareness about the ongoing housing and infrastructure needs settlements by policymakers, and their tendency to underprioritize and poorly understand the dimensions of settlement rehabilitation [16].

In many situations, neither public bodies nor formal markets are able to respond to rapid urbanization quickly enough to with sufficiently priced housing to pre-empt informal settlement by channeling housing demand into formal pathways. New residents arrive to cities more rapidly and with less ability to pay than formal-sector arrangements can offer.

Most public jurisdictions do not have the planning and financing tools to deal with the rapid urbanization, or the tools in place are not sufficiently responsive to the reality on the ground [17]. For example, Tellman, et.al. [3] note that in Mexico City, "regularization is a complex, multi-institutional process, dependent on land tenure type and location, taking anywhere from 5 to 20 years. Many communities are never regularized due to this long and convoluted process. Informal urbanization has outpaced regularization" [3]. The World Bank notes that globally, "narrowly-focused, neighborhood-level slum upgrading interventions, while generally effective, have fallen short of addressing the magnitude and scope of expanding informality and slums" and that the output of informal settlement regularization projects has been minimal when compared to the growing slum population worldwide [18].

2.4. Financing Land Regularization and Acquisition through Private Sector-Led Approaches Has Been Hypothesized to Address Many of the Weaknesses in Models of Informal Settlement Regularization

In low-income countries the financial sector frequently filters out the informal urban population from access to its products and services such as housing mortgages due to perceived high risks and low returns [15]. The financial sector perceives risks of non-payment, delinquency, and default because land and housing of the informal settlement itself cannot be used as a collateral, household savings are often small or non-existent and incomes are irregular. The financial sector also perceives low returns because of the low relative property values and long payback horizons compared to other market segments.

In theory, incentives and legal measures provided by the public sector should be able to bridge the financing gap created by these risks and induce private sector involvement in housing provision. However, the track record of such financial strategies, both those aimed at existing informal settlements and at new social housing, is weak due to misalignment of incentives by developers and government, and land cost itself [19]. The experience of Brazil's Minha Casa Minha Vida program, for example, which has provided a variety of public credits, subsidies, and guarantees for private developers and contracted over 3 million housing units since 2009, shows high delinquency of repayment (28 percent), abandonment of housing due to undesirable locations and moral hazard in the management of the program resulting in high costs for land and construction [20].

Strategies such as loan guarantees or social funding, modeled on microfinance ideas, at the lower end of the market have been called for by observers of these issues [15]. For example, Nzau and Trillo [19] conclude their review of informal settlement upgrading with the observation that, while public-sector driven attempts to provide decent housing to slum residents in developing countries have either failed or achieved minimal output when compared to the growing slum population, these countries exhibit vibrant real estate markets that may hold the potential to bear the costs of regenerating informal settlements.

Based on these conclusions about the inadequacy of public sector led responses to informal settlements' needs for regularization, Nzau and Trillo [19] call for private-sector or public-private partnerships and focus on the need for engaging private sector developers by attracting their finances and expertise, in other words, fit-for-purpose land financing. They note that in many urban areas the real estate market is vibrant and highly dynamic and may hold the potential of bearing the costs of regenerating slums through the value increase from regularization and upgrading which can be shared by owner-occupants, private sector actors and the public sector [19]. However, they maintain that little has been explored in the way public–private-based approaches to develop a market-driven processes for informal settlement regularization. They speculate that this lack of experiences may be due to the limitations of the social construct or social cohesion in informal settlements which could also be understood as coordination failures [19]. Similarly, UN has also called for participatory and inclusive approaches that explore new innovative and effective financing avenues for land regularization and financing of land and infrastructure acquisition. [21]

Ward et al. [16] also call for new financing mechanisms (mortgages, etc.) to facilitate property sales for informal settlement upgrading, including buy-outs by inheritors and other stakeholders. In this call for action, governmental and private sector actors are envisioned to support provision of micro credits for home improvement and rehabilitation, linked to financial assistance to promote clean titles among stakeholders in order to leverage loans and financing [16].

3. Fit-for-Purpose Solutions

3.1. Taken as Whole This Review of the Literature on Informal Settlement Regularization and Upgrading Identifies a Need for Fit-for-Purpose Solutions Which Bundle Land Regularization and Land Financing Together in a Package to Resolve the Needs of Residents

The above literature demonstrates the need to solve critical land financing imperfections in the relationship of informal settlements to private land markets and property rights. The discussion above identifies a need or gap for private sector-led or public-private partnerships to leverage land markets and land values in the regularization and upgrading of informal settlements. The identification of the need to provide this type of service within a number of constraints linked to affordability and tenure security can be viewed as a search for fit-for-purpose land regularization bundled together with fit-for-purpose land financing. This implies the need to establish new categories of land market transactions which are based on the ability of occupants of informal settlements to pay, which recognize the 'sunk capital' or 'fact-on-the-ground' of existing occupation, and the potential for security of tenure and value increases to support a stream of payments for land regularization and financing services.

However, in most situations where urban land value is expected to increase over time, the notional expected market price will still be above the reservation price or maximum willingness to pay of low-income occupants. This situation has been referred to as the "fundamental financing problem of the poor" [22]. Therefore, a negotiated, "adjusted discount price" may be required to buy out a prior owner whose land is occupied by an informal settlement. Creating the conditions to manage and consolidate these types of market transactions via coordination of the informal settlement residents and negotiation with the underlying land owner, can be viewed as fit-for-purpose land financing.

Fit-for-purpose land financing for informal settlements is a natural extension of fit-for-purpose land administration. Fit-for-purpose land administration means applying the spatial, legal, and institutional methodologies that are most fit for the purpose of providing secure tenure for all, usually meaning they are low-cost, technologically accessible, and precise enough for the needs of the land boundaries in question [23]. In the context of informal settlements fit-for-purpose land administration implies bringing together the advantages of formality (legal security, ability to access infrastructure and public services), with the advantages of informality (affordability, accessibility, adaptability).

As the above discussion has demonstrated, provision of the mechanisms for financing of land acquisition, and infrastructure bundled together with land regularization may be one of the missing elements for addressing regularization and upgrading of informal settlements at a wider scale. Such mechanisms potentially can create 'bridges' between informal settlements and formal property markets by connecting the low-income market segment of informal, self-built housing with formal property markets and financial institutions, and bring the settlement into a position to receive and pay for municipal infrastructure.

Financing of land acquisition and infrastructure is often a key missing element of informal settlement upgrading, and it is therefore valuable to identify cases in which land regularization and land financing are bundled together as a fit-for-purpose response [19]. The idea of bundling regularization and financing together appears to be an important innovation to enable the informal settlement regularization within a reasonable time and at affordable costs. The paper now turns to the relevance of these concepts in Brazil, and focuses on an innovative, private sector social enterprise which puts these principles into action.

3.2. The Need for Fit-for-Purpose Bundled Land Regularization and Land Financing Is Acute in Brazil

Addressing the financing element is central to Brazil's challenge to regularize tenure and upgrade urban services for informal settlements. As Brazil urbanized, going from 37 percent urban in 1950 to 86 percent urban in 2018, informal settlements of self-built housing filled the gap for affordable housing and access to urban space for large segments of the population. Informal settlements in Brazil are typically comprised of self-built housing constructed by semi-organized groups on unoccupied land, either on the urban periphery or in less favorable building sites within urban agglomerations, such as the slopes of hills. Although some informal settlements in Brazil occupy public land, most are built on land with an underlying private owner. The fact of underlying private ownership makes regularization of informal settlements in Brazil financially demanding due to requirements for expropriation with compensation to the private owner, as well as payment of the costs of preparing urbanization plans and legal procedures to title regularized parcels. Before the informal parcels and structures can be regularized (i.e., parcels and occupants recognized through regular titles of ownership and the settlement incorporated into municipal service arrangement for roads, water, and sewerage), the underlying private land must be acquired from the original owner with due compensation, and an urbanization plan accepted by the municipal government.

These situations generally create long-term impasses with high costs to residents, underlying landowners, and local governments. Residents of informal settlements on private land and the underlying owners in Brazil often become trapped in a type of ad hoc unresolved arrangement in which residents occupy land but do not have access to land titles, while owners hold legal title but have no ability to use or dispose of the land.

Evictions of informal settlement residents by underlying private landowners are also costly and contentious. Landowners must undergo costly and lengthy legal processes to evict informal settlement dwellers. Even if successful in legally recovering the land, the underlying private owners still have to cover the costs of reconfiguration, infrastructure, and overdue taxes. Meanwhile, informal occupants live under the threat of forced evictions, without formal infrastructure and in social exclusion from education, health, and postal services. Local government receives no tax collection on the informally occupied areas and is burdened by low-quality urban surroundings.

In the next sections, we examine how these issues are being resolved through a private-sector-led experience in Brazil, with the aim of analyzing its fit-for-purpose qualities which could make it relevant for replication in a wider set of contexts, especially for the financing of land acquisition and infrastructure.

4. The Terra Nova Case

4.1. An Innovative Private Sector Social Enterprise Called Terra Nova Is Bundling Land Regularization and Financing for Informal Settlement Upgrading in Brazil

The Brazilian private social enterprise Terra Nova takes a new approach to solving the financing problem of informal settlements in Brazil which overcomes the low-income land financing gap and can be viewed as a fit-for-purpose bundling of land regularization and land financing for informal settlement regularization and upgrading.

Terra Nova began as a family-owned business begun in 2001 by two brothers, Andre and Daniel Albuquerque, with experience in dispute resolution. Originally, the business worked as an advocate with community associations to lobby authorities to invest in infrastructure and services for informal settlements. This evolved into mediating agreements between landowners and associations representing residents in which the land would be expropriated by municipal or state government and compensation paid to the underlying private owner by Terra Nova in installments based on streams of payments from residents, under an agreement ratified by a court order. Terra Nova would restructure the settlement's spatial plan to permit municipal infrastructure, document and map each individual parcel, supervise execution of contracts and manage payments, earning revenue from a retained share of the installment payments made by the residents. As individual residents completed payments, full ownership title was provided to them. Municipal governments in turn provided infrastructure to the regularized settlements. Legal title and infrastructure raised property values and made the properties marketable.

The value of informal settlement regularization is high. It has been estimated that the value of all private land under dispute in Brazil was R$15 billion, with the potential of the land value to reach R$45 billion once areas were regularized and received infrastructure [24].

The evidence from Terra Nova's experience suggests that a private entity can help resolve this problem by creating a coordination mechanism that is able to negotiate and mediate between informal settlement residents and private owners to reach a solution to the financing problem. The historical performance of the enterprise demonstrates that this approach is feasible ('feasible' here is defined as having the ability to achieve the objective of regularization over a sustained period of time and an expanding base of settlements without external subsidy at levels of affordability which enable low-income residents to pay for the regularization process themselves at a level which is manageable within their household budgets) from both financial and social points of view and is able to overcome the fundamental financing problem in two critical ways. From the residents' payment side, it solves the problem of coordination of individual residents by aggregating residents' payments. It solves the affordability problem by stretching the period of repayments to 7–10 years, which allows for monthly payments which do not restrict consumption beneath residents' subsistence threshold. From the land price side, discounted purchase prices are arranged which reduce the sales price below prevailing market rates, to levels which are likely closer to the owners' true reservation price for the occupied site.

Terra Nova typically enters when the ad hoc unresolved legal status generates a conflict over the land, creating an opportunity to broker an agreement between the parties to purchase and compensate the land from the underlying private owner. Terra Nova simultaneously coordinates the residents to enter into contracts with Terra Nova to pay for the land compensation and the regularization procedure, and with the underlying private owner to accept a stream of payments for the overall compensation. Terra Nova then maintains both the collection and the payment accounts over a period of time, using local and federal regulation and government titling mechanisms to fully legalize and include the settlements into formally provided municipal services.

As an intermediary, Terra Nova negotiates a compensation price at which landowners will agree to sell (typically at a discount of 20–40 percent below comparable market prices) and informal residents can afford to buy. The purchase is made by low monthly installment payments over a 7–10-year period adjusted to affordability determined by a social evaluation of the settlement. Local authorities provide the verified land title to Terra Nova after acquisition of the land and Terra Nova provides title to residents after the full stream of payments is completed.

With this fit-for-purpose model for regularization and financing of informal settlements as a bundled package, Terra Nova empowers community associations and facilitates authorities to invest in basic infrastructure and services for the underserved areas. Since 2001, Terra Nova has received judicial approval to regularize 30 areas with about 20,000 occupants in total and 3 million square meters of housing stock. It has facilitated the installation of 21 km of water and energy networks, 12 km of sewage networks and 19 km of paved roads. The enterprise has also facilitated the construction of four schools, eight daycare centers, five health care centers, five social assistance centers, and 18 recreational facilities. Since 2012, residents' payments had grown eightfold, and net revenues of the enterprise reaching R$1.9 million in 2018 [24].

4.2. Terra Nova's Model Adapts Traditional Land Purchase Arrangements for the Low-Income Population in a Fit-for-Purpose Manner in Which Payment Requirements Are Adapted to the Capacities of Residents, and the Price of Compensation to the Original Landowners Is Also Negotiated

In this way Terra Nova's model can be viewed as a type of fit-for-purpose land financing, distinct from both traditional commercial financing, and from subsidized public financing.

The initial step is to reach an agreement with the informal settlement population to carry out the regularization process. This is facilitated by community neighborhood associations discussing the regularization plan and terms with residents. It is necessary to have the agreement of a minimum of 75 percent of the households in a settlement to ensure the financial stability of the compensation arrangement. A parcel map and upgraded land use map for the post-regularization street and social infrastructure lay-out is prepared. Usually, the process of reaching agreement with the community members takes 30–60 days of meetings and, in some cases, door-to-door discussions with individual residents. In some large communities, factions among residents, conflicts of interest, and organized groups may oppose the regularization project. In these cases, Terra Nova either has to withdraw or facilitate a leadership change among the community.

Working with the local neighborhood association, Terra Nova's staff carries out a social assessment to estimate the payment capacity of the residents, prior to negotiation with the landowners about compensation. In the social assessment the household is interviewed to determine its sources of incomes, the stability of those sources (most often in the informal sectors), and the household's expenditures. Checks are made with local references to ensure the veracity of the information. The target for the monthly installments to be paid by residents is approximately 30 percent of the one individual's minimum salary. Depending on the capacity to pay and the negotiated land price, these monthly instalments are typically paid for 7–10 years (at IPCA + 4% or ~9% p.a.), agreed to in an individual contract made directly between Terra Nova and the resident Currently, the average instalments are R$ 250–350 per month per household.

At the same time, Terra Nova makes an assessment of the land value and enters into a negotiation with the landowner for a compensation arrangement. Typically, this is a discounted market price, to be paid over a 7-to-10-year period as a compensation contract by Terra Nova to the landowner. The municipal government then expropriates the land from the landowner. In most of the settlements regularized to date, the original landowner receives 60–65% of the monthly payments from the overall stream of payments from residents. Terra Nova receives 30–35% for its staff and services. Lawyers and community-owed Social Funds receive 5%. The median land value for contracted parcels is R$ 7400 (constant 2010 Reals) and the median lot size, which includes shared common areas such as walkways, is 175 square meters.

Terra Nova's property titling process, the duration of the payment process is based upon 10 years of affordable monthly payments for a favela resident to acquire the property title but may be advanced to as few as seven years depending on the payment capacity of residents and the cost of the land acquisition.

Along with this focus on adaptability and affordability, other fit-for-purpose factors are present in the regularization and financing model. The project identification and selection process is rigorous and restrictive; only when residents, the landowner, and the municipal government are all in agreement can the process be carried out. This means that strong relationships with local bodies and municipalities have to be maintained at all times. Terra Nova uses only highly experienced negotiators with deep expertise in legal matters to work out a win-win solution for all stakeholders. Terra Nova has to accurately assess the payment capacity of residents. It also employs a strong technological backbone to track individual payments in from residents and payments out to landowners and has a robust delinquency assessment [24]

5. Evidence of Performance

5.1. Materials and Methods

This section presents evidence about the performance of the approach over time from the case study of Terra Nova. The evidence presented aims to show two basic performance indicators of the viability of the social enterprise over time: (1) sufficient payments with low-enough delinquency rates to maintain the business in operation and achieve formalization for landholders; and (2) increases in land value of the regularized land parcels.

The dataset used for this study represents all financial transactions and obligations between Terra Nova and residents from 2000 to 2019. There are a total of 8193 contracts (which correspond to 7103 lots and 7414 occupants) in the dataset. We also analyze a subset of lots pertaining to Paraná state (4527 lots, 5388 contracts, and 4990 occupants) since Terra Nova has the longest history in this area. We do not consider transactions in 2020 given the unique circumstances of the COVID-19 pandemic and its economic impact in Brazil.

Several types of financial transactions and obligations are considered which include the following: contract entry fee (2% of total obligations), administrative fees (3%), and instalments (95%). We do not distinguish between these types of financial flows, but rather we compare each title holder's complete vector of 'obligations' versus 'payments' to run our analysis on the reliability of Terra Nova's credit model. All financial data is in real Brazilian Reals, deflated using Consumer Price Index (base year 2010) available from the World Bank.

Besides financial transactions and obligations, the dataset offers basic demographic information on title holders (year of birth, sex, profession, marital status) but no data on other members of each household. Occupant identifiers are anonymized, and geographic information is limited to city and neighborhood. The data also contains lot-level information such as lot size (meters squared), size of shared space with neighboring lots (such as common walkways), and price per square meter (assessed at a fixed value for each community on a yearly basis).

5.2. Empirical Results on Payments and Parcel Value

This section presents evidence about the capacity of residents to make payments for land regularization. Regular payments by residents are the critical element of the revenue stream of the social enterprise.

We analyzed financial transactions and obligations between Terra Nova and residents from 2000 to 2019 in the state of Paraná, which is where Terra Nova started its operations and is home to 64% of Terra Nova's total lots. We discuss the utility of Terra Nova's business model from two perspectives. One, from the point of view of the business, we ask, "Do title holders meet their financial obligations?" We verify that occupants overwhelmingly tend to meet their financial obligations, which suggests the business model is sustainable. In fact, we find that many occupants complete payments and do so ahead of schedule, indicating a strong willingness to become landowners under Terra Nova's structure.

Two, from the point of view of the occupants, we ask, "Do real land prices increase over time?" We find that the real price of land (as assessed by independent appraisers hired by Terra Nova) on average increases over time. This asset appreciation can be expected to render future benefits for the occupants in terms of job security, credit worthiness, and the ability to make long-term economic investments, as well as improved mental and physical health as municipal amenities are improved for regularized settlements by municipal governments.

5.2.1. Descriptive Data

By 2019, Terra Nova had projects in three states: Paraná (64% of total lots), São Paulo (35%), and Minas Gerais (1%). In Paraná, Terra Nova accumulated 5388 unique contracts and 4990 title holders, covering 4527 lots. The median age of title holders in Paraná is 50, with a minimum age of 19 and a maximum age of 99 years. Terra Nova did not record sex for around half of the title holders, however 23% report as female and 27% report as male. Around 40% of title holders are single (never married), 40% are married or living with a spouse, 15% are divorced or separated, and 5% are widowed.

Each of the regularized parcels is issued a contract by Terra Nova which sets out the terms for repayment in installments and titling. These contracts are the basis for the data we analyze. The number and year of origination of the contracts included in the dataset are depicted in Figure 1, as well histograms describing the distributions of contract length, real price of the parcel, the area of the property and the age of the title holders. The median term length of the contracts is 48 months, with a minimum of zero months (the case for 206 contracts in which land was bought upfront) and a maximum of 600 months (representing two outlier contracts). The median land value is R$ 7400 (constant 2010 Reals) for the parcel and the median parcel size, which includes shared common areas such as walkways, is 175 square meters.

5.2.2. Empirical Questions

Question I: Do Title Holders Meet Their Financial Obligations?

As depicted in Figure 2, over the twenty-year period studied (2000–2019), obligations owed by title holders in Paraná state totaled R$ 23.8 million (constant 2010 Reals), while the same title holders paid a total of R$ 20.9 million (constant 2010 Reals). The unpaid amount (R$ 3.1 million) represents 12.9% of the total financial obligations over the period. The unpaid amount, in this aggregate measure, includes all lengths of delinquency, ranging from a few days to many months of lateness. Roughly 25% of the volume of late payments is in arrears by one month or less, while less than 1% is in arrears by six months or more. Overall, Terra Nova occupants have shown a high capacity to pay on their obligations and the organization has operated successfully from a financial standpoint over the time period.

As of 2019, most delinquent title holders owe a relatively small sum (see Figure 2). Less than 2% of title holders owe more than R$ 8000. Of the title holders with an outstanding obligation, over 81% owe less than R$ 2000 (constant 2010 Reals). This evidence suggests that most occupants are reluctant to fall behind on their payments and few assume tremendous debt (see Figure 3).

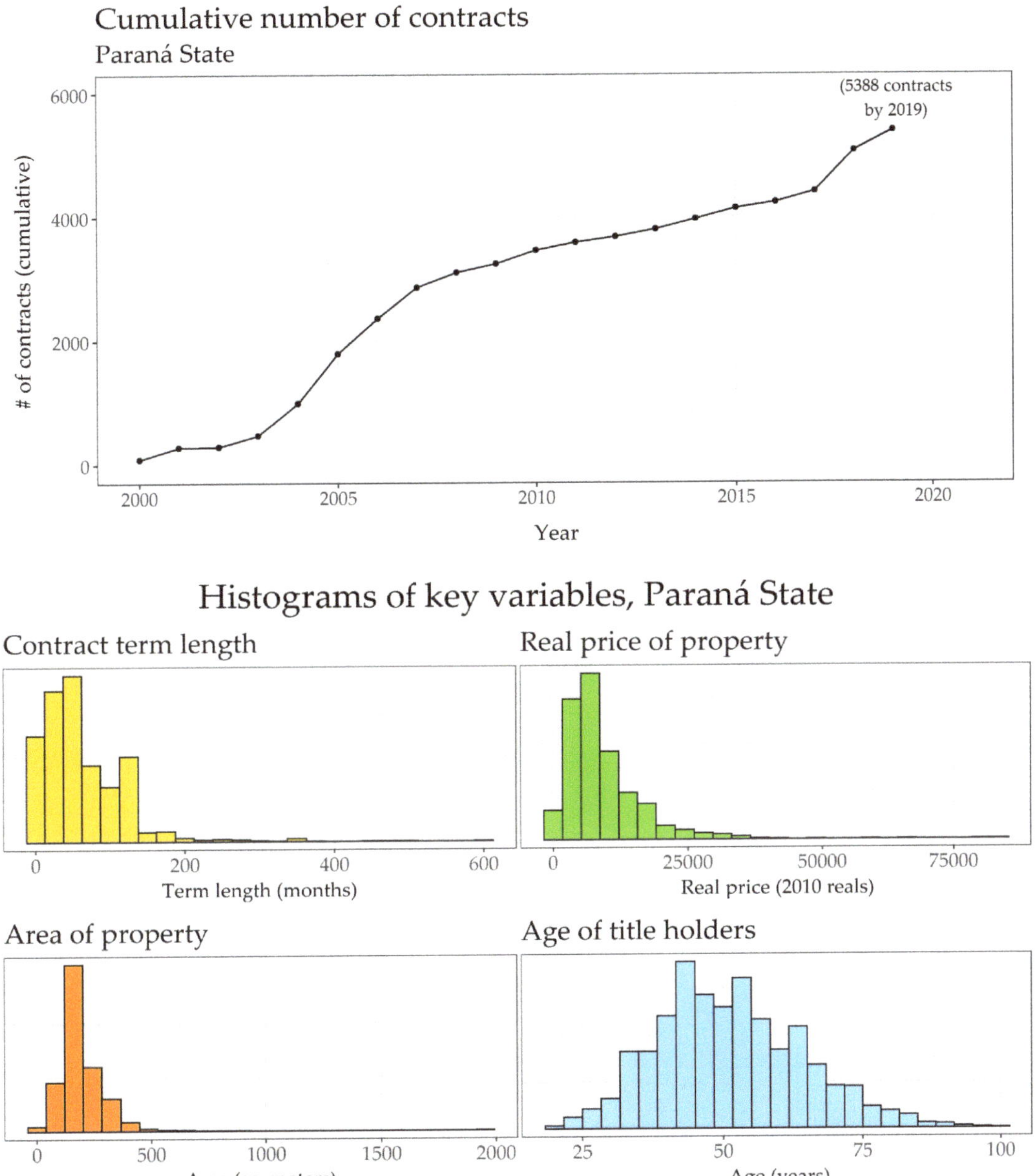

Figure 1. Cumulative number of contracts for parcel regularization in the dataset and descriptive statistics about contracts in the state of Paraná. This graph shows the growth of parcels being regularized by Terra Nova from 2000 to 2019. The histograms show the distribution of these parcel regularization contracts in terms of contract length, real price of the property in constant 2010 reales, parcel area, and age of the title holders.

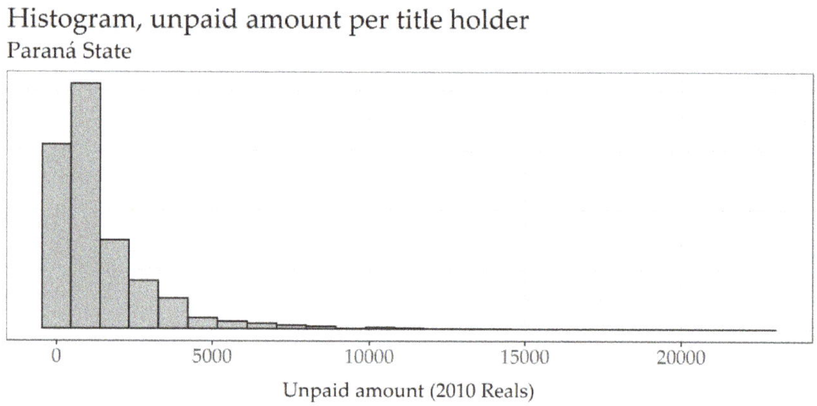

Figure 2. Histogram of unpaid amount per title holder.

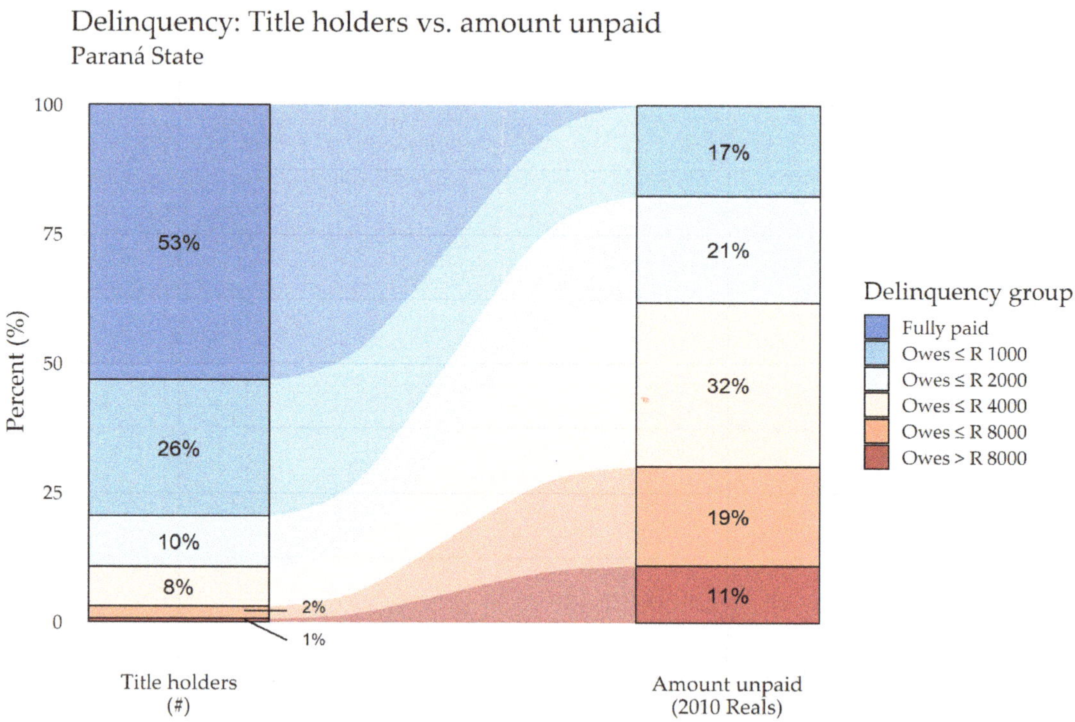

Figure 3. Delinquency: Title holders vs. amount unpaid. The left-hand side of the figure depicts the distribution of title holders stratified according to their level of delinquency. The right-hand side depicts the total amount of delinquent payments in constant 2010 Reals. Eighty-nine percent of title holders in delinquency is for less than 2000 Reals, and only 1 percent of delinquency is greater than 8000 Reals.

Furthermore, many occupants have paid ahead of their obligations: 44% of occupants have paid more than is due as of 2019, totaling around R 300,000 (constant 2010 Reals) in early payments to Terra Nova. This reinforces the contention that occupants are eager to become landowners and are making great efforts to meet their financial obligations in this regard.

Question II: Do Real Land Prices Increase Over Time?

For occupants, there are numerous benefits of land ownership; the most basic concern is to avoid repossession by the hitherto legally recognized landowner. However, there are also plentiful economic benefits of owning land, especially land that is increasing in value.

In the process of negotiating contracts with landowners and occupants, Terra Nova hires outside appraisers who assess the value of land each year within each settlement. Land is appraised at a fixed rate (value per square meter) across an entire settlement at regular intervals. Thus, assessed land values can change over time within a settlement. We analyze real land prices over time at the project level in Paraná state.

Figure 1 below presents the change in real median price for the largest ten projects in Paraná by number of lots. 'Years active' refers to years during which new contracts are made. For most projects that are active for more than one year, the change in the median real price per square meter is positive; moreover, the increase is greatest for projects that are active for many years (such as Jardim União and Vila Nova). In the one case (Vila União) where the median price decreases, the decrease is relatively small (R$ -4).

The median price increase is most prominent for Jardim União and Vila Nova, the two projects with the longest history with Terra Nova (see Figure 4 below). For Jardim União, the real mean price per square meter was just R$ 49 in 2004 rising to R$ 111 in 2017, a change of R$ 62. For Vila Nova, the real mean price per square meter was just R$ 14 in 2005 rising to R$ 72 in 2018, a change of R$ 60.

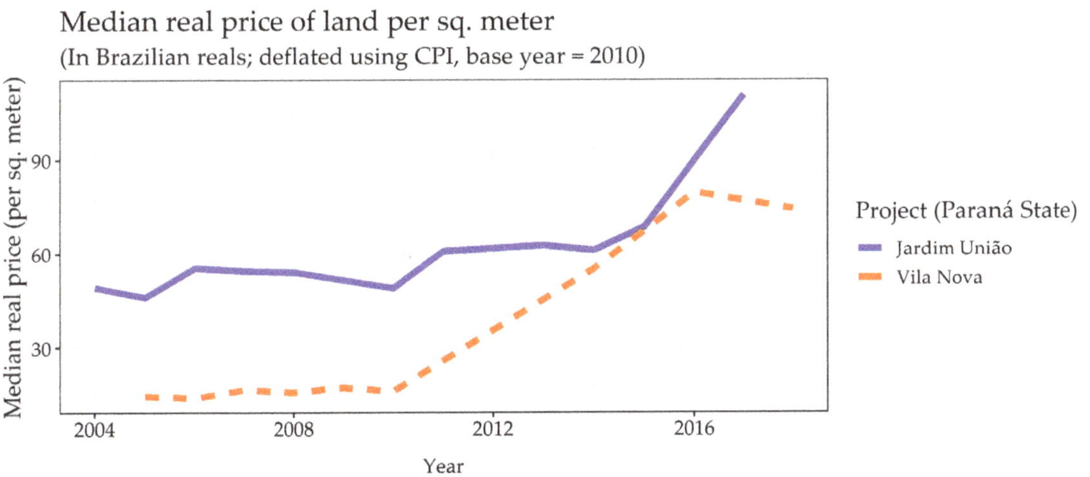

Figure 4. Evolution of land price in Terra Nova regularized settlements from 2004–2018.

5.3. Qualitative Evidence of Impact: Pictures and Perceptions

It is also relevant to view the operation of the social enterprise from a qualitative perspective. Below in Figure 5 we present several before-after images of settlements regularized by Terra Nova, and a summary of preliminary findings of an ongoing qualitative impact assessment in one settlement which was regularized by Terra Nova.

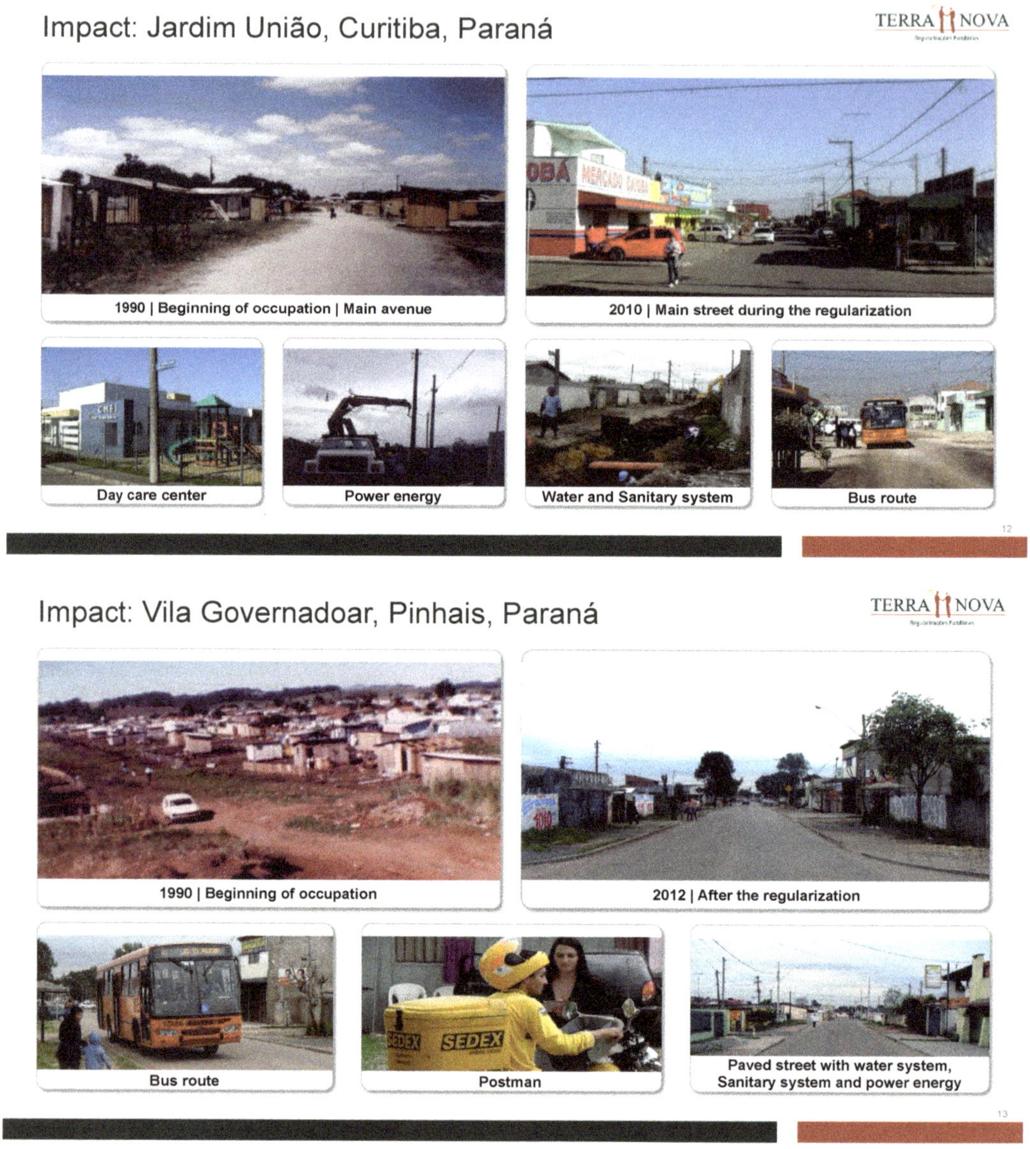

Figure 5. Photographs of Impact of Terra Nova Regularization in Curitiba, Parana, and Pinhais, Parana.

5.4. Residents Perceptions of Impact

Residents' subjective perceptions of the value of the regularized land are positive. According to initial reports from qualitative studies conducted by the Getulio Vargas Foundation in 2020 in communities regularized by Terra Nova, residents perceive a variety of significant benefits from the regularization. Benefits cited by residents included relief from fear of eviction, gains in the value of fixed assets, individual feelings of achievement, access to energy, clean water and sewage services, access to postal codes, public education and health services, and changing life expectations for children. Residents also expressed that payment for the land to achieve full ownership was a major household budget priority, even during the period of economic hardship imposed by the COVID-19 pandemic.

6. Discussion: Bundling Fit-for-Purpose Land Administration and Land Financing Together

Terra Nova demonstrates a case of a feasible, fit-for-purpose social enterprise model for informal settlement regularization and financing which bundles land regularization and land financing together. The case fills a gap in the practice of the regularization of informal settlements which has been remarked on in the literature and thus deserves to be discussed more widely and potentially to be replicated. The relevance of the case is high in Brazil, where most informal settlements are located on land with an underlying private landowner who must be compensated in order to relinquish rights to the occupants. The case may be relevant for a wider set of applications as well in which the land value and willingness-to-pay for it by residents can be leveraged to finance infrastructure for the informal settlement in a bundled package with land regularization.

The presentation of these findings has certain limitations. The data used was not originally intended for analytical use and may have gaps and inconsistencies for that reason. Only a single case is presented and there is no evidence that the approach or model presented has been replicated directly elsewhere. The results are all based on performance prior to the COVID-19 pandemic which has been reported to have reduced the incomes of many residents of informal settlements in Brazil.

7. Conclusion: An Example of Fit-for-Purpose, Private-Sector Led Land Regularization and Financing

Terra Nova's model demonstrates feasibility in Brazil and has the potential for wider replication. It combines fit-for-purpose, socially-embedded procedures for land regularization among informal residents, land owners, and municipal governments in an affordable package which is broadly sustainable for both residents and the social enterprise delivering the service. Terra Nova's example of fit-for-purpose land regularization and fit-for-purpose land financing addresses the coordination failure among residents, the market failure in the residential land market for low-income groups, and the governance failure in public regularization and social housing.

Further developments to scale up which Terra Nova is now developing include the packaging of infrastructure provision into the financing package (i.e., land and infrastructure together), creation of compensation funds which can be flexibly utilized to compensate multiple landowners and securitizing streams of payments to leverage of investment capital to increase the number of settlements which can be regularized at any given time by expanding operating capital and workflow. In the current circumstances of limited public sector activity for informal settlement regularization and upgrading in Brazil, private-sector led regularization and financing is likely the only model which is feasible at the scale of the problem. Continued success of Terra Nova's example in Brazil could pave the way for replication in other countries in Latin America and potentially in other regions where informal settlement has long been identified as a widespread and critical issue for social wellbeing and asset-building for low-income groups.

Author Contributions: M.C. contributed the conceptualization of the paper, the main drafting of the literature review, discussion and conclusions. S.C. contributed the preparation of the data set and the main drafting of the empirical results section and interpretation of results. E.B. contributed to the overall conceptualization of the paper, drafting of the sections on informal settlements in Brazil and the qualitative impact results. All authors have read and agreed to the published version of the manuscript.

Funding: Support for this research was provided in part by a grant from the Omidyar Network (No. 00033443-3). Funding for the publication costs of this article has kindly been provided by the School of Land Administration, University of Twente in combination with Kadaster International, the Netherlands.

Institutional Review Board Statement: This study is exempt from IRB approval because it utilizes de-identified data provided by the social enterprise for purposes of improving the efficacy of its programs through additional analysis.

Informed Consent Statement: The study utilizes de-identified data provided by the social enterprise.

Data Availability Statement: Data is available upon request from qualified researchers through the authors.

Conflicts of Interest: The authors declare no conflict of interest.

References

1. UN Habitat. Pro-Poor Action for Climate Change in Informal Settlements. 2018. Available online: https://unhabitat.org/pro-poor-climate-action-in-informal-settlement (accessed on 26 July 2021).
2. Portes, A.; Castells, M.; Benton, L. (Eds.) The World Underneath: The Origins, Dynamics and Effects of the In-formal Economy. In *The Informal Economy: Studies in Advanced and Less Developed Countries*; John Hopkins University Press: Baltimore, MA, USA, 1989.
3. Tellmann, B.; Eakin, H.; Janssen, M.; de Alba, F.; Turner, B. The role of institutional entrepreneurs and in-formal land transactions Mexico City's urban expansion. *World Dev.* **2021**, *140*, 105374. [CrossRef]
4. High Level Forum on Sustainable Development. 2018. Available online: https://sustainabledevelopment.un.org/content/documents/197282018_background_notes_SDG_11_v3.pdf (accessed on 26 July 2021).
5. UN Habitat. A Practical Guide to Designing, Planning and Implementing City-wide Slum Upgrading Programs. 2014. Available online: https://www.ohchr.org/Documents/Issues/Housing/InformalSettlements/UNHABITAT_A_PracticalGuidetoDesigningPlaningandExecutingCitywideSlum.pdf (accessed on 26 July 2021).
6. Fernandes, E.; Regularization of Informal Settlements in Latin America. Policy Focus Report. *The Lincoln Institute of Land Policy.* 2011. Available online: https://www.lincolninst.edu/sites/default/files/pubfiles/regularization-informal-settlements-latin-america-full_0.pdf (accessed on 26 July 2021).
7. Huchzermeyer, M.; Karam, A. *Informal Settlements: A Perpetual Problem?* 1st ed.; University of Cape Town Press: Cape Town, South Africa, 1 April 2007.
8. Huchzermeyer, M. The struggle for in situ upgrading of informal settlements: A reflection on cases in Gauteng. *Dev. South. Afr.* **2009**, *26*, 59–73. [CrossRef]
9. Huchzermeyer, M. Pounding at the Tip of the Iceberg: The Dominant Politics of Informal Settlement Eradication in South Africa. *Politikon* **2010**, *37*, 129–148. [CrossRef]
10. UN Stats SDG 11. 2016. Available online: https://unstats.un.org/sdgs/report/2016/goal-11/ (accessed on 26 July 2021).
11. Durand-Lesserve, A. Informal settlement and the Millenium Development Goals: Global Policy Debates on Property ownership and security of tenure. *Glob. Urban Dev.* **2006**, *2*, 1–15.
12. UN. The Sustainable Development Goals Report. 2020a. Available online: https://unstats.un.org/sdgs/report/2020/The-Sustainable-Development-Goals-Report-2020.pdf (accessed on 26 July 2021).
13. Habitat III. The New Urban Agenda. 2016. Available online: https://habitat3.org/the-new-urban-agenda/ (accessed on 26 July 2021).
14. Millington, K.A.; Cleland, J. *Counting People and Making People Count: Implications of Future Population Change for Sustainable Development*; K4D Helpdesk Report; Institute of Development Studies: Brighton, UK, 2017. Available online: https://www.ids.ac.uk/publications/counting-people-and-making-people-count-implications-of-future-population-change-for-sustainable-development/ (accessed on 26 July 2021).
15. Wekesa, B.; Steyn, G.; Otieno, F. A review of physical and socio-economic characteristics and intervention approaches of informal settlements. *Habitat Int.* **2011**, *35*, 238–245. [CrossRef]
16. Ward, M.P.; Jiménez, E.; di Virgilio, M.; Camargo, A. (Eds.) Políticas de vivienda en ciudades Latinoamericanas: Una nueva generación de estrategias. In *Housing Policy in Latin American Cities: A New generation of Strategies and Approaches for 2016 UN-Habitat III*; Routledge: London, UK, 2015.
17. Cities Alliance. Informal Settlement. 2021. Available online: https://www.citiesalliance.org/themes/informal-settlements (accessed on 26 July 2021).
18. World Bank. Systems of Cities: Harnessing Urbanization for Growth and Poverty Alleviation. 2020. Available online: https://www.worldbank.org/en/topic/urbandevelopment/publication/urban-local-government-strategy (accessed on 26 July 2021).
19. Nzau, B.; Trillo, C. Affordable Housing Provision in Informal Settlements through Land Value Capture and Inclusionary Housing. *Sustainability* **2020**, *12*, 5975. [CrossRef]
20. Acolin, A.; Hoek-Smit, M.C.; Eloy, C.M. High delinquency rates in Brazil's Minha Casa Minha Vida housing program: Possible causes and necessary reforms. *Habitat Int.* **2019**, *83*, 99–110. [CrossRef]
21. UN. Cities of Opportunity: Partnerships for an Inclusive and Sustainable Future. In Proceedings of the Fifth Asia-Pacific Urban Forum, Bangkok, Thailand, 20–25 June 2011. Available online: https://www.unescap.org/sites/default/files/report.pdf (accessed on 26 July 2021).

22. Binswanger, H.P.; Elgin, M. Reflections on land reform and farm size. In *Agricultural Development in the Third World*; Eicher, C.K., Staatz, J.M., Eds.; The Johns Hopkins University Press: Baltimore, MA, USA, 1990.
23. Enemark, S.; McLaren, R.; Lemmen, C. *Fit-for-Purpose Land Administration—Guiding Principles for Country Implementation*; GLTN/UN-Habitat: Nairobi, Kenya, 2016.
24. Harvard Business School. *Terra Nova: A Social Business Trying to Unlock Land Rights for the Urban Poor in Brazil—Case Study*; Harvard Business School: Boston, MA, USA, 2019.

land

Article

Exploring PPPs in Support of Fit-for-Purpose Land Administration: A Case Study from Côte d'Ivoire

Ana García-Morán [1,*], Simon Ulvund [1], Eva-Maria Unger [2] and Rohan Mark Bennett [2,3]

1 Meridia, Landmapp B.V., Mauritskade 63, 1092 AD Amsterdam, The Netherlands; simon@meridia.land
2 Kadaster, The Netherlands Cadastre, Land Registry and Mapping Agency, 7311 KZ Apeldoorn, The Netherlands; eva-maria.unger@kadaster.nl (E.-M.U.); rohanbennett@swin.edu.au (R.M.B.)
3 Swinburne Business School, Swinburne University of Technology, Hawthorn, VIC 3122, Australia
* Correspondence: ana.garciamoran@meridia.land

Abstract: Public-private partnerships (PPPs) may facilitate the implementation of fit-for-purpose land administration (FFPLA); however, the approach can be compromised when funding for land registration is insufficient or donor projects end. This paper aims to introduce a new form of PPP to the literature on FFPLA, further extending the discourse and options available on PPPs for FFPLA. A background review finds that whilst PPPs have had long standing application in land administration, there is room to explore approaches that seek increased involvement of non-conventional land sector actors. A case study methodology is applied to analyse recent developments of FFPLA in Côte d'Ivoire that includes a partnership between the government and a consortium of private sector companies. Results describe the novelty, challenges, opportunities, and success factors for the approach, when compared to existing forms of PPPs. It is found that the innovative partnership approach may create novel avenues for financing FFPLA in developing countries and for more active forms of participation of the private sector in improved land tenure governance. The model potentially creates sustainable buy-in from private sector corporations, who whilst not conventionally closely undertaking land administration efforts, rely intrinsically on it to achieve corporate social responsibility objectives.

Keywords: fit-for-purpose land administration; land and resources rights; land tenure security; pro-poor land recordation; land governance; public-private partnerships; corporate social responsibility; poverty reduction; business driven solutions; social enterprises

Citation: García-Morán, A.; Ulvund, S.; Unger, E.-M.; Bennett, R.M. Exploring PPPs in Support of Fit-for-Purpose Land Administration: A Case Study from Côte d'Ivoire. *Land* **2021**, *10*, 892. https://doi.org/10.3390/land10090892

Academic Editors: Stig Enemark, Robin McLaren and Christiaan Lemmen

Received: 9 May 2021
Accepted: 18 August 2021
Published: 25 August 2021

1. Introduction

Secure land rights are intrinsic to achieving the 2030 agenda and specifically the land related Sustainable Development Goals (SDGs)[1] [1]. The need for secure land rights demands that existing land governance approaches are adapted and gives rise to calls for more responsive or responsible land administration [2]: conventional land administration approaches have been seen to be too inefficient and not sustainable in delivering formal land documentation [3]. The various inefficiencies present in existing land administration approaches need to be overcome in order to achieve the 2030 agenda [4].

In this vein, a marked momentum towards innovative land administration can be seen. Global institutions, in support of national and more local governments, promote developments reflecting the continuum of land rights [5,6]. The International Federation of Surveyors (FIG) supports the adoption of fit-for-purpose land administration (FFPLA). This philosophy counters dogmatic legal and administrative techniques and calls for nationally or locally appropriate methods [3,7]. FFPLA solutions are intended to be flexible, affordable, and achievable: they present opportunities for the application of innovative and suitable technologies [8]. This approach allows flexibility in land administration design, and upgrades or improvements in terms of data quality, or reductions to time and costs.

Key issues for FFPLA include sustainable funding models and overcoming land agency inertia, amongst others. Even though global development institutions—most notably via the World Bank—assist and fund reforms and programmes, implementation may remain dependent on the recipient government's human resources and operational budget allocated for execution and long-term maintenance. The responsibility to deliver on donor projects objectives can be too overwhelming for governments in this regard.

Existing research has explored land administration interventions in the frame of partnerships mostly funded by donors, where the private sector actors involved tend to come from within the land sector, as service providers, and their involvement is often merely focusing on technical delivery [9–11]. It is therefore worthy to continue exploring novel ways in which broader private sector actors, such as food and agricultural sectors, who may often deeply rely on land tenure security, can engage in the land administration initiatives through partnerships.

In this regard, the paper begins from the premise that private sector companies in sectors such as agriculture and food production are now increasingly incentivised to engage in land administration. This is particularly relevant to those transitioning towards sustainable supply chains and having the required evidence surrounding those to ensure the achievement of corporate social responsibility ambitions. Private sector corporations often have a close relationship to certain commodity-producing countries and governments. This creates a context and opportunity that allows for new flexible types of public-private partnership financing and implementation of FFPLA.

In response, the aim of this paper is to introduce a novel PPP to the FFPLA literature; one that further examines alternatives for the participation of private actors from outside the land administration sector. Rather than building solely from theory or literature, a case from practice in the context of Côte d'Ivoire is used to illustrate the approach. Due to the recency of the case, and constraints on presenting any longitudinal impacts, the paper focuses on presenting the drivers, requirements, and possible design options of such a PPP. Despite this limitation, what is presented is considered to be of high interest to the FFPLA discourse.

The remainder of the paper is structured as follows. First, a background section unpacks the needs and challenges in adopting the FFPLA approach and previous experiences of land administration PPPs. Second, the case study method applied is explained. Third, the results of the case study are described based on analysis of the context, differences of the case when contrasted with other land administration PPPs, and other novel aspects. Fourth, the key learnings, challenges and opportunities—including the possibility for generalisation—are discussed in order to lay the ground for future innovations to PPPs implementing FFPLA.

2. Background

2.1. Sustainability and FFPLA

It is necessary to provide a brief background on FFPLA in terms of how it intends to contribute to the broader sustainability agenda, but, also to unpack the issues relating to the sustainability of FFPLA initiatives in their own right, particularly with regards to financing.

Recording the relationships of people to land unveils complexities that are intrinsic to each local social system. For those involved in the design and execution of land administration, registering land rights is a cross-cutting disciplinary challenge, from technical implications through to social dimensions. Acknowledging that tenures come in different forms, and can exist beyond legal or statutory prescriptions, new tools and approaches are needed to facilitate the recording of so-called 'customary' or 'traditional' rights. Attention must be paid to ways that the complexities and pluralities of existing customary rights can be addressed in land administration systems [12]. It is argued that land administration systems should seek to follow emerging global policies and guidelines, as advised in the Framework for Effective Land Administration (FELA), a guidance document which serves as a global reference for land administration policy [13].

In comparison to conventional land administration approaches, which tend to be generalised and technically standardised, the FFPLA approach focuses on fitting characteristics into a specific context, for a purpose, and to meet the needs of all people, including vulnerable groups. On this, FELA refers to FFPLA because it aligns to the continuum of land rights, with the objective of providing 'security of tenure through recognition of legitimate rights' and recording the corresponding evidence of rights on a national register that is publicly accessible [14]. Especially, with the adoption of technology, FFPLA can contribute improvements to processes and enable the necessary functionalities for the recordation of the continuum of land rights that conventional methods tend to lack. This is in addition to, for example, recognition of broader land-related social elements; cost-effectiveness, efficient and interoperable capabilities; expedited data collection and automation; and bottom-up processes (although the FFPLA approach is usually implemented top-down), among others [15–17].

When implementing FFPLA, a multi-stakeholder and multi-disciplinary approach can potentially translate into more complexity in technical processes [16]. It is therefore important to carefully discern the roles and responsibilities of actors, funding, partnerships, and other FELA aspects. Implementers need to achieve 'balance' between global technical norms and local land governance arrangements. An issue complicating innovation is often found at the financial level: developing innovative approaches, adapting them to unique contexts and involving multiple stakeholders across all levels, can all help to spiral out costs as in previous experiences in Canada and Malaysia (addressed later) [9].

Sustainable funding mechanisms remain a major barrier for governments in developing countries to both introduce and sustain land administration efforts. The land administration community should always remain open to exploring novel models of collaboration between stakeholders; ones that may open new funding opportunities to deliver FFPLA. In this regard, the private sector can benefit from FFPLA as a vehicle for delivering on the Voluntary Guidelines on the Responsible Tenure of Land, Fisheries and Forests (VGGT), that supports the SDGs and corporate social responsibility objectives [8]. In these guidelines, the private sector can adopt voluntary roles such protecting human and legitimate tenure rights of local and indigenous communities, creating partnerships, and preventing conflict [8]. Additionally, understanding that sectors are interdependent, there may be opportunities for a deeper engagement of the private sector in FFPLA for generating shared value between partners and supporting SDGs delivery [18,19]. The literature provides an understanding of how land administration partnerships allow private sector actors, mainly from within the land sector to assist governments. However, alternative models bringing corporate actors into FFPLA are still rare and underexplored, at least in the published literature.

Therefore, in summary there is an opportunity to explore cases within FFPLA where private sector actors play a key role—even as drivers of FFPLA—and facilitate recordation and documentation of the land rights of smallholders[2] [20] within supply chains, as part of their broader social corporate responsibility and sustainability actions.

2.2. PPPs in Land Administration

Historically, the public sector has been responsible for owning (on behalf of citizens), operating and delivering most land administration services. Large-scale operations require large investments to not only cover the cost of land documentation, but, other overhead expenditures (e.g., technical assistance, training etc.) [9,21]. Developing countries have often struggled to afford land administration systems without external financial support, in comparison with developed countries, where land administration systems are well established and often profitable through enforced land related taxes, fees, and other charges [22].

For decades, the international community[3] has been providing financial support to governments to assist their endeavours to register land and to develop or modernise land administration methods and systems [23–25]. On this, the World Bank has supported land reforms, policy and land registration projects in developing countries [26]. Not only financial institutions, but, also different actors other than governments can play a substantial role too. A PPP can be considered an approach involving shared responsibilities through contractual cooperation between the public and private sectors to provide a product or service, often sharing risks, responsibilities and remuneration [27–29].

Able to be classified as different types (addressed later in the paper), PPPs involve different sectors, not only land, and have been utilised in developing countries to maximise use of public resources. Also, PPPs allow other non-traditional actors and international development organisations to take part in solution development.

There already exists recorded cases of PPPs dedicated to traditional land administration activities, originated by governments and/or donors with land service operators [30]. Learnings include that PPPs can facilitate private investment, cost-sharing, efficient risk-allocation, and efficient use of public funds. In terms of management, PPPs can enable expertise-based efficiency, and strengthen the public sector capacity for delivery, and allow the private sector to share innovations and new technologies securely [10]. PPPs may also bring flexible and customer-oriented land registration services and may improve procedures of land registration [27]. Additionally, PPPs are said to bring higher levels of service delivery, cost-effectiveness and reduction of investment risks for the parties involved [8,27]. However, the success or failure of a PPP can be marked by any resultant increase in levels of corruption, and any increase of costs of services as a result of a PPP [31,32].

Over the years, PPPs for conventional land administration focused on reform and infrastructure developments, and also focused on Land Administration Systems (LAS) [28]. In most cases, private specialised firms executed projects directly through contracting out services and focusing on creating land registries in both developing and developed countries [8,10]. The following paragraphs show some of these experiences.

Recent land administration PPPs in Canada and Australia, relating to the management and storage of transactions, are characterised by long-term relationships between governments and operators for periods of over 35-40 years duration, as well as the interest of reformulating processes and building automated systems with IT infrastructures [8,10]. Some of the challenges identified in those two cases are: the risk of customer needs being poorly integrated into the registry system; unexpected delays and increased implementation costs; risky public data management; overly long-term periods of registries operated by private parties (it may be difficult for the government to resume control of the system successfully immediately without the experience, capacity or human resources required); a structural lack of transparency; vulnerability to corruption within related sector, and dependency to private operators interests [8,10]. Besides, it is important to consider if long-term relationships have implications on the property market.

Cases in the Philippines, India and Malaysia have focused on progressing digitisation of processes and e-Government [8–10]. Showing that difficulties to digitise registries, even with participation of private sector, may originate from the government's lack of operational capacity; failure of private parties to ensure fully functioning systems; lack of expertise by private sector parties in land registry/automation; and thereby, unexpected long delays to delivery. In the case of Malaysia, the creation of e-Tanah (electronic land administration and management system) provides important lessons. After an initial failure, the private sector re-invested and eleven years were necessary before e-Tanah showed results[4].

On the other hand, New Zealand seems to have adopted an innovative approach in which collaboration with the private sector consists of procuring a suitable service model and paying the provider for its use [8].

Other experiences of a less dominant, but, complementary participation of private sector parties in diverse FFPLA projects worldwide, have depended heavily on donor funding. In Mozambique, for example, registration projects have been mainly funded by international donors focused on supporting local communities in registering their land rights. Mapping, delimitation and registration processes were supported by private land service providers contracted by the government with donor funding. Since 2011, land rights registration has been conducted with five different complex programmes but most of the donors involved have been discontinuing their support. Despite the potential opportunities, implementation challenges included lack of capacity, funding to develop a database of community landholdings, difficulties in maintaining the register, and the weak integration of local communities' data into governmental datasets [33,34]. In the cases of Canada and Malaysia, for example, risks have been jointly mitigated between parties through re-investment, sustaining allocation of resources and agreeing on an extension of the project duration [9].

These past experiences provide not only lessons on challenges—such as limited resources and weak contract/contractor management—but, also indications of success factors for private sector implementers. Examples include knowledge of global and regional contexts, requirements, processes, products and activities; capacity in different disciplines and capacity required for land administration; organisational stability despite context changes; risk management capacities; and direct supervision over contractors and collaboration flexibility [35].

That said, it is still somewhat unclear if and to what extent the FFPLA approach has been adopted in previous land administration PPPs. FFPLA PPPs seem new, or the FFPLA literature often does not describe the use of PPPs in FFPLA explicitly, and it is still necessary to seek to understand how FFPLA and PPPs can be most suitably combined to identify differences with the traditional PPP models.

Overall, no cases led by commodity sector industries (such as the cocoa sector) are observed in the literature. Exploring other collaboration formulas may contribute to broader and more innovative avenues for providing financial solutions and added capacity to FFPLA.

3. Methodology

Fundamentally, the exploratory work underpinning this paper can be considered to follow the interpretivist, if not pragmatic, research paradigm, and subsequently a qualitative research methodological approach could be applied. Specifically, in order to respond to the aims of the paper, that is, to further explore and expand upon knowledge relating to the use of PPPs in FFPLA, and to introduce a new type of PPP, a case study approach was applied. Specifically, a case study of the 'Côte d'Ivoire Land Partnership' [36], was undertaken, and the potential innovations to PPPs in land administration were explored. The case study initiative was selected because of the familiarity of the authors with the case. This potential for bias needs to be acknowledged. The case study largely included collection and analysis of secondary data sources, although the author experiences also informed the critique. The case study was exploratory in nature, rather than explanatory or confirmatory. The aim was to explore the involvement of non-conventional private sector actors in innovative FFPLA partnerships—in terms of how they might engage actively, or not, in securing land rights for vulnerable smallholders in developing countries.

Regarding the case study selection, Côte d'Ivoire was considered ideal since it is actively seeking to develop its agricultural markets and also its land administration system at the same time. The country has a high cocoa production as well as a current lack of alternatives for formalisation of land rights different to donor-led initiatives. The case, the Côte d'Ivoire Land Partnership (CLAP), was setup by the partners in CLAP, primarily through Meridia (explained below). Since July 2019, three phases were introduced before an early scale implementation: country scoping, feasibility, and design and testing (to be finalised in mid-2021). The early scale implementation is intended to run from April 2021 until the end of 2023. Therefore, it is acknowledged that results presented here, with regards to implementation success and impact, can only be considered preliminary. However, aspects relating to drivers, requirements, and design inputs are considered to provide insights on how a PPP in FFPLA could look like.

A review of existing evidence in the literature presenting FFPLA and relevant PPP references, guidelines and previous experiences, as well as private documentation of Meridia's FFPLA experiences, and the authors' experiences informed the data collection.

Then, data was analysed and contrasted with the existing literature to analyse the results systematically. Including the specific context of Côte d'Ivoire, partners, rationale and nature of the PPP model, how it adopts FFPLA and the challenges and limitations involved were explored.

In terms of data collection, to commence the case study work, an initial literature review based on documents and reports—both publicly available and privately sourced—was used to build the background context on PPPs and the case country. Secondary data was then collected from the documented experiences of involved private sector actors, that were involved in Côte d'Ivoire between 2019 and 2021. These materials were made available to the authorship team: it is acknowledged that some members of the authorship team have direct affiliation with Meridia, the Dutch-based private land documentation firm, implementing the case explored.

With regards to data analysis, extracting details on the country context, the partners involved, the rationale of the PPP, the type and design of partnership models, the adoption of FFPLA by the PPP and implementation challenges were considered. It is important to note that the paper does not examine in-depth the funding/finance issues. Novel aspects were also analysed with existing FFPLA frameworks in mind [10,37,38].

After compiling the data into results, these were presented systematically. This paid attention to PPP design characteristics, the functioning of the case, and the generalisation of findings to other FFPLA interventions.

4. Case Study Results

In the following sub-sections, the case of Côte d'Ivoire, describing a variation of a conventional land administration PPP from the cocoa sector—in the implementation of land registration for rural right holders—is described. First, an overview of land administration is provided. Then, a description of the novelty in terms of design, requirements, and implementation, based on the early learnings from an on-going pilot, are presented. Subsequently, discussion undertakes a comparison and contrast with other more traditional partnership models.

4.1. Land Administration in Cote d'Ivoire

Until 2019, rural land comprised 71% of the country and occupied 48.75% of the total population. Agriculture has been a major driver for economic growth since 1950s. Similarly, cocoa production has been mobilising people within and from outside the country. The country has seen decades of conflicts and political tension [39,40].

Conflicts between customary land holders and more recent migrants (both nationals and/or foreigners, or allochtones and/or allogènes in French—respectively) are mostly related to disputed land rights [41]. Without secure over ownership or use rights, tensions between social groups are, and were, constantly increasing. Besides, land governance is complex because customary and statutory tenures can coexist: land has been traditionally held by customary law, but the State administers land and property through formal registration processes. Findings from similar settings in other geographic locations indicate that FFPLA could be relevant to the Ivorian context: FFPLA has the same characteristics in both land-related conflict contexts and non-conflict contexts [42].

The Ivorian Government has set up reforms, updated policies and introduced specialised institutions that facilitate the advancement of rural land registration. The Rural Land Law, created in 1998 and amended in 2019, recognised rights acquired before the law took place, taking customary rights into account while promoting the formalisation of these and other types of rights [39,43]. The law promotes the conversion of customary rights into statutory rights, and originally established a 10-year period for rural land registration and delivery of land certificates to traditional owners before unregistered land would return to the State's ownership. Also, modifications to the constitution in 2016 established the right to property as a guarantee for all Ivorians and vested ownership of rural and agricultural lands to the State, public entities and Ivorian citizens.

Rural land registration in the country can be done through three different types of formal documents recognised by the government: land titles, land certificates and land use contracts[5] [44].

Land registration involves an interaction of multiple actors across the national, sub-prefecture and village level. As part of the institutional setting, the Ivorian government created AFOR (Agence Foncière Rurale—AFOR), or Rural Land Agency, in 2016. AFOR is a decentralised institution dedicated to implementing the Rural Land Law and is in charge of land registration operations. Just as in other cases, decentralisation can be considered as effective, but also challenging for the implementation of FFPLA [45].

At any rate, the Ivorian government has not been able to register existing customary rights over rural land, at least, at scale. In 2019, only around 0.5% of the targeted land certificates have been delivered[6] [46]. The government must ensure efficiency in the implementation of its land registration programme before unregistered land can be retrieved by the state. However, challenges have overruled recent efforts. The commonly expensive cost of issuing land certificates, without funding from donors, are not affordable for landowners, including cocoa farmers, considering that the income of a cocoa household from all sources is around 2400 EUR per annum [47]. People used to pay between XOF 980 k—1.4 M (equivalent to USD 1747–2621 or EUR 1498–2247) when applying for a certificate individually.

In early pilot free registration programmes[7] the cost of land certification to the government was between 500,000 XOF and 1 million XOF (equivalent to around 760 to 1500 EUR) for a 3-hectare parcel in the south of the country [39].

AFOR has been working to improve documentation processes at the same time as prioritising land certificates and land use contracts [48]. Currently, AFOR—with support from the World Bank—is implementing the Land Policy Improvement and Implementation Project for Cote d'Ivoire (PAMOFOR). AFOR is being assisted in implementing the country's land policy, establishing national geodetic infrastructure, developing the capacities necessary to operate land registration, modernising the land information system, and testing the roll out of systematic registration and documentation under participatory, streamlined, simplified and less costly processes [49].

AFOR developed the national land administration system (Système d'Informations Foncières de l'AFOR—SIFOR) and started a free registration project. According to AFOR's Manual of Operations, the national registration process involves institutions from all levels including grassroot village land governance institutions [46]. At the community level, village land management committees (Comité Villageois de Gestion Foncière Rurale—CVGFR) are in charge of community-based management of traditional land. These committees are formed by the traditional chief as well as other community leadership and group representatives. These committees are supervised by sub-prefectures, sub-prefectural rural land management committees (Comité Sous-Préfectoral de Gestion Foncière Rurale—CSPGFR) and AFOR, and are supported by land surveyors—especially in management of land owner-tenant agreements contracts. Given the pressing timeline, limited capacities and human capital available to implement the Rural Land Law it appears that, although supported by the World Bank through PAMOFOR[8], the government still requires support in order to meet the rural land registration deadline.

4.2. *The Cote d'Ivoire Land Partnership (CLAP)*

4.2.1. Drivers and Motivation

The fragile tenure security of cocoa farmers including migrants, in conjunction with the sustainability issues of the country related to agricultural activities (deforestation, land pressure), led to attention by private transnational companies to securing rural land tenure, in support of their own objectives. Since 2019, The Hershey Company, based on an earlier positive experience in Ghana with USAID[9], engaging on the sensitive topic of affordable and acceptable land titling documents, brought together companies to promote the creation of the Côte d'Ivoire Land Partnership (CLAP).

With this antecedent, CLAP focused on developing affordable land documentation at scale for Ivorian smallholder farmers by aiming to develop a similar cost-efficient service approach—thus accelerating land registration in cocoa farming areas, as well as helping the government maximise public resources in other areas across the country as needed. Then, documents were to be facilitated and subsidised for supply chain farmers to register their own rights to land or crops depending on their tenure situation. The only commitment needed from farmers was to contribute to the cost of the document by paying a small part (up to 20%).

4.2.2. Partners and Relationships

The partnership was originally formed by cocoa industry leaders—The Hershey Company, Unilever and Cocoa Horizons Foundation, the Ivorian government through AFOR, the Foundation of the German Cocoa and Chocolate Industry, and Meridia. While the private sector actors provide funding and service delivery, the local government enabled a political environment for interventions, provided in-kind contributions, collaborated in the execution of projects and operated the land information system/registry to which CLAP service connects with.

The design also involves donor agencies and civil society (see Figure 1): International development agencies provided funding and technical assistance needed at the social and sustainability levels, and civil society organisations address needs at the community level.

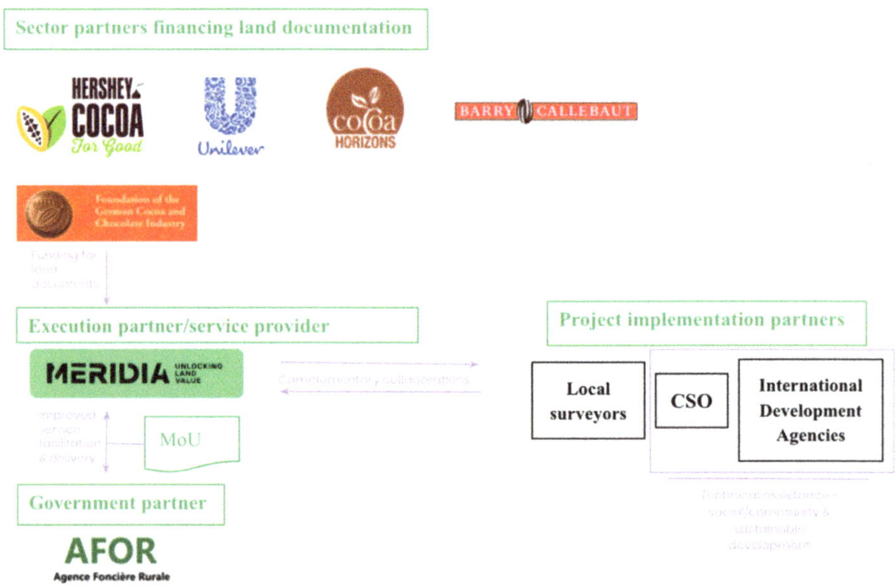

Figure 1. Partners in CLAP.

4.2.3. Initiation and Timeline

In 2020, CLAP conducted a feasibility study; one recommended step for the establishment of a PPP for land administration [50]. During eight months, the partners explored the land tenure and social local contexts in the country's farming areas; the legal, regulatory and institutional frameworks and their suitability for CLAP tenure projects under a commercial model; a solution to scale-up land rights registration specifically focused on cocoa farming areas; and a financial model adjusted to the context. Specifically, the feasibility study components included the understanding of specific issues of the Ivorian rural land context, alignment of stakeholders and identification of actors relevant to the partnership and its objective, social/community acceptance (including demand and affordability from farmers), identification of vulnerable groups and types of access to land, legal and institutional structures (from government to community levels), legal operational compliance, validation of a business case for investment by the private sector in the land rights of cocoa farmers within supply chains, and adaptation of systems and fit-for-purpose technologies.

Then, in the design phase, from mid-2020 to mid-2021, the partnership with support from the Dutch government, focused on testing the service model through a first pilot in one area in the south of the country[10]. The on-going pilot was accepted by the communities targeted, the rest of the stakeholders subsequently committing to continuing the partnership and running an early scale project until 2023.

In terms of duration, CLAP has been conducted through sub-projects and by stages without foreseeable permanent system leasing with the government. In the first stage, the PPP aimed to operate an early scale registration project from 2021 to 2023. The short period of the first project may be an opportunity for improvements and adjustments that long-term agreements in other PPPs do not easily enable.

4.2.4. Ownership, Management, and Finance

CLAP focused on service delivery and improvement of processes. Compared to other PPPs in the sector, no relevant partnership type was found to align fully with CLAP.

Starting with the distribution of responsibilities, in CLAP, these differed from those in the most common and accepted types for land registration services discussed in existing scholarship and thematic texts [8,24]. The land administration system, to which the CLAP projects integrated, is fully designed, built, owned and operated by the Ivorian government. For this reason, CLAP does not fit either in the types of 'Design Build' or 'Operation and Maintenance Contract'. Additionally, since the private sector actors do not only finance the intervention, but, also bring the registration service, CLAP cannot be categorised under 'Finance Only'. Moreover, the registry, asset or facility are not built, leased or transferred, so CLAP is not 'Design, Build, Finance and Operate', 'Operate and Transfer', 'Lease, Develop and Operate', 'Build Lease, or 'Build, Own, Operate and Transfer'. Whilst CLAP does relate partially to 'Build, Own, Operate', since the private sector partners were in charge of the end-to-end process, those actors do not take part in ownership transfer to the public sector, or permanent service, or administrative operations.

Whilst Meridia was (and is) responsible for collection, management, cleaning, processing and submission of data to the system, AFOR validates data and issues the land documents. No asset has been transferred between the public and private sectors.

CLAP cocoa sector members provided direct investment in land documentation. The private land firm was responsible for improving and innovating processes and the overall service delivery based on the FFPLA approach. The firm held a contract with the local government and both sectors cooperated to improve the service, execute projects, reduce administrative costs and increase efficiency of rural land registration. The government also facilitated engagement with other authorities.

Based on this formula, it can be recognised that the private sector engagement in CLAP was high, moving from management or operational contracts, leases or concessions, to a more hybrid PPP types between 'partial divestiture of public assets', 'joint venture' and 'full private sector divestiture' [10]. However, it cannot be considered as any of those categories solely, and can be best described as a new model that fits the purpose.

When compared with other cases, this case varies in terms of funding. CLAP projects included two main components: land documentation delivery and technical assistance. The financial model was a mixed set up where land registration was paid by the private sector actors and users (80/20% of the costs of documents, respectively)[11]. Revenue from land documentation went to the private land service provider and local private surveyors sub-contracted. These payments for land documentation represented 60% of the total project cost. Then, contributions from donors leveraged the other 40% to cover technical assistance, development-focused actions needed, capacity-building for community land governance institutions, and other project costs. For example, the contribution was dedicated to organisations from the civil society supporting activities in the community and participatory processes, and any work required for context exploration and attention to social issues in the communities. In general, capacity-development was financed by both private sector actors and donors. The government held a contract with the private land service provider (as manager of the partnership and representing all partners) to work together in improving services for rural communities and contributing internal capacity for joint technical activities and final issuance of land documents. Overall, this scheme can be replicated in any other sector or context.

4.2.5. Data, Innovation, Processes and Technologies

Two central aspects found in the technical approach are worthy of mention: adaptability and replicability. As the approach followed the configuration of existing legitimate rights it can be adapted to register different tenure types as long as the registration approach complies with the legal requirements and regulations set by the law. The tools applied can be adapted to the accuracy required and integrated to other systems—including the national land administration system and existing government data. Use can also be made of the National Reference System, parcel monuments and use of Continuously Operating Reference Stations (CORS). Also, different alternatives of registration and 'smart' surveying tools are available, particularly for use in mass operations where digitisation, technology and processes interrelate. Those tools can be used by professionals and non-professionals. Community members, including women, can collect data, if trained technically. However, just as with other digital tools, digital illiteracy in combination with technology can imply a first layer of exclusion for some individuals. Data goes through seamless data processing and certified local surveyors and authorities can use the tools (backoffice) to fully validate data once submitted to the land administration system. Then, authorities approve applications and issues the respective land documents.

4.2.6. Gender and Vulnerable Group Sensitivity

Specifically, in relation to tenure insecurity, the CLAP model entailed attention to vulnerable groups through affordability approaches such as tailored subsidies. The finance model vested the major part of the cost in companies' payment. It aimed to first reduce the cost of a sole land certificate by more than 10 to 15 times: further funding was used to subsidise up to 80% of the total cost of documents to make those affordable for smallholders within the cocoa supply chains.

Then, as higher volumes drive the cost down at scale, it was (and is) expected that smallholders' contributions will increase over time, without depending mainly on external funding. If the model is successful, the outcome would be an alternative for affordable documentation filling the gaps after PAMOFOR concludes.

The partnership can contribute to addressing the Ivorian Government's financial gap, reducing costs along lands involved in cocoa supply chains, and bringing efficient land administration service delivery [23]. In return, some of the benefits for the companies engaging in this can be an overall strengthening of the land security of farmers within their supply chains.

The method also facilitates joint documentation for families to secure not only the farmer, who is often a male, but also secure the rights of female spouses (or other members of the family) by also recording their name in the land documents. In addition, sensitization activities in the process included discussions with women and specific vulnerable groups (e.g., youth, migrants), with a view to seeking to make their voices heard.

4.2.7. CLAP, PPP, FFPLA, and FELA

As explained in Figure 2 below, the PPP analysed enabled the creation of a FFPLA solution for Côte d'Ivoire. This solution then allowed for alternative service delivery. Both the PPP model and the service may evolve over time informed by the early scale efforts.

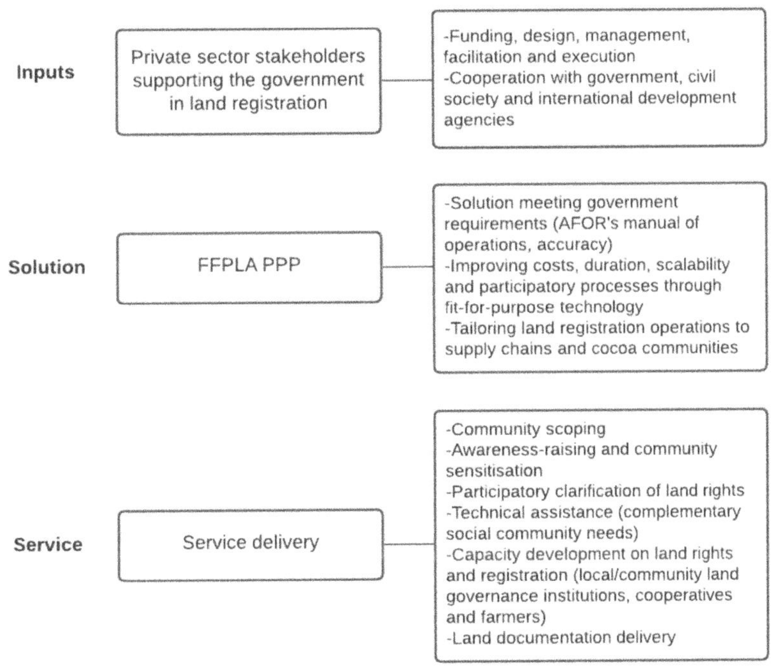

Figure 2. PPP to service delivery processes.

Table 1 presents the detailed key distinguishing features of the CLAP FFPLA PPP against the nine strategic pathways (SP) of FELA. FELA's nine strategic pathways provide an overarching guide for an overall summary of key features.

Table 1. CLAP as a FFPLA PPP.

FELA SP1: Governance, Institutions and Accountability	
Actors and Responsibilities	1. Private sector: 1a. Industry actors (The Hershey Company, Unilever and Cocoa Horizons Foundation) Role: Funding for land registration and partnership governance 1b. Private land service operator (Meridia) Role: Design, partnership governance, management and overall execution (in collaboration with local private surveyors) 1c. The Foundation of the German Cocoa and Chocolate Industry Role: Funding

Table 1. *Cont.*

Actors and Responsibilities	2. Public sector: 2a. Government (AFOR) Role: Political enablement, in-kind contributions, execution, application files approval, issuance of documents and registry operation 2b. International development and cooperation government agencies (donor countries) Role: Funding for technical assistance, development-oriented activities and execution 3. Beneficiaries: 3a: Smallholders (Landowners [12]) Role: Accepting the project by applying for land documents; allowing land occupants apply for land documentation; collaboration in implementation (clarification of land rights); and partial payment for land documents 3b. Smallholders (occupants/users under agreement with landowners [13]) Role: Accepting the project by applying for land documents; and partial payment for land documents 4. Civil Society Organisations Role: Advisory, strong implementation support
FELA SP2: Policy and Legal	
Alignment and Legitimizing Aspects	- Support to the government in facilitation of implementation of the Rural Land Law - Support to the government in facilitation of implementation of the national programme for rural land registration and other related policy.
FELA SP3: Financial	
Financial and Business Model	Hybrid model - Large investment payment by companies to the private land administration firm to design and execute land registration projects, and to cover a major part of the cost - Small partial payment from smallholders to the private land administration firm - Leveraged funding from donors for implementing complementary technical assistance in CLAP intervention areas in order to maximise efforts Payment: Joint payment to the private land service provider
FELA SP4: Data	
Approach and Mindset	- Data collection and mapping approach is designed to be scalable - Set up model for replication in other specific contexts - Recognition of land tenure at multiple levels and based on traditional/customary rights

Table 1. *Cont.*

FELA SP5: Innovation	
Technology Orientation	- Cost and time reduction by implementing technological innovations - Adaption of innovations to the Ivorian context (software and digital tools for robust data collection, accuracy and requirements
FELA SP7: Partnerships	
Other Key Actors	1. Community leaderships and local village land management committees (CVGFRs) Role: Adoption of interventions, collaboration in operations (clarification of land rights and elaboration of land use contracts) and officialization of document grantees 2. Cooperatives Role: Collaboration in farmer organisation, communication, and coordination of operations
PPP Constructions	Model: Hybrid Not within existing common models described in literature. Similarities with Private Finance Initiative (PFI) and Joint venture/Partial divestiture of public assets [14]. Level of private sector participation: High - Design, development, finance, integration and operation by the private sector—no transfer, integration with the LAS only
FELA SP8: Capacity and Education	
Multi-level and Multi-sector	- Capacity-development for community leaderships and local surveyors, community members (including women) and village-level governance institutions, conducted by the land service provider and donors.
FELA SP9: Advocacy and Awareness	
Modern Approach	- Gender and Vulnerable Group Sensitivity - Attention to women and vulnerable groups in the awareness-raising and sensitization processes [15] - Facilitation of joint documentation (a family tenure view)

5. Discussion

Having presented the core results from the Côte d'Ivoire case study, a discussion is undertaken with regards to (i) the potential benefits and advantages of the CLAP model (with respect to other PPPs), (ii) the potential challenges with the approach, (iii) the generalisation/adaptability of the model more generally; and (iv) the future ahead for the specific case.

5.1. Potential Benefits and Advantages

CLAP serves as a novel reference for more innovative PPP models to transit from traditional land administration PPPs to FFPLA PPPs, and enable more active engagement of the private sector in improving land tenure governance from social responsibility and sustainability perspectives.

CLAP is a relatively new partnership, for which early implementation started in 2021. Despite no further empirical evidence being able to be offered at this point, the potential benefits of the model can be already perceived in the creation of a FFPLA PPP framework, utilising the FELA, that merges efforts and objectives from a non-common group of actors, who do not naturally interact in the land space (at the funding and execution levels). It provides broader support to the government without it needing to depend on donor funding solely.

For governments, this type of partnership may allow for operational flexibility. While CLAP focuses on cocoa farming areas, the government can still orient public resources and donor funding to other areas. Although, it is yet unknown how both services will co-exist, the commercial approach by CLAP tests an alternative and appears an affordable choice for smallholders.

CLAP as a PPP entails cooperation between the public sector, both through the local government and governments, from donor countries and the private sector. The partnership is formalised through contracts[16], and; cooperation with non-governmental organisations, and offers the possibility to cooperate with other different civil society institutes and foundations.

Other novelties are found in the focus on a certain commodity value chains; re-thinking the financial streams for funding land administration, applying the FFPLA approach in developing countries; the sector-oriented attention to the poor and social needs, and; the interaction of non-conventional actors when driven by the private sector (outside traditional land sector actors), supported by government, compatible with international development agencies, and managed by a land rights documentation private company.

CLAP as a FFPLA PPP adds purpose and flexibility to land registration services, provides possibilities for multiple context-specific developments for scaling, and drives service delivery through technological approaches. It further facilitates attention to wome—e.g., facilitation of joint documentation with the name of women recorded on land documents—and vulnerable groups, as mandated by corporate sustainability ambitions.

5.2. Challenges and Limitations

At the social and political levels, any lack of commitment from government, community leadership would challenge land tenure interventions under a PPP like CLAP. The same applies to cooperatives, since they are at the heart of value-chain implementations.

At the financial level, funding is still an aspect to pay attention to. The finance availability in this FFPLA PPP may be still influenced by the number of companies engaging in the partnership and their contributions and the extent of donor funding leveraging. Similar to the situation of the governments' lack of funding, companies alone may not be able to allocate the necessary budget for systematic registration either. Without leveraging sufficient co-funding from other actors, it may be difficult to navigate how the FFPLA-focused services can be delivered systematically in the communities. Though it should be again noted, any financial risks are mostly shared by the private sector partners, in terms of investment and revenue, and by the beneficiaries in terms of sharing the service cost. Similarly, the level of community acceptance and demand for land tenure interventions may enable or limit projects under this type of PPP financially and socially.

It also must be recognised that dependence on private funding by private companies could also potentially present drawbacks, and needs to be further explored in future discussions.

On the other hand, continuing cooperation with donors is important to guarantee funding for technical assistance, but, the risk of 'competing' with other future free documentation programmes for donor funding must be considered.

At the technical level, as Côte d'Ivoire is still setting up its land administration system, CLAP may still depend to a certain extent on the government progress if traditional methods dominate the national approach. In order to sustain the efficiency of the process, projects would rely on technology and digital processes to overcome any delays of the land administration system being developed in PAMOFOR. Even if the system is not fully in place, implementation can continue as long as the private land firm provides the service, manages the hybrid methods, and maintains data digitally, until the integration with the national system can be done.

Overall, improvements to CLAP processes will need to be a constant. Also, due to the early stage of the CLAP pilot project, issues such as the sustainability of the model and relationship to the national efforts cannot be fully assessed yet. A deeper evaluation of the PPP may be possible only in later stages.

5.3. Generalisation and Adaptability

Whereas CLAP has been developed for Côte d'Ivoire, the aim is for replicability to other contexts. Transfer to other contexts is possible as long as social, legal and institutional environments allow it, and technological innovations are adjusted to the local context. Nevertheless, not only the advantages but also challenges at different levels may be transferred too.

In the end, the blueprint developed by CLAP will inform future innovations and partnerships about opportunities and success elements. It is worth to continue investigating how CLAP develops, from a pioneering model to potentially a mature partnership.

Similarly, it may be relevant for other cases to monitor the outcomes and evidence gathered in the following years as the CLAP land registration project progresses. The flexibility of CLAP can be expected to allow more actors to join the PPP and so develop expansion possibilities. In a future scenario, two important steps for CLAP would be important: to ensure a future integration with other value chains through a more systematic land registration, as well as to continue working with the government and donors to fulfil existing gaps. The CLAP model also allows the opportunity for donors to experiment with innovative collaborations with the private sector in land administration, outside the traditional PPPs.

It is yet unknown if and how the broader industry or broader sectors may embrace a FFPLA PPP. However, considering the current interest and local acceptance of the early scale project by cocoa farming communities, the government, the private sector actors and generally all the stakeholders involved, the authors argue that a FFPLA PPP such as CLAP may lead to strengthening actions by the private sector across different sectors to support the achievement of the SDGs. Promising, as well, is the realisation of the potential of multi-stakeholder cooperation through partnerships (SDG 17) and the impact of the private sector in securing land rights.

5.4. Looking Ahead

Whilst a longitudinal study will be required to further assess the long-term validity and added value of the CLAP model, at this moment, the most important preliminary key success factor to highlight is political will and a long-term commitment from all parties. Especially the political-will of customary landowners, local village land management committees and the local government and community leaderships can be considered a preliminary key success factor for the model. Even though the duration of this CLAP model is relatively still short (3 years), in comparison to other cases (with a 10 year or longer period), this type of industry-based partnerships requires a long-term commitment from all actors involved and continuous budget allocation. This support will depend directly on the focus on land rights within the corporate sustainability agendas, which can be considered as a relatively pioneering focus in this contemporary period, but is likely to increase in the future.

This FFPLA PPP provides the private and public sectors with an alternative to document land for farmers or the agricultural sector. Especially relevant for private companies is to integrate this type of effort within the supply chains, and to explore the potential of securing investment in farmers and land, the ability to strengthen existing sustainability/social programming, and the potential of obtaining an overall return on investment.

Overall, this PPP type offers one way of involving other sectors to land registration processes within the country, or experimenting in other contexts. This can further benefit other government entities such as those responsible for taxation, infrastructure planning and development, rural development and agriculture.

There is also an opportunity that the scaling up of the project volume drives the cost of documentation down, as dependencies on private investment are reducing. Hence, documents continue to be affordable to smallholders and less subsidies will be required. This aspect reduces the possibility of costs increases later on, which often represents a disadvantage for PPPs.

6. Conclusions

This paper sought to introduce a new form of PPP to the FFPLA literature, by expanding the discourse on PPPs in land administration and focusing specifically on their role in FFPLA initiatives, and more specifically exploring the role of non-conventional land administration sector actors. These actors, for example, global food and agricultural corporations, are typically not directly involved in land administration interventions, but, for corporate social responsibility objectives, are intrinsically reliant upon it. Closer embedding of these actors into land administration interventions were suggested as having the potential to unlock the financial capacities of those organisations, making them available for national government-led land administration initiatives, that often lack sustainable funding mechanisms.

The work therefore commenced from the premise that increased corporate-style partnerships may also further facilitate the implementation of FFPLA—because, anecdotally, major funding is required by governments to implement FFPLA—and donor interventions would not be sufficient after projects end. The background review confirmed this notion, finding that whilst public-private partnerships have long standing use in land administration, there is room to explore innovation of partnerships with an increased involvement of non-conventional private sector actors in certain roles and activities.

The case study methodology was applied to review the recent developments of FFPLA in Côte d'Ivoire through a partnership between the government and a consortium of companies. The results showed that the CLAP approach brought novelty to PPPs in FFPLA in terms of actors, roles, the financing model, and the specific use of the FFPLA technical approach. That said, it was recognised that a longitudinal study would be needed to fully assess the validity and sustainability of the approach. Preliminary key success factors, like all land administration interventions included the importance of high-level political will (from government and communities) and long-term stability surrounding land access and availability for the corporate actors. Another factor that would protect land tenure interventions is the government's capacity to incorporate the demand and applications from the partnership projects into their national registry in order to sustain the efficiency of the process.

In terms of generalisation possibilities and future developments, it was found that the innovative partnership may create new avenues for financing FFPLA in developing countries and for more active forms of participation of the corporate. Results from the case study are considered preliminary and even though achievements and impact cannot be fully defined until the approach is scaled up, the type of partnership studied can be considered novel in terms of design, requirements and implementation. The model potentially creates a long lasting and sustainable buy-in from private sector corporations, who whilst not conventionally closely tied to land administration efforts, rely intrinsically on it to achieve corporate social responsibility objectives.

Author Contributions: Conceptualization, A.G.-M., S.U. and E.-M.U.; methodology, A.G.-M. and E.-M.U.; formal analysis, A.G.-M., E.-M.U. and R.M.B.; investigation, A.G.-M. and S.U.; writing, A.G.-M., S.U., E.-M.U. and R.M.B.; original draft preparation, A.G.-M., E.-M.U. and R.M.B.; data collection, A.G.-M. and S.U.; writing—review and editing, E.-M.U. and R.M.B. All authors have read and agreed to the published version of the manuscript.

Funding: The funding for the publication costs for this article were kindly provided by the School of Land Administration Studies, Faculty ITC from the University of Twente, in combination with Kadaster International, The Netherlands.

Institutional Review Board Statement: Not applicable.

Informed Consent Statement: Not applicable.

Acknowledgments: The authors acknowledge the journal editors and guest editors of the Special Issue "Fit-for-Purpose Land Administration—Providing Secure Land Rights at Scale" for their support and the granting of funding covering the fees related to the publishing of this paper.

Conflicts of Interest: The first and second authors mentioned at the beginning of the manuscript declare their direct involvement in the PPP case presented in this paper. The funders of this article had no role in the design of the study; in the collection, analyses, or interpretation of data; in the writing of the manuscript, or in the decision to publish the results.

Notes

1. Specifically, SDGs: 1 No Poverty, with the target 1.4; 2 Zero Hunger, with the target 2.3; 5 Gender Equality, with the target 5.a, and; 15 Life on Land with the target 15.2.
2. The term 'smallholder' varies across countries under different criteria. The definition used in this paper for reference is "..a farmer (producing crop or livestock) practicing a mix of commercial and subsistence production . . . , where family provides the majority of labour and the farm provides the principal source of income".
3. 'International community', refers to global governance and multilateral entities such as the World Bank, and international development and cooperation agencies such as the United States Agency for International Development (USAID).
4. Including improvement of the delivery of land administration services, introduction of innovations such as online applications and payments, simplification of procedures and better customer experience, reduction of the time for transactions, increasing revenue for the government agency in charge, and contributing to efficient e-Government platforms.
5. Land titles are the final formal document proving the permanent conversion of land rights to property rights. Land certificates prove registered ownership rights over land and plots, and are the document which can be converted to titles. Land use contracts are community-managed formalised agreements between customary owners and occupants. Indeed, tenancy contracts are the most common agreement that have allowed non-landowner cocoa farmers to access to land and their application may be key in conflict prevention in Côte d'Ivoire.
6. 6421 out of 1,500,000 land certificates aimed by the government.
7. Referring to the DP3 and DP4 projects partly funded by the European Union debt swap programmes (Devis-Programme, or DP).
8. PAMOFOR also includes support to the government in aspects such as institutional capacity building and institutional systemic improvements.
9. In Ghana, costs of land titling documents were reduced from $400 to $100 EUR, the process for farmers was made easier and the delivery time shortened from 18 months to 8 weeks.
10. The region where the service model has been tested is Lôh-Djiboua. The service design phase was still running and conversations around the onboarding of one donor to the partnership was in process when this paper was written.
11. For the private sector actors, return on investment was expected in the productivity and sustainable sourcing dimensions across their supply chains. For the cocoa farmers, return on investment was expected to be figuratively in the form of sense of land security and any positive behavioural change that led to their welfare. Unless farmers who own land and receive documentation decided to use the document for immediate land transactions, for a direct economic return.
12. Referring to cocoa farmers who are landowners or to people owning land occupied by cocoa farmers. Both types of landowners can be beneficiaries of the project.
13. Most cocoa farmers are migrants who do not own the land they occupy. Therefore, in that situation they are land occupants/users.
14. In this FFPLA PPP, the private sector brings funding for covering 80 percent of the total cost. Smallholders pay the 20 percent left and the Government provides in-kind contributions at the same time while working with the land administration firm to reduce administrative costs.
15. These two processes mentioned are implemented by organisations from the civil society subcontracted for the projects. These activities are independent to any structural efforts by the government (through PAMOFOR) to communicate the national implementation of the Rural Land Law and rural land registration policies.
16. On behalf of the private partners in CLAP and as the main intervention executer, Meridia signs contracts with AFOR for land tenure projects under the partnership. The first contract of the PPP was the on-going pilot (at the time of writing) conducted in two villages of the country to test the service.

References

1. Musinguzi, M.; Enemark, S. A Fit-For-Purpose Approach to Land Administration in Africa: Supporting the global agenda. *Int. J. Technosc. Dev.* **2019**, *4*, 69–89. Available online: https://vbn.aau.dk/ws/portalfiles/portal/318235502/2019_issue_Vol_4_Musinguzi_Enemark.pdf (accessed on 8 May 2021).
2. Unger, E.M. Responsible Land Administration in Disaster Risk Management: Approaches for Modelling Integrated Governance and Policy Transfer of People, Land and Disasters. Ph.D. Thesis, University of Twente, Enschede, The Netherlands, 26 September 2019. [CrossRef]
3. Enemark, S.; Bell, K.C.; Lemmen, C.H.J.; McLaren, R. *Fit for Purpose Land Administration, No. 60*; International Federation of Surveyors (FIG), The World Bank: Copenhagen, Denmark, 2014.
4. Zevenbergen, J.A.; De Vries, W.; Bennett, R.M. (Eds.) *Advances in Responsible Land Administration*; CRC Press: Boca Raton, FL, USA, 2016. [CrossRef]
5. Zevenbergen, J.A. *Handling Land: Innovative tools for Land Governance and Secure Tenure: E-Book*; UN-HABITAT, International Institute of Rural Reconstruction (IIRR), Global Land Tool Network (GLTN): Nairobi, Kenya, 2012; 152p. Available online: https://www.local2030.org/library/76/Handling-land,-Innovative-Tools-for-Land-Governance-and-Secure-Tenure.pdf (accessed on 23 April 2021).
6. Asiama, K.O. Responsible Consolidation of Customary Lands. Ph.D. Thesis, University of Twente, Enschede, The Netherlands, 17 April 2019. [CrossRef]
7. Bennett, R.; Masli, E.; Potel, J.; Unger, E.M.; Lemmen, C.H.J.; de Zeeuw, K. Cadastral Entrepreneurs Recognizing the Innovators of Sustainable Land Administration. In Proceedings of the FIG Working Week 2019: Geospatial Information for a Smarter Life and Environmental Resilience, Hanoi, Vietnam, 24–26 April 2019. Available online: http://www.fig.net/resources/proceedings/fig_proceedings/fig2019/papers/ts02g/TS02G_bennett_masli_et_al_9950.pdf (accessed on 8 May 2021).
8. FAO. *Voluntary Guidelines on the Responsible Governance of Tenure of Land, Fisheries and Forests in the Context of National Food Security*; Food and Agriculture Organization (FAO): Rome, Italy, 2012; 40p. Available online: http://www.fao.org/3/i2801e/i2801e.pdf (accessed on 21 June 2020).
9. Bell, K. *Global Experiences with Public Private Partnerships for Land Registry Services: A Critical Review*; World Bank: Washington, DC, USA, 2017. [CrossRef]
10. Anand, A.; Halid, N. *Public-Private Partnerships in Land Administration: The Case of e-Tanah in Kuala Lumpur, Malaysia*; World Bank: Washington, DC, USA, 2020. [CrossRef]
11. World Bank. *Public-Private Partnerships in Land Administration: Analytical and Operational Frameworks*; World Bank: Washington, DC, USA, 2020. Available online: https://openknowledge.worldbank.org/handle/10986/34072 (accessed on 2 July 2020).
12. Lavigne Delville, P.; Deininger, K.; Augustinus, C.; Enemark, S.; Munro-Faure, P. Registering and Administering Customary Land Rights: Can we deal with complexity. In *Innovations in Land Rights Recognition, Administration and Governance*; Deininger, K., Augustinus, C., Enemark, S., Munro-Faure, P., Eds.; World Bank: Washington, DC, USA, 2010; pp. 25–43.
13. UN-GGIM. Framework for Effective Land Administration: A reference for developing, reforming, renewing, strengthening, modernizing, and monitoring land administration. In Proceedings of the Tenth Session of the United Nations Committee of Experts on Global Geospatial Information Management (UN-GGIM), New York, NY, USA, 26–27 August–4 September 2020. Available online: https://ggim.un.org/meetings/GGIM-committee/10th-Session/documents/E-C.20-2020-29-Add_2-Framework-for-Effective-Land-Administration.pdf (accessed on 3 June 2021).
14. Enemark, S.; Mclaren, R.; Lemmen, C. *Fit-For-Purpose Land Administration: Guiding Principles for Country Implementation*; United Nations Human Settlements Programme (UN-Habitat), Kadaster International, Global Land Tool Network (GLTN): Nairobi, Kenya, 2016; p. 118. Available online: https://gltn.net/download/fit-for-purpose-land-administration-guiding-principles-for-country-implementation/ (accessed on 12 November 2020).
15. Lemmen, C.; Unger, E.M.; Bennett, R. How Geospatial Surveying is Driving Land Administration: Latest innovations and developments. *GIM Int.* **2020**, *34*, 25–29.
16. Lengoiboni, M.; Richter, C.; Zevenbergen, J. Cross-cutting challenges to innovation in land tenure documentation. *Land Use Policy* **2019**, *85*, 21–32. [CrossRef]
17. Ulvund, S.; Unger, E.M.; Lemmen, C.; Van den Berg, C.; Wits, T.; Utami, R.D.; Lukman, H. Inclusive and Gender Aware Participatory Land Registration in Indonesia. In Proceedings of the 2019 World Bank Conference on Land and Poverty: Catalyzing Innovation, Washington, DC, USA, 25–29 March 2019. Available online: https://research.utwente.nl/en/publications/inclusive-and-gender-aware-participatory-land-registration-in-ind (accessed on 12 November 2020).
18. Chandler, D. *Strategic Corporate Social Responsibility: Sustainable Value Creation*; SAGE Publications: Thousand Oaks, CA, USA, 2019.
19. Porter, M.E.; Kramer, M.R. The Big Idea: Creating Shared Value. How to Reinvent Capitalism—and Unleash a Wave of Innovation and Growth. *Harv. Bus. Rev.* **2011**, *89*, 62–77.
20. Narayanan, S.; Gulati, A. *Globalization and the Smallholders: A Review of Issues, Approaches, and Implications*; Markets and Structural Studies Division [Discussion Paper], 50; International Food Policy Research Institute: Washington, DC, USA, 2002.
21. Pagiola, S. *Economic Analysis of Rural Land Administraton Projects*; World Bank: Washington, DC, USA, 1999.
22. Williamson, I.; Enemark, S.; Wallace, J.; Rajabifard, A. *Land Administration for Sustainable Development*; ESRI Press Academic: Redland, AB, Canada, 2010.

23. Van Tulder, R. *Business and the Sustainable Development Goals: A Framework for Effective Corporate Involvement*; Rotterdam Schoool of Management, Erasmus University: Rotterdam, The Netherlands, 2018.
24. Bell, K.C. *Focusing on Innovation and Sustainability in Rural and Urban Land Development: Experiences from World Bank Development Support for Land Reform*; World Bank: Washington, DC, USA, 2011.
25. De Zeeuw, K.; Lemmen, C. Securing Land Rights for the World. In Proceedings of the FIG Working Week 2015: From the Wisdom of the Ages to the Challenges of the Modern World, Sofia, Bulgaria, 17–21 May 2015. Available online: https://www.fig.net/resources/proceedings/fig_proceedings/fig2015/papers/ts01b/TS01B_de_zeeuw_lemmen_7502_abs.pdf (accessed on 20 November 2020).
26. World Bank. *Public-Private Partnerships: Reference Guide Version 3*; World Bank: Washington, DC, USA, 2017. Available online: https://openknowledge.worldbank.org/handle/10986/29052 (accessed on 3 August 2020).
27. Burns, T.; Wright, F.; Rickersey, K. How to Conceptualize a PPP for Land Administration Services: Understanding the private sector and commercial feasibility. In Proceedings of the 2020 World Bank Conference on Land and Poverty: Institutions for Equity and Resilience, Washington, DC, USA, 16–20 March 2020.
28. Kalantari, M.; Paez, D.; Anand, A.; Burns, T.; Rickersey, K. A Review of Public-Private Partnerships in Land Administration. In Proceedings of the 2019 World Bank Conference on Land and Poverty: Catalyzing Innovation, Washington, DC, USA, 25–29 March 2019. Available online: https://www.conftool.com/landandpoverty2019/index.php/04-11-Kalantari-625_paper.pdf?page=downloadPaper&filename=04-11-Kalantari-625_paper.pdf&form_id=625&form_version=final (accessed on 13 November 2020).
29. Tang, L.; Shen, Q.; Skitmore, M.; Cheng, E.W. Ranked Critical Factors in PPP Briefings. *J. Manag. Eng.* **2013**, *29*, 164–171. [CrossRef]
30. Datta, A. Public-Private Partnerships in India: A case for reform? *Econ. Political Wkly.* **2009**, *44*, 73–78. Available online: http://www.jstor.org/stable/25663449 (accessed on 13 November 2020).
31. Galilea, P.; Medda, F. Does the Political and economic Context Influence the Success of a Transport Project?: An analysis of transport public-private partnerships. *Res. Transp. Econ.* **2010**, *30*, 102–109. [CrossRef]
32. Vives, A.; Paris, A.M.; Benavides, J.; Raymond, P.D.; Quiroga, D.; Marcus, J. *Financial Structring of Infrastructure Projects in Public-Private Partnerships: An Application to Water Projects*; Inter-American Development Bank: Washington, DC, USA, 2006.
33. Norfolk, S.; Quan, J.; Mullins, D. *Options for Securing Tenure and Documenting Land Rights in Mozambique: A Land Policy & Practice Paper*; LEGEND, Natural Resources Institute, University of Greenwich: Chatham, UK, 2020. Available online: https://images.agri-profocus.nl/upload/post/Moz_Policy_Paper1587375947.pdf (accessed on 12 November 2020).
34. Monteiro, J.; Inguane, A.; Oliveira, E.; Weber, M.; Palomino, D. Mapping Community Land In Mozambique: Opportunity and challenges for combining technology with good land governance. In Proceedings of the 2018 World Bank Conference on Land and Poverty: Land Governance in an Interconnected World, Washington, DC, USA, 19–23 March 2018.
35. Endo, V.; Pantoja, E.; Trevino, L. Contracting Out Services for Land Regularization: The new roles of the private sector in land administration. In Proceedings of the 2018 World Bank Conference on Land and Poverty: Land Governance in an Interconnected World, Washington, DC, USA, 19–23 March 2018.
36. Yin, R. *Case Study Research: Design and Methods*, 4th ed.; SAGE: Washington, DC, USA, 2009.
37. Enemark, S.; McLaren, R.; Lemmen, C. Building Fit-For-Purpose Land Administration Systems: Guiding principles. In Proceedings of the FIG Working Week 2016: Recovery from Disaster, Christchurch, New Zealand, 2–6 May 2016; International Federation of Surveyors: Copenhagen, Denmark.
38. GLTN. *Handling Land: Innovative Tools for Land Governance and Secure Tenure*; UN-Habitat, Global Land Tools Network and IIRR: Nairobi, Kenya, 2020. Available online: https://cepa.rmportal.net/Library/natural-resources/Handling%20Land%20-%20Innovative%20tools%20for%20land%20governance%20and%20secure%20tenure.pdf (accessed on 26 November 2020).
39. Boone, C.; Bado, P.; Dion, A.M.; Irigo, Z. Push, Pull, and Push-Back to Land Certification: Regional dynamics in pilot certification projects in Côte d'Ivoire. *J. Mod. Afr. Stud.* **2020**. draft.
40. François, K.N. Social constraints of rural land management in Côte d'Ivoire and legal contradictions. *Int. J. Asian Soc. Sci.* **2012**, *2*, 310–319. Available online: http://www.aessweb.com/pdf-files/IJASS,%202(3),PP.310-319.pdf (accessed on 10 December 2020).
41. Colin, J.P.; Kouamé, G.; Soro, D. Outside the Autochthon-Migrant Configuration: Access to land, land conflicts and inter-ethnic relationships in a former pioneer area of lower Côte d'Ivoire. *J. Mod. Afr. Stud.* **2007**, *45*, 33–59. Available online: https://www.jstor.org/stable/4486719?seq=1 (accessed on 10 December 2020). [CrossRef]
42. Augustinus, C.; Tempra, O. Fit-for-Purpose Land Administration in Violent Conflict Settings. *Land J.* **2021**, *10*, 139. Available online: https://www.mdpi.com/2073-445X/10/2/139/htm (accessed on 13 January 2021).
43. AFOR—La Consolidation des Droits Concedes. Available online: http://www.foncierural.ci/index.php/la-consolidation-des-droits-concedes (accessed on 18 April 2021).
44. Ruf, F.O. "You Weed and We'll Share": Land dividing contracts and cocoa booms in Ghana, Côte d'Ivoire and Indonesia. *CIRAD/GTZ Surv.* **2010**. [CrossRef]
45. Ho, S.; Choudhury, P.R.; Haran, N.; Leshinsky, R. Decentralization as a Strategy to Scale Fit-for-Purpose Land Administration: An Indian Perspective on Institutional Challenges. *Land J.* **2021**, *10*, 199. [CrossRef]
46. Koffi, M. *Présentation du Projet d'Amélioration et de Mise en Œuvre de la Politique Foncière Rurale de Côte d'Ivoire—(PAMOFOR)*; AFOR, World Cocoa Foundation: Abidjan, Côte d'Ivoire, 2020.

47. Bymolt, R.; Laven, A.; Tyzler, M. *Demystifying the Cocoa Sector in Ghana and Côte d'Ivoire*; The Royal Tropical Institute (KIT): Amsterdam, The Netherlands, 2018.
48. AFOR. *Manuel de Sécurisation Foncière: Opérations Intégrées*; Ministere de L'Agriculture et du Developpement Rural: Abidjan, Côte d'Ivoire, 2019.
49. World Bank Côte d'Ivoire-Land Policy Improvement and Implementation Project (No. PIDISDSC18852, 1–14). 3 October 2017. Available online: http://documents1.worldbank.org/curated/en/823571509717469081/pdf/ITM00184-P157206-11-03-2017-1509717464849.pdf (accessed on 18 April 2021).
50. Paez, D.; Burns, T. Conceptual Design of a Private Investment Scorecard for Land Administration in Developing Countries. In Proceedings of the 2018 World Bank Conference on Land and Poverty: Land governance in an Interconnected World, Washington, DC, USA, 19–23 March 2018.

land

MDPI

Article

The Amazon Forest Preservation by Clarifying Property Rights and Potential Conflicts: How Experiments Using Fit-for-Purpose Can Help

Bastiaan Reydon [1,*,†], Mathilde Molendijk [1], Nicolas Porras [1,‡] and Gabriel Siqueira [2]

1 Kadaster International Cadastre, Land Registry and Mapping Agency (Kadaster), P.O. Box 9046, 7300 GH Apeldoorn, The Netherlands; Mathilde.molendijk@kadaster.nl (M.M.); nicolas.porras@landinpeace.com (N.P.)
2 Institute of Economics, State University of Campinas, Campinas 13083-857, Brazil; G210058@dac.unicamp.br
* Correspondence: bastiaan.reydon@kadaster.nl; Tel.: +31-088-1832200
† This author is in Brazil.
‡ This author is in Colombia.

Citation: Reydon, B.; Molendijk, M.; Porras, N.; Siqueira, G. The Amazon Forest Preservation by Clarifying Property Rights and Potential Conflicts: How Experiments Using Fit-for-Purpose Can Help. *Land* **2021**, *10*, 225. https://doi.org/10.3390/land10020225

Academic Editors: Stig Enemark, Robin McLaren and Christiaan Lemmen

Received: 28 December 2020
Accepted: 19 February 2021
Published: 23 February 2021

Abstract: The burning and the deforestation of the Brazilian Amazon forest, which has been recently highlighted by the international press and occurs mostly on public or undesignated land, calls for an in-depth examination. This has traditionally been the main way to grab land, speculate, and simultaneously prove ownership by its occupation. The absence of mapping, registration, and an effective regulation of land property in Brazil, particularly in the Amazon, plays an important role in its deforestation. Recent estimations, besides others, show that the amount of land in this condition is around 200 million ha, near enough $\frac{1}{4}$ of the national surface. This article, besides examining the Brazilian deforestation characteristics, provides evidence that clear landholders' rights diminishes deforestation, and that proposals based on concrete cases of participatory clarification of land rights in forest regions using fit for purpose (FfP) methodology promote forest preservation. The article finishes with an example of a land rights clarifying case from small, medium, large, and traditional population landholders. The case is important to illustrate that it is possible to clarify land rights in a FfP way and how that increases the security of landholders, diminishing the pressure on the land and thus reducing the potential deforestation.

Keywords: Amazon; deforestation; Fit-For-Purpose land administration; participatory mapping

1. Introduction

The burning and deforestation of the Brazilian Amazon forest, which has recently been highlighted by the international press, play an important role in the global climate equilibrium and on global greenhouse gas (GHG) emissions, an important aspect of the UN's Sustainable Development Goals (SDG). With Bolsonaro's administration, there was an evident escalation of deforestation in Brazil, which stimulated the discussion around it[1]. Along with the recent dismantling of environmental policies, the government proposed important changes in the legislation regarding land regularization that could increase the possibility to grab land, and thereby also incentivize deforestation (see Kluck (2020) [2] for details). Not only because of the parliament and pressure from social movements, but it was possible to avoid further damage[2], given a coordinated effort that also raised awareness regarding the legal undefinition of land as an important driver of the current deforestation. Due to this, many seek concrete solutions for the regularization of these

[1] The article in Science by Escobar (2020) [1] shows that since the beginning of Bolsonaro's government, forest protection policies have diminished and deforestation has increased.
[2] There were many public discussions, technical publications, and political motions in Brazil regarding the Provisional Measure (*Medida Provisória*) no. 910 and later Law Proposal no. 2633 during the year 2020.

troubling issues in the Amazon region. This article aims to provide a solution for these as it proposes the use of a participatory fit-for-purpose (FfP) approach to clarify land rights in the Amazon region.

In the literature, there is a perception that deforestation usually occurs when land is grabbed or bought to be used immediately or in the long run, as evidenced by Reydon (2011) [3]. Deforestation occurs as it creates revenues from logging, crops, cattle ranching, land appreciation, and ultimately, deforestation is necessary to prove or assure ownership. The absence of a cadaster, as efficient registration, and an effective regulation of land property in Brazil, especially in the Amazon region, contributes to deforestation as an attractive venture, as shown by Reydon et al. (2019) [4] and others.

Since the beginning of the 2000s, mainly based on the Constitution of 1988 and other specific policies, Brazil has created numerous protected areas for its indigenous people and for environmental purposes, summing about 205.8 million hectares. The indigenous reservations and the protected areas represent 24.2 % of the Brazilian surface[3] and are the ones that mostly protect the forests, and after the New Forest Code of 2012 (*Código Florestal*), there were expectations for the diminishing of deforestation.

With the Forest Code was created the CAR (Cadastro Ambiental Rural), a land use georeferenced mapping system, to monitor the forested areas in private properties. It is an opensource dataset that made possible many studies on the deforestation and patterns of forests maintenance on private properties. Two important examples are Alix-Garcia et al. (2017) [6] and L'Roe, J et al. (2016) [7], who, in different ways, showed that this cadastral system plays an important role in monitoring the deforestation on private properties.

On the other hand, Moutinho et al. (2016) [8] showed that much of the deforestation occurs on public land or undesignated land[4], but could not be evidenced only by the CAR data, as its dataset is focused on presumed ownership of georeferenced areas. Based not only on evidence from Brazil, but Robinson et al. (2014) [9] also showed, based on international literature, that clear land rights have a decisive role in preserving forests, especially in Latin America.

As most studies on deforestation have dealt with private properties, the main aim of this article is to emphasize that, besides the efforts of reducing deforestation on private properties, there is an urgent need to clarify land rights in general, but more intensively on public and undesignated land. This article will not only present the characteristics of deforestation and the consequences of the lack of land administration in Brazil, but will also show a concrete example of actions that clarified land rights and avoided conflict around land in forested areas in the Amazon region.

Therefore, this article will be divided into four items:

(a) Deforestation in the Brazilian Amazon: quantification, importance, and characteristics;
(b) Evidence of the relation between deforestation and lack of clear property rights in Brazil and in the Amazon region;
(c) Why good land administration reduces deforestation; and
(d) A case using a participatory fit-for-purpose approach to help clarify land rights in a forested region.

By presenting a case study, it is expected to highlight methodologies that can clarify landholder ownership and other traditional population landholder rights, but also contribute to diminish potential conflicts over undesignated public land. As a result, the experiences were conducted to find ways to improve the legislation and the institutional settings, so that land rights can be clarified in an easier/affordable way and can help maintain the Amazon rainforest.

[3] For details, see the amount of land for each category in Sparovek et al. (2019) [5].

[4] Sparovek et al. (2019) [5] estimates, based on all Brazilian available spatial mapping efforts, that the country has 196,056 million ha with no clear destination as they are undesignated or unregistered land that are the most vulnerable for grabbing and deforestation.

2. Amazon Rainforest Deforestation

Figure 1 demonstrates that the deforestation in the Amazon region in recent years is of 69 to 11.1 thousand km^2 a year[5], based on satellite images. This is a lot less than previous decades, but is still a high level of deforestation for a biome like the Amazon, considering its biodiversity and its role in regulating the world's climate and especially rainwater, which is extremely important for good agriculture productivity without irrigation in the Central and Southern regions of Brazil[6].

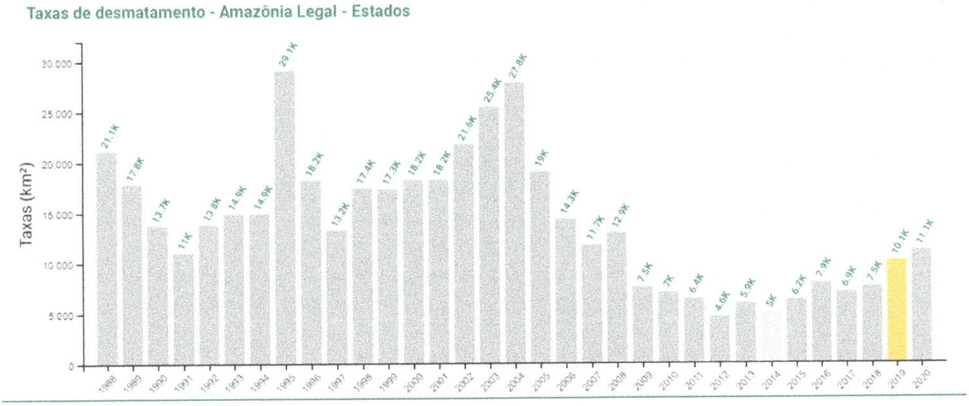

Figure 1. Annual deforestation in the Legal Amazon (Km^2 a year) [11].

Numerous studies[7] have evaluated the causes of deforestation in the Brazilian Amazon. One of them, by Moutinho et al. (2016) [8], lists the six main factors as follows: (a) The growth acceleration plan (PAC) and infrastructure constructions; (b) growth in the demand for commodities (meat and grain); (c) unsustainable policy on rural settlements (Agrarian Reform); (d) inadequate application of the Forest Code; (e) lobbying by agribusiness in the National Congress; and (f) land ownership ambiguities and the existence of undesignated public forests. In more general terms, Margulis (2003) [12] states that the main drivers of deforestation are:

a. Increase in profits linked to the use of land in the Amazon;
b. Accessibility of public policies and loans for the region;
c. Installation of infrastructure for access to frontier areas; and
d. Phases of GDP growth.

While not disagreeing with the aforementioned conclusions, Reydon et al. 2019 [4] stresses that the mechanism of Amazon deforestation is the product of the traditional form of continuous expansion of the agricultural frontier in Brazil, with the occupation of (private or public) virgin lands, the (il) legal extraction of timber, the introduction of extensive livestock farming[8] and, subsequently, the development of a more modern

[5] In the beginning of the 2000s, it was around 25 million hectares. That drop represented a substantial improvement, caused mainly by very strong command and control policies.

[6] Foley, J.A. et al. (2007) [10] shows this in a very clear way.

[7] Moutinho et al. (2016:2) observes [8] that: "A vast body of literature discusses the principal drivers of deforestation in the Brazilian Amazon (Nepstad et al., 2001; Kaimowitz et al., 2004; Fearnside, 2005; Etter et al., 2006; Scouvart et al., 2008; Boucher et al., 2011; Guerra, 2014; Nepstad et al., 2014; Azevedo-Ramos et al., 2015). There is still, however, no consensus concerning which intervention was the most effective in prompting the dramatic reduction in deforestation in the region since 2005."

[8] Reydon (2011) [3] shows that the main driver of the transformation to livestock farming is, on one hand, the large amounts of vacant land to be grabbed, linked to the possibility of introducing livestock at low cost, rendering deforestation an unbeatable capital appreciation strategy. A survey conducted by the National Institute for Space Research (INPE) showed that 62.2 % of the near 720,000 km^2 clearance of forest, was occupied by pastureland.

agriculture and livestock sector[9]. These economic activities exercise the role of generating income, legitimizing short-term occupation by new squatters, virtually without the need of many resources[10]. In the long run, the land remains with more intensive livestock farming or, if there is demand, it will be converted to grain farming or other agricultural activities.

However, what is important for occupation or deforestation is the existence of an expectation that there will be demand for this land[11], to be used at some point in the future, causing its price to rise significantly. The closer it is to being used productively, the higher the land value appreciation.

The macro policies as the turnarounds and the changing governments in Brazil also influenced the deforestation levels. After 2014, the federal government's macro policies changed toward economic austerity-oriented policies. In 2016, there was a major turmoil with the impeachment[12] that resulted in institutional instability and the deposition of Dilma Rousseff. From 2016 onward, the austerity pattern on macro policies became more intense and in relation to the deforestation, also coupled with a conservative push against social and environmental policies, the result is a weakened institutional capacity, especially through cuts in environmental, social, and science-related governmental branches [13,14]. After Bolsonaro's election in 2018 with an anti-environmental, anti-indigenous people and pro-deforestation rhetoric, the area deforested in the Amazon increased again, as shown by Figure 1. Based on scattered information, the increase in deforestation in 2020 is still higher, and it is happening in unowned land and in indigenous reservations, which is also caused by mining activities.

3. Undefinition of Land Rights and Deforestation: Some Evidence

To diminish deforestation in the Amazon biome, besides the more general policies[13] that must impact the Amazonian region as a whole, there is a need to fine tune the policies associated with the land ownership and responsibilities. Therefore, it is important to have an overview of what kind of properties are deforesting in order to establish the effective policies.

The only existing information are estimations based on satellite images. As can be seen in Table 1 adapted from Moutinho et al.(2016) [8], most deforestation in 2016 happened on private properties, then rural settlements with 35.5% and 28.7%, respectively. To diminish the deforestation on this type of land, the main policies in place are the Forest Code and other specific policies such as law 13,465, which will not be discussed here. Between 2012 to 2015, the summed deforestation of the categories 'land with no information', 'federal', and 'state lands' was always around 37% of the total deforestation. This is a typical kind of land over which there is no control, as the federal and state governments do not have clear cadasters of their land. The reason for the fall of this participation in 2016 to 25.1% is still unknown, but it might be that all private landholders that registered at CAR started to be private properties. Furthermore, what is important to highlight from this table is that private landowners deforested more in 2016, reaching 2462 km^2, 35.5% of the total. It is expected that all of the deforestation on private properties is legal, but that is still not possible to confirm, as Forest Code (2012) authorizes only 20% of private properties to be deforested for productive use in this region.

9 Reydon et al. (2019) [4] also argues that speculation with land in general and the conversion of forest into pastureland are important drivers of the deforestation.

10 It is frequently these same occupiers who make use of slave labor.

11 This is the result of increases in the prices of beef, soy, or even of reports that Brazil is going to be the largest alcohol producer in the world. In recent times, these factors have converged, causing the demand for land to grow even more as well as its price, encouraging deforestation.

12 Or coup, depending on one's point of view.

13 Most studies agree that command and control policies played an important role in the diminishing of the deforestation at the beginning of the 2000s.

Table 1. Deforested area in the Amazon by land title category from 2012 to 2016. Adapted with permission from [8] Moutinho et al. (2016).

Agrarian Category	2012		2013		2014		2015		2016	
	(Km^2)	(%)	(Km^2)	(%)	(Km^2)	(%)	(Km^2)	(%)	(Km^2)	(%)
Indigenous Lands	168	3.8	170	3.2	71	1.6	62	1.2	88	1.3
Federal Conservation Unit	175	4	187	3.5	120	2.8	184	3.5	201	2.9
State Conservation Unit	117	2.7	175	3.3	174	4	233	4.4	322	4.6
Environmental Protection Areas	124	2.8	228	4.3	202	4.6	245	4.7	207	3.0
Rural Settlements	1239	28.3	1518	28.7	1269	29.2	1437	27.3	1986	28.6
Private properties	986	22.5	1009	0	883	20.3	1113	21.2	2462	35.5
Federal Public Lands	574	13.1	743	14.1	584	13.4	670	12.7	855	12.3
State Public Lands	15	0.3	31	0.6	0	0	7	0.1	59	0.9
No information	982	22.4	1222	23.1	1047	24.1	1306	24.8	758	10.9
Grand total	4381	100.0	5282	100.0	5350	100.0	5256	100.0	6938	100.0

The other important information from Table 1 is that the smallest amount of deforestation happened on indigenous land, and in all kinds of conservation units and protected areas. It is clear that the main effort to maintain the Amazon forest is related to clarifying property rights: giving out titles to private owners, establishing clear boundaries for indigenous territories, protected areas, among other types of land use, but also to have a good, mapped cadaster of it all to enforce the Forest Code and its protection rules.

It is insufficient to know what kind of land has been deforested to avoid further deforestation; there is also a need to understand the amount of land that is under risk for each of these types. One study by Azevedo-Ramos, and Moutinho [15] stated that: "what is not widely known is that 70 million hectares ha of that public land—an area nearly twice the size of Germany—remains undesignated." (p. 125). They estimated the amount of land that is yet to be designated and that there should be a specific policy for its protection and avoidance against its deforestation. However, the ownership, possession, or responsibility over those 70 million ha were not clearly defined by the authors. In Figure 2, it can be seen that these areas are mostly in rather accessible areas where deforestation can happen easily. However, this definition of public forests is not very precise, as it comes from the cadaster of the *Serviço Florestal Brasileiro*. Mostly, there are people in those areas and there is a need to know what the real agrarian situation is, and if that land is under risk of deforestation.

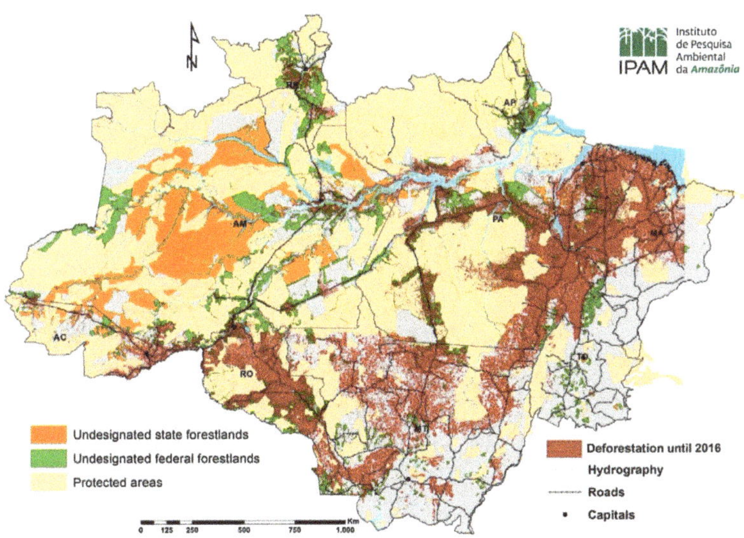

Figure 2. Public forests of the Brazilian Legal Amazon and its 70 million hectares of undesignated forestlands [15].

Sparoveck et al. [5] were able to estimate the different kinds of properties existing in the total area of Brazil using other sources and satellite images, as demonstrated by Table 2. This table integrates all information existing in the country and makes it possible to start understanding the real agrarian situation of Brazil. One important result that has come out of this is that 196,056 million ha have no clear allocation as they are undesignated or unregistered land[14].

Table 2. Area and number of units of Brazilian land tenure categories [5].

Land Tenure Category	Area (ha)	%	Number	%
Indigenous Reserves	112,412,239	13.2%	600	0.0%
Conservation Unit[15]	93,403,026	11.0%	1337	0.0%
Community Territory	1,779,373	0.2%	815	0.0%
Military	3,006,965	0.4%	104	0.0%
Rural Settlement	41,736,096	4.9%	7547	0.2%
Undesignated Lands	54,599,607	6.4%	22,016	0.5%
Total Public Land	306,937,306	36.1%	32,419	1%
Private property from CAR[16]				
Small	83,400,520	9.8%	3,805,698	79.0%
Medium	42,077,338	4.9%	167,537	3.5%
Large	48,366,589	5.7%	34,779	0.7%
Private property from SIGEF[17]				
Small	12,700,175	1.5%	206,070	4.3%
Medium	41,551,394	4.9%	110,830	2.3%
Large	134,531,227	15.8%	62,677	1.3%
Private property from Terra Legal Program	9,830,630	1.2%	116,854	2.4%
Quilombola Territory	3,117,971	0.4%	378	0.0%
Total Private Land	375,575,843	44.2%	4,504,823	94%
Unregistered land	141,454,569	16.6%		
Transportation network, Urban area and Water bodies	26,310,500	3.1%	280,692	5.8%
Total Brazil	850,278,218	100.0%	4,817,934	100%

To have an idea of its location, the best way is to look at Figure 3, which demonstrates that most of these areas are in the Amazon region. Thus, what can be concluded from the previous information is that there is a strong need to clarify land ownership, legitimize occupants, and build a good land administration system to enforce those limits. This combination could protect the forest and make deforestation much more difficult.

To understand the large amounts of undesignated land, Reydon et al, (2019) [4] showed that land with no information or is public land plays an important role because since the Land Law of 1850, it has been legally defined that whatever land that is not private and registered at the registration offices is State land. By doing so, Brazilians developed a 'habit' of grabbing this kind of land and later obtaining documentation for it (by any means necessary) because it was easier and possible to do so. In the article by Reydon at al. (2019) [4], this phenomenon also helps to explain how the Brazilian Land Administration

14 Public lands that have not been designated to a final use. The findings of Sparoveck et al. (2019) [5] differ from the 65.5 million ha of undesignated forest lands in the Amazon found by Azevedo-Ramos and Moutinho (2018) [15] due to the hierarchy rules adopted, where forest type B have a low level of priority and are classified as other categories. Forests type B include Federal or State lands covered with forests whose final designations have not been decided yet. They are under the administration of the Brazilian Forest Service (SFB). Additionally, Sparoveck et al.'s (2019) [5] estimations are relative to the whole country, but surely most of the undesignated and unregistered land are in the Amazon region.

15 We excluded APAs from the conservation unit category. APA (area of environmental protection) is a type of conservation unit of sustainable use which may occur in areas of public or private domain that allow human occupation and economic activities including intensive agriculture. Its creation does not imply expropriation of private land ownership. It sums 44 million ha. Its inclusion would confuse interpretation of land ownership and overlaps as it necessarily coincides with other land tenure categories.

16 Cadastro Ambiental Rural (Rural Environmental Registry).

17 Sistema de Gestão Fundiária—INCRA (Land tenure management system from INCRA).

System was built and why its malfunctioning is at the core of many Brazilian problems, especially that of deforestation.

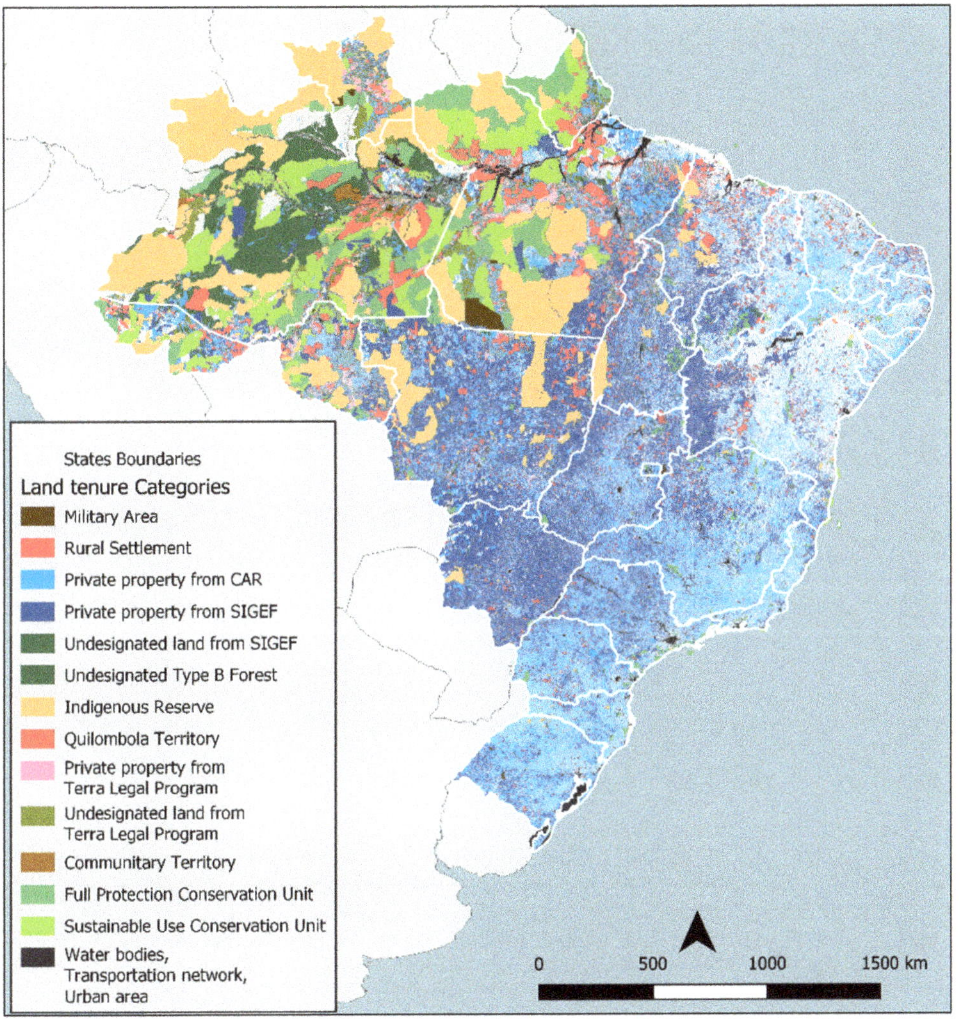

Figure 3. Reprinted with permission from Sparoveck et al. (2019) [16] from the Atlas of Brazilian Agriculture.

It is also important to highlight that as the system does not work properly, the main efforts have been directed to create land ownership cadasters. In the same article, Reydon et al. (2019) [4] explained CNIR as the official mapped landownership cadaster from INCRA and *Receita Federal*, which has SIGEF as its operational system, and that CAR was created by the Environmental Ministry with the Forest Code and is a self-mapped cadaster of land users.

This is the reason that the issue of land registration, as per the legislation discussed above, is so important in the country. If the registration of possessions is possible, with large amounts of land 'available' due to its legal uncertainty, this will act as an incentive for land grabbers to officialize land holdings.

The experience of Terra Legal, a federal government program to clear ownership on registered national public land, achieved important results between 2009 and 2017 [17], which was based on mapping, the titling of small landholders, and transferring larger amounts of land to indigenous and environmental reservations. As this was done, part of the public land was registered at the registration offices, so there would be no future controversy around it [18]. However, as it was shown, the amount of land that is still unregistered is still very large, and those areas are suffering larger amounts of pressure for occupation, conflicts, and deforestation. Therefore, the clearing of ownership rights in these areas is very important to avoid further similar issues.

4. Why Good Land Administration Reduces Deforestation

In Reydon et al. [4], it was possible to show that the lack of a good land administration system in Brazil is an immense problem and plays an important role in deforestation.

When analyzing the Brazilian Amazon rainforest, Araujo et al. (2009) [19] found that insecure property rights had a positive impact on deforestation in the period of 1988–2000 and, therefore, guaranteeing clear and secure rights on rural land ownership could decrease or avoid future deforestation.

In the same way, Assunção et al. (2015) [20] calculated the avoided deforestation in the same region through coordinated public policies between various levels of government and showed that in the period of 2004–2009, about 59% of the predicted deforestation was avoided for this reason.

Finally, a meta-analysis study, made with more than 118 published articles, by Robinson et al. (2014) [9] contributes to the conclusion that land tenure security is significantly associated with lower rates of deforestation, adding that this occurs regardless of the specific land access regime (tenure, property, customary systems, etc.). What this study shows, based on international literature, is that the clarification of land rights is very important as the main instrument to stop deforestation.

In Brown et al. (2016) [21], through a different approach, the authors analyzed the effects of land occupation on deforestation in Brazil, reaching the conclusion that in an environment of the low security of property titles, and with policies that value land deforestation over forested land, occupation has a direct influence on deforestation including occupations in a given municipality that can affect deforestation in adjacent areas.

More than one analyzed study deals with the forms of governance of forests and indigenous communities and their impacts on deforestation. Blackman et al. (2017) [22] analyzed one of these cases in Peru and concluded that the titling of these communities reduced deforestation by two thirds in a period of two years after the program.

Fernandes (2018) [23] showed, with systematic review methodology[18], that improvements in land governance had an observable positive impact on economic development, more specifically on the economic aspects of (a) production, productivity, and access to credit; (b) diminishing of poverty; and (c) dynamization of land markets. However, it also has large impacts on women's rights over land and on environment protection (deforestation and erosion decrease).

In Reydon et al. [4] (pp. 12), the authors concluded that:

"It is clear that, in order to combat deforestation in the Brazilian Amazon, it is necessary to move forward with improvements in the Land Administration/Governance System. The land cadaster must be completed and integrated with the CAR. The regularization of ownership, along the lines of the Terra Legal program, must be continued and expanded to include state-controlled public lands. Once a reliable land cadaster is in place, land taxes can be improved, which in turn diminish speculation and improve control over forested lands and associated environmental crimes. Finally, it is important to stress that, while better land governance is a necessary precondition for reducing deforestation, it is not in itself sufficient. Land governance and environmental protection must be structured under long term compromises and insulated against the politics of the day. "

[18] Systematic review methodology consists of: "[...] studies which synthesize [sic] all the existing high-quality evidence using transparent methods to give the best possible, generalizable statements about what is known" (Waddington et al. 2012, p. 360) [24].

In conclusion, Reydon et al. [4] proposed that what is needed to diminish rainforest deforestation, mainly in the Amazon, but also in other regions, is:

a. an improvement in the land administration/governance system, mostly the integration of cadaster and registry systems (CNIR, SIGEF, SINTER, CAR, and others), so that every landholder can be identified and localized[19].
b. The regulation of ownership similar to Terra Legal has to be continued and amplified to public land that are under national and states' responsibility.

Without secure property rights, farmers cannot obtain access to investment loans or public benefits and are almost invisible to the government. Without land administration reflecting the realities on the ground, land governance is difficult: the government cannot promote sustainable planning, which happens mostly at the cost of the natural environment and vulnerable groups. The main existing argument is that the areas are too big and with the existing technology, it is not possible to identify all landholders' rights in the Amazon region. The next section presents a case in Brazil where the small landholder's regulation process was quick, affordable, scalable, and successful. Another example of a case in Colombia is also mentioned, as it used the FfP method to clarify land rights and avoid conflicts to enhance the possibility of forest preservation.

5. Materials and Methods

5.1. Method—A Participatory Fit-For-Purpose (FfP) Approach in Areas under Deforestation Pressure

Forests are frequently located in different arrangements of property rights from communal rights to private and public ones, among others; so it needs a strong intervention to solve eventual controversies and recognize all those rights in a legitimate way. To achieve this, this article will show how to identify and cadaster landholders using the participatory fit-for-purpose methodology, which was very clearly defined by Ennemark et al. [25] (p. 6) as:

"the approach used for building land administration systems in less developed countries should be flexible and focused on citizens' needs such as providing security of tenure and control of land use, rather than focusing on top-end technical solutions and high accuracy surveys." A fit-for-purpose approach includes the following elements:

- Flexibility in the spatial data capture approaches to provide for varying use and occupation.
- Inclusive in scope to cover all tenure and all land.
- Participatory in approach to data capture and use to ensure community support.
- Affordable for the government to establish and operate, and for society to use.
- Reliable in terms of information that is authoritative and up-to-date.
- Attainable in relation to establishing the system within a short timeframe and within available resources.
- Upgradeable with regard to incremental upgrades and improvement over time in response to social and legal needs and emerging economic opportunities.

A country's legal and institutional framework must be revised to apply the elements of the fit-for-purpose approach. This means that the fit-for-purpose approach must be enshrined in law, it must still be implemented within a robust land governance framework, and the information must be made accessible to all users.

Considering this, the FfP method should be quick, affordable, and as accurate as possible so that it can solve concrete land ownership conflicts. To do so, the methodology must be flexible to fit different institutional settings and the technical demands of each location. These determinations must be respected in real scenarios, but its results will enable discussion on the viability of those standards and settings. Furthermore, the methodology must consider all different types of rights and legitimate occupants, therefore, the participation of all neighboring parties are central to it. Due to the involvement of all parties, any conflict

[19] For more information on the existing Brazilian Cadasters and Registering System, see Reydon et al. (2019) [4] and Reydon et al. (2017) [17].

resolution is much easier, once the final data are plotted for all participants to see and understand the nature of the overlap or if there are any illegitimate claims. This is why an affordable technology, with good-enough accuracy, can provide much good and spur development, especially in regions with weak or unclear land rights.

An example of the use of FfP methodology with good results that can be an inspiration for the issues in the Amazon region was conducted in Santa Teresita del Tuparro, an indigenous protected area located in Cumaribo, Vichada, Colombia[20]. It has been constituted as a special reserve since 1983, with an area of 180,000 ha, but most of its boundaries are determined by natural boundaries such as rivers, or dirt roads. More critically in the southern area of the reserve, there was an imminent conflict regarding overlapping claims.

Due to this factor, the pressure over the land and the risk of conflicts increased largely and for that, a FfP intervention was necessary. Therefore, the parcels in the current cadaster were used to identify the conflict and the actors in the area. Even though it did not show the actual reality, it was used to understand the dispute between the Santa Teresita del Tuparro indigenous groups and the adjacent parcels of farmers (colonos). The mapping of the conflicting claims was done using a FfP approach to clear the conflict by understanding the origin of the overlap.

The problem relied on the titles given in the past by the state and their poorly defined boundaries, which made the local parties understand that the land dispute was not being caused by them. The dispute could be solved by showing the indigenous people and the farmers the maps that where self-measured by them, which made them confident of the results and the official cadaster that was being used.

The fact that there were real and accurate data on the perceived limits helped to show the exact part of the land that the dispute and what the problem was about. After this, their rights were clearer and the communities were engaged in solving their overlapping boundaries in a peaceful way by recognizing their rights among their peers and neighbors, with the certainty that these areas will be respected after this process.

As shown before, land rights play an important role to reduce deforestation in the long run, especially in undesignated land in the Amazon region. In this section, a case study that applied this methodology will be detailed to illustrate and reinforce the potential of this method to solve complex land issues associated with deforestation in a participatory way. The next case shows a successful experience that will contribute to similar situations in the Amazon region. The case was the 'Tangará da Serra' in Brazil, demonstrating that it is possible to clarify property rights and provide formal titling to small land holders in a quick and low-cost way. It helps to show how it is possible to assure different land rights and peacefully solve conflicts over land using the FfP participatory approach.

The costs and timeframe necessary to regulate these situations are also central aspects of this methodology. Due to this, it is important to find experiences that would allow estimations of real costs and the minimum timeframe necessary, but also to extrapolate the results and conclusions to a national perspective. Once it is understood or has identified land holders, communal land, state lands, private possessions, or any other arrangement, it is necessary to estimate the feasibility of the current legislation, procedures/regulations, and expansive accuracy standards, especially considering the goal of having all land holdings identified and mapped within an updated national cadastral system. Not only for legal reasons, but the correct definition of all kinds of land rights, from community based to legitimate possession, are necessary for communities to thrive.

The importance of the formal recognition of land rights and the institutional capacity to enforce them has already been discussed, but there are few innovative ways to solve such complex issues. For example, in Brazil, a very controversial law at the time (Law 13,467 of 2017, the "Land Regularization Law") was enacted that, among some widely criticized aspects, eased the administrative regulation process that empowered registry offices to regularize properties in situations where there were no conflicts or disputes over land or

[20] See Molendijk et al. (2020) [26] for a complete description of the case.

boundaries in a much simpler configuration than the usual judicial process. The main innovation presented in the new legislation is the need of a formal agreement between neighbors for their shared boundaries that must be registered within the property deed (including the size, shape, and borders of the property). This formal agreement on the boundaries between neighbors gives the registry office enough security to go through the process with the certainty that property rights are being formalized with consent, thus speeding up the process without compromising any of the parties. This was also one of the reasons why a case in Brazil was chosen in order to evaluate the gains and consequences of this legal change.

5.2. Materials and Equipment's

To implement this theoretical approach, different test cases were carried out using GNSS receptors (Global Navigation Satellite System) from Trimble®R1 and R2, which are usually simpler than those used to georeference properties; in exchange, it has been used as an accessible and affordable technology, with adequate accuracy for mapping rural plots. The usage of this simple technology by the local population can be seen in Figure 4.

Figure 4. Images from the pilot case in Tangará da Serra.

Different countries and regions have different legal and technical demands for recognizing rights over land including different legal settings that must also be respected. For example, INCRA in Brazil, which is the national regulator of accuracy standards regarding these matters, requires a maximum error for georeferencing a property of 0.5 m to be certified. Other regulations may vary the level of precision, but the main argument brought by the FfP methodology is that many rights and conflicts could be solved by using a less precise survey[21], but still 'good enough' to speed up the regulation process, otherwise it may take decades and millions of dollars to do so.

Considering the Brazilian context, it is also important to highlight Law no. 10,267 enacted in 2001, which determines, among other relevant things, that all private properties should be certified by the INCRA under these accuracy standards. The same law sets a time limit for rural property owners to seek certification based on the size of their properties: the largest properties have a stricter timeframe, which has long been overdue, and smaller properties (up to 100 hectares) are supposed to expire by 2023. The compulsory certification

[21] Considering the Brazilian legal standards, a minimum horizontal accuracy of 0.5 m is required for 'artificial limits', three meters for 'natural limits', and up to 7.5 m in 'inaccessible limits'.

processes, determined by INCRA itself, demands an expensive georeferencing/mapping of the property because it must be done by a verified and contracted engineer with specific precision standards even in difficult terrain. Considering this institutional framework, a fit-for-purpose approach becomes extremely necessary, so it is better to be pro-active and start collecting evidence from the field of the advantages of a FfP methodology than to push forward this public issue regarding informality for years to come. Surely as soon as the properties are defined, the owners will have to comply with the Forest Code, therefore, maintaining 80% of their areas covered with native vegetation.

6. The Tangará Case: From Georeferencing to Titling under Six Months

As shown for the Brazilian Amazon, property rights are fragile or absent. This not only leads to serious conflicts over land and increases deforestation, but also hampers economic growth. Evidence shows that informal rights outnumber formal land rights in Brazil, for both urban and rural areas. Formalizing land rights can be very time consuming (over 20 years for a conflict resolution over a parcel within the judicial system is not an exception) and costly (around R$ 30.000 or USD 9.138[22])[23], which makes it a challenging task, especially for smallholder communities that are social and economically vulnerable.

To address these challenges of land regularization, a participatory FfP approach was developed in Tangará da Serra, in the state of Mato Grosso, Brazil during the year 2018. The initial goal of this pilot project was to develop a test case for the implementation of a fit-for-purpose methodology adapted to Brazil's legal and institutional framework, which could contribute to a viable model for a country-wide regularization process. Furthermore, an area in the Amazon region was chosen to test that reality. Therefore, the project was carried out by the Dutch Cadaster and Land Registry (Kadaster) and the State University of Campinas (UNICAMP), in collaboration with INCRA (the responsible national institution for rural land administration, cadaster, and certification), the local registry office ("cartório"), a local lawyer to represent the case, and a responsible engineer to certify the final maps.

The project consisted of a FFP approach on a small rural community with a selection of farmers, who could not afford to go through the usual process individually, starting with the social mapping of smallholder's properties that did not have any legal documents that officially stated their rights over their specific piece of land. This lack of formality is due to the time spent and financial burden that is required to seek regularization, and therefore, most of their parcels remained unregistered and unmapped.

The original focus of the project was to test a method for georeferencing properties (compliant with the accuracy standards demanded by the INCRA) as a viable solution for mapping tenure in adverse possession for smallholders in Brazil. By using a much more affordable technology and a social mapping strategy with the community including public inspection where neighbors sign to agree on the location of their boundaries, the intention was to optimize the timeframe and reduce costs at the maximum. As will be shown, this experience was very useful for all kinds of identification and land holding clarification.

First, Tangará da Serra is an interesting case because all types of land tenure and land use can be found there: commercial farms, small farmers, informal tenure, indigenous lands, natural areas, and state land, and all of them could be addressed in this pilot. Second, it is situated in the Amazon region where most land related problems are such as the overlapping of rights, the absence of clear rights, the invasion of public land, and others. Third, as all of the institutional stakeholders in Tangará da Serra had a keen interest in participating in this project, it was very feasible to prove the advantages of the FFP method with this arrangement. Especially considering the institutional complexity and limitations of the land registry as one of the top bottlenecks of securing land rights in Brazil, since in

22 Using an approximate exchange rate (from 15 June 2017) of 1 USD being 3,283 BR$, according to the Brazilian Central Bank (https://www.bcb.gov.br/conversao).

23 Reydon (2010) [27] estimated the cost for georeferencing the whole country based on the regularizing experience at the municipality of São José das Pontas in the state of Pará.

Tangará da Serra they were partners, a strong case could be made in an integrated approach to improve land administration in the region.

In Figure 5, the mosaic of properties that are available on the cadaster from the state of Mato Grosso and the municipality of Tangará da Serra is visible. It offers a scattered view on the existing land situations and people relations: a discontinuous map of tenure, showing 'islands' of formalized properties within an uncertain background. In Mato Grosso, as in Brazil as a whole, there is still the need for a national complete georeferenced cadastral map (with continuous coverage of the whole country instead of the current 'patchwork' map of land tenure), registered, and integrated with all of these smaller plots and also those that still have not obtained formal recognition of their rights.

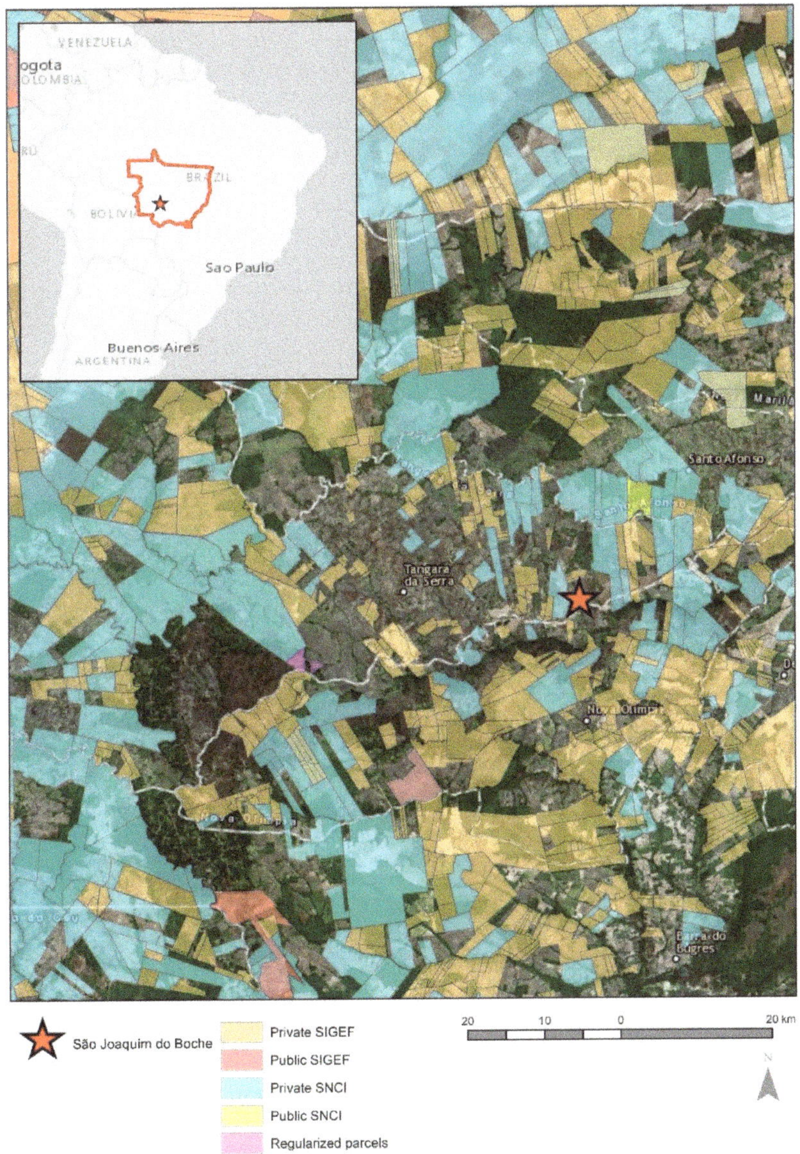

Figure 5. Land tenure mosaic of the Tangará da Serra region and intervention case location.

Tangará da Serra provides us with a unique opportunity of work in Mato Grosso and in Brazil due to its pioneer initiative of mapping all known properties that are already registered and by other secondary information from Brazil's official cadaster system that are within the municipality. Therefore, we were able to identify an unregistered community that could be formalized by the fit-for-purpose methodology. Then, the community of "São Joaquim do Boche", in the rural area of the municipality was chosen to be part of this initiative, with 60 known parcels that include formal georeferenced properties, informal georeferenced plots, and a vast majority of informal, not georeferenced plots. From those, we excluded the already registered plots and used the georeferenced ones for a precision survey comparison between thee FFP methodology and an official survey that had been done by a regular technician.

This test case started at the end of 2017 and within four months of coordinated work, 52 rural properties were regularized in January 2018 (with no costs to the smallholders, while complying with the current legal standards of the state of Mato Grosso and Brazil), as it can be seen on Figure 6. Although we managed to reduce the legal costs in the project to the minimum, there were still many costly procedures that are officially required during the regularization process such as the georeferencing by a verified engineer, the legal assistance provided by a lawyer, and the costs regarding the registry office practices. Nevertheless, the intentions on reducing costs allowed us to understand one important bottleneck for a national wide regularization, especially for smallholders, where many will not be able to pay for the process individually. As the Brazilian state is currently facing an enormous deficit, the public service is also not likely to cover these costs. Therefore, there is an urgent need to look for FfP solutions, leading to a fast, affordable, and complete land administration.

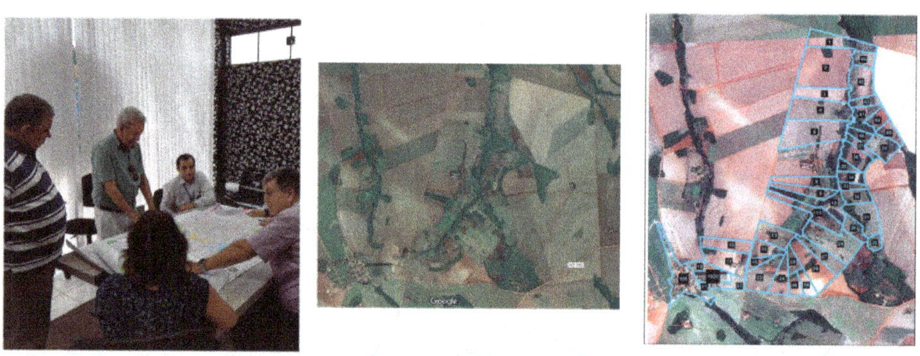

Figure 6. Images of the fit-for-purpose experience in Tangará da Serra.

The technical results obtained from this experience presented an average of error (using the R2 device, the most sophisticated one) of 0.31 meters, which is below the required standard. Furthermore, INCRA's normative requires a certified engineer to design the georeferenced map of the property (which is very costly), although an interpretation of the law allowed us to prepare the maps using the FfP approach and a hired engineer was only required to validate the maps, optimizing his work and the cost–benefit relation for the process of the regularization of smallholders. For private landowners, besides providing the legal security rights, it proved that social mapping helps smallholders to have considerably more secure tenure rights, which prevent conflicts.

The results from this project have already led to significant changes in national policies and made policy makers aware of the problem and complexity of the current requirements.

Very much inspired by the technical results obtained by the FfP social mapping approach, an update in the National Cadastral System (SIGEF) was proposed and named as "SIGEF 2.0." The new proposed system will integrate the social mapping of adverse possessions of land use, confronting the obtained information with the already certified land properties (public and private) and (if not in conflict) certifying informal rural households. Through this, the INCRA will promote a "good-enough" tenure on land, securing land holders that are in areas where there are no conflicts of rights with any other claims in a complete mapping of the current situation.

7. Conclusions

As the burning and deforestation of the Brazilian Amazon forest and other rainforests in the world play such important roles in the global climate equilibrium and on the emission of greenhouse gases, its control plays a very important role in the SDG's agenda.

This article started showing that the deforestation in the Brazilian Amazon started growing again in 2019 and that it occurred mostly on public land or undesignated land. Based on recent studies, it was possible to show that these areas sum up to near 200 million ha, about 25% of the Brazilian surface. Not all of it is in the Amazon region or with forest cover, but in the Amazon region, this is the type of land that is mostly possible to be grabbed, deforested, and used for speculative reasons.

Furthermore, the article presented evidence, based on international literature, that clear property rights are essential to the preservation of primary forests all around the globe. The article concluded by showing one participatory land rights clarification case in the Amazon region using fit-for-purpose methodology to help forest preservation. From the study, it is expected to mainstream this methodology to help to clarify the ownership of small landholders, the rights of traditional population landholders as well as diminish potential conflicts over undesignated public land. It also aimed to find ways to improve the legislation and the institutional settings to make the clarification of property rights easier and thus ultimately help maintain the Amazon rainforest.

Author Contributions: Conceptualization, B.R. and M.M.; methodology, B.R., M.M. and G.S.; software, N.P.; investigation, N.P. and G.S.; data curation, B.R., M.M., N.P. and G.S.; writing—original draft preparation, B.R.; writing—review and editing, B.R. and G.S.; project administration, B.R. and M.M.; supervision, B.R. and M.M.; funding acquisition, M.M. All authors have read and agreed to the published version of the manuscript.

Funding: The funding of the publication costs for this article has been kindly provided by the School of Land Administration Studies, University of Twente, in combination with Kadaster International from the Netherlands.

Institutional Review Board Statement: Not applicable.

Informed Consent Statement: Not applicable.

Data Availability Statement: Not applicable.

Acknowledgments: Acknowledge the contribution from the registrar José de Arimatéia for making this project viable and the good will of the people from Tangará da Serra.

Conflicts of Interest: The authors declare no conflict of interest. The funders had no role in the design of the study; in the collection, analyses, or interpretation of data; in the writing of the manuscript, or in the decision to publish the results.

References

1. Escobar, H. Deforestation in the Brazilian Amazon is still rising sharply. *Science* **2020**, *369*, 613. [CrossRef] [PubMed]
2. Kluck, E.G.J. Terra e Floresta em foco: Percursos recentes da regularização da terra e anistia a crimes ambientais. *Rev. ANPEGE* **2020**, *16*, 675–705.
3. Reydon, B.P. O desmatamento da floresta amazônica: Causas e soluções. *Política Ambient.* **2011**, *1*, 143–155.
4. Reydon, B.P.; Fernandes, V.B.; Telles, T.S. Land governance as a precondition for decreasing deforestation in the Brazilian Amazon. *Land Use Policy* **2019**, *94*, 104313. [CrossRef]

5. Sparovek, G.; Reydon, B.P.; Pinto, L.F.G.; Faria, V.; de Freitas, F.L.M.; Azevedo-Ramos, C.; Gardner, T.; Hamamura, C.; Rajão, R.; Cerignoni, F.; et al. Who owns Brazilian lands? *Land Use Policy* **2019**, *87*, 104062. [CrossRef]
6. Alix-Garcia, J.; Rausch, L.; L'Roe, J.; Gibbs, H.K.; Munger, J. Avoided deforestation linked to environmental registration in the Brazilian Amazon. *Conserv. Lett.* **2017**, *11*, e12414. [CrossRef]
7. L'Roe, J.; Rausch, L.; Munger, J.; Gibbs, H.K. Policy intersections, land registration, and forest clearing patterns in the Amazo-nian state of Para. *Land Use Policy* **2016**, *57*, 193–203. [CrossRef]
8. Moutinho, P.; Guerra, R.; Azevedo-Ramos, C.; Kapuscinski, A.R.; Frumhoff, P.C. Achieving zero deforestation in the Brazilian Amazon: What is missing? Zero deforestation in the Brazilian Amazon. *Elem. Sci. Anthr.* **2016**, *4*, 000125. [CrossRef]
9. Robinson, B.E.; Holland, M.B.; Naughton-Treves, L. Does secure land tenure save forests? A meta-analysis of the relationship between land tenure and tropical deforestation. *Glob. Environ. Chang.* **2014**, *29*, 281–293. [CrossRef]
10. Foley, J.A.; de Fries, R.; Asner, G.P.; Barford, C.; Bonan, G.; Carpenter, S.R.; Chapin, F.S.; Coe, M.T.; Daily, G.C.; Gibbs, H.K.; et al. Global Consequences of Land Use. *Science* **2007**, *309*, 570–574. [CrossRef] [PubMed]
11. Observação da Terra—A Estimativa da Taxa de Desmatamento por Corte Raso Para a Amazônia Legal. Available online: http://terrabrasilis.dpi.inpe.br/en/home-page/ (accessed on 15 January 2021).
12. Margulis, S. *Causes of Deforestation of the Brazilian Amazon*; The World Bank: Washington, DC, USA, 2003; Volume 22.
13. Petherick, A. Funding: Austerity bites deeply. *Nat. Cell Biol.* **2017**, *548*, 249–251. [CrossRef]
14. Magalhães, A.L.B. Brazil: Biodiversity at risk from austerity law. *Nat. Cell Biol.* **2017**, *542*, 295. [CrossRef]
15. Azevedo-Ramos, C.; Moutinho, P. No man's land in the Brazilian Amazon: Could undesignated public forests slow Amazon deforestation? *Land Use Policy* **2018**, *73*, 125–127. [CrossRef]
16. Sparovek, G.; Reydon, B.P.; Guedes Pinto, L.F.; Faria, V.; de Freitas, F.L.M.; Azevedo-Ramos, C.; Gardner, T.; Hamamura, C.; Rajão, R.; Cerignoni, F.; et al. Types of land tenure in brazil: The first estimate from available geo-referenced information. In *Proceedings of the 20th Annual World Bank Conference on Land and Poverty, Washington, DC, USA, 25–29 March 2019*; The World Bank: Washington, DC, USA, 2019.
17. Espinoza, J.; Bastiaan, P.R.; Cabral, R.; Siqueira, G.P.; de Siqueira, E.; Moreira, O.; Nascimento, M. Public land allocation and the Terra Legal regularization program in the Brazilian Amazon. In *Proceedings of the 20th Annual World Bank Conference on Land and Poverty, Washington, DC, USA, 25–29 March 2019*; The World Bank: Washington, DC, USA, 2019.
18. Reydon, B.P.; Fernandes, V.B.; Bueno, A.P.S.; Moreira, G.L. Casos de regularização fundiária bem-sucedidos que promovem a boa governança de fundiária. In *Governança de Terras: Da Teoria à Realidade Brasileira*, 1st ed.; Reydon, B.P., Fernandes, V.B., Bueno, A.P.S., Siqueira, G.P., Eds.; FAO/SEAD: Brasília, Brazil, 2017; pp. 193–207.
19. Araujo, C.; Bonjean, C.A.; Combes, J.-L.; Motel, P.C.; Reis, E.J. Property rights and deforestation in the Brazilian Amazon. *Ecol. Econ.* **2009**, *68*, 2461–2468. [CrossRef]
20. Assunção, J.J.; Gandour, C.; Rocha, R. Deforestation slowdown in the Brazilian Amazon: Prices or policies? *Environ. Dev. Econ.* **2015**, *20*, 697–722. [CrossRef]
21. Brown, D.S.; Brown, J.C.; Brown, C. Land occupations and deforestation in the Brazilian Amazon. *Land Use Policy* **2016**, *54*, 331–338. [CrossRef]
22. Blackman, A.; Corral, L.; Lima, E.S.; Asner, G.P. Titling indigenous communities protects forests in the Peruvian Amazon. *Proc. Natl. Acad. Sci. USA* **2017**, *114*, 4123–4128. [CrossRef] [PubMed]
23. Fernandes, V.B. A Relação Entre a Governança de Terras e o Desenvolvimento Econômico: Da Teoria a uma Revisão Sistemática de Resultados Empíricos. Ph.D. Thesis, State Universtity of Campinas, Campinas, Brazil, 2018.
24. Waddington, H.; White, H.; Snilstveit, B.; Hombrados, J.G.; Vojtkova, M.; Davies, P.; Bhavsar, A.; Eyers, J.; Koehlmoos, T.P.; Petticrew, M.; et al. How to do a good systematic review of effects in international development: A tool kit. *J. Dev. Eff.* **2012**, *4*, 359–387. [CrossRef]
25. Enemark, S.; Bell, K.C.; Lemmen, C.H.J.; Mclaren, R. *Fit-for-Purpose Land Administration*, 2nd ed.; Enemark, S., Ed.; International Federation of Surveyors and the World Bank: Copenhagen, Denmark, 2014; Volume 60, p. 44.
26. Molendijk, M.; Reydon, B.P.; Spijkers, P.; Morales, G.J.; Siqueira, G.P.; Becera, L. "Fit for purpose": Solutions on land boundary conflicts between Colombian indigenous land reserves and neighboring cattle ranchers. In Proceedings of the 21st Annual World Bank Conference on Land and Poverty, Washington, DC, USA, 16–20 March 2020; The World Bank: Washington, DC, USA, 2020.
27. Reydon, B.P.; Costa, A.S. O custo da regularização das propriedades rurais brasileiras uma estimativa. *Rev. Política Agríc.* **2010**, *19*, 50–64.

land

MDPI

Article

Quality Assurance for Spatial Data Collected in Fit-for-Purpose Land Administration Approaches in Colombia

Luz Angela Rocha *, Jonathan Montoya and Alvaro Ortiz

Facultad de Ingeniería, Universidad Distrital Francisco José de Caldas, Bogotá 11021-110231588, Colombia; jmontoyac@correo.udistrital.edu.co (J.M.); aeortizd@udistrital.edu.co (A.O.)
* Correspondence: lrocha@udistrital.edu.co

Abstract: The Fit-For-Purpose Land Administration (FFPLA) approach uses flexible techniques under basic regulations, avoiding complicated systems and aiming to fulfill the objective of land tenure security for all. In addition, a land administration system should evolve, starting as a simple system in rural areas and gradually evolving into a more complex system in more populated areas where requirements and quality increase progressively. The system can develop to a precision system. Implementing the FFPLA methodology in Colombia has allowed processes to be developed for data capture in the field using real-time technology and efficient methods for information management. These processes are under quality control by applying technical specifications in alignment with the FFPLA principles. This article presents the results of creating a FFPLA quality assurance model, which includes the application of the ISO 19100 family of technical standards based on the product's life cycle and quality model concepts. Furthermore, the article documents essential aspects for controlling the quality of the parcel boundary data collected in the field, using direct and indirect methods to measure the applicable spatial data quality elements (logical consistency and positional accuracy) preserving FFPLA principles.

Keywords: land administration; fit for purpose; spatial data quality; spatial data quality assurance

Citation: Rocha, L.A.; Montoya, J.; Ortiz, A. Quality Assurance for Spatial Data Collected in Fit-for-Purpose Land Administration Approaches in Colombia. *Land* **2021**, *10*, 496. https://doi.org/10.3390/land10050496

Academic Editors: Stig Enemark and Christiaan Lemmen

Received: 9 February 2021
Accepted: 22 April 2021
Published: 7 May 2021

1. Introduction

New policies in Colombia include the use of a multipurpose cadaster that allows for the formalization and registration of land and property rights in rural areas where cadaster systems and data are outdated. Many projects are being developed in Colombia that aim to update the Colombian cadaster, especially in rural areas. One of them is known as the 'Modernization Land Administration Project', the aim of which is the collection of cadastral data (physical, legal, and economic) as a basis for public policy design and implementation of a Land Management System, in which is essential the production of quality information according to the standards established by the cadaster and mapping agency (IGAC) [1]. Therefore, the possibility of implementing a fast, simple, and low-cost methodology for land administration, such as a Fit-For-Purpose approach, allows the right holders to register the boundaries of their parcels using a smartphone application connected to a GPS receiver. Using this methodology, landholders walk along the borders of their land to survey their parcels by collecting points, creating a polygon as a spatial representation of their land parcels. Then, in a public inspection process in a community meeting, the owners or possessors mutually agree on the boundaries of their parcels. This is related to an efficient process of conflict resolution and the parcels can be formalized by the government organizations in charge of this mission in the country [2].

In the Colombian context, the Fit for Purpose Land Administration methodology has been implemented in several pilot projects with the participation of governmental organizations involved in the land administration process and in join efforts with academic institutions such as ITC of Twente University and the Universidad Distrital Francisco José de Caldas to evaluate this methodology in the Colombian context through a Memorandum of Understanding with the Cadastre, Land Registry and Mapping Agency (Kadaster) of the Netherlands [3].

The implementation of Fit for Purpose Land Administration in Colombia is restricted by rigid rules regarding data acquisition procedures for cadastral information, especially in terms of the quality of the data. For this reason, it is necessary to determine how to control and measure the data quality for the data collected with the FFPLA methodology in order to ensure the correct application of the technical specifications for cadastral surveys in the country, and thus to show that it is not only about the question if the procedure of this methodology is applicable in Colombia, but also about ensuring the quality of the data.

This paper aims to show the results of a project that had as its primary purpose the creation of the Fit for Purpose Land Administration quality assurance model, which includes the application of ISO 19100 technical standards for spatial data and the concepts of the product life cycle and the FFPLA data quality model. In order to validate the quality assurance model, the quality elements of Positional Accuracy and Logical consistency were measured in order to demonstrate how well a dataset conforms to the product specifications and to determine whether FFPLA data are of sufficient quality for the Colombian multipurpose cadaster.

2. Materials and Methods

The Fit for Purpose Land Administration methodology is based on gradual quality upgrades, which must guarantee total coverage of cadastral data with adequate quality to be included in the Colombian Multipurpose Cadastral System. This implies that the capturing and recording of land information must be maintained permanently until gradually reaching the highest levels of quality; in other words, the methodology involves the following five levels of quality [4]:

(a) Quality Level 1: Recording of Land Interest;
(b) Quality Level 2: Recording of Land Claim;
(c) Quality Level 3: Recording of Recognition;
(d) Quality Level 4: Registered Land Right;
(e) Quality Level 5: Published Land Right.

In Colombia, the Fit for Purpose Land Administration approach was focused for the Land Titling process; for this reason, we considered starting with a basic quality level, which means quality levels one to four, and then making continuous improvements to reach optimal quality with level five [4].

2.1. FFP Quality Assurance Model

The Fit for Purpose (FFP) quality assurance model, as shown in Figure 1, integrates the methodology's production flow, the product's life cycle (stage 1 to stage 5), and the quality evaluation specifications (Q1 to Q5); therefore, this model involves the following stages in the Colombian context [5]:

- Stage 1: The people involved must understand and become familiar with the FFPLA methodology and the activities in which the respective technical specifications must be applied.
- Stage 2: The main activity is focused on exploring and analyzing existing information in different databases of governmental entities (municipalities) in the area to be measured.
- Stage 3: Field work preparation and data collection are the activities carried out at this stage. Likewise, the technical specifications for the creation of the property data must be defined (Q1, Q2, and Q3 requirements, according to the level of quality).
- Stage 4: Data storage and post-processing are then carried out, applying the technical specifications to accomplish the quality requirements.
- Stage 5: This stage involves activities such as public inspection, FFP validation and product elaboration, quality evaluation, and quality reporting (Q4 and Q5 requirements, according to the level of quality).

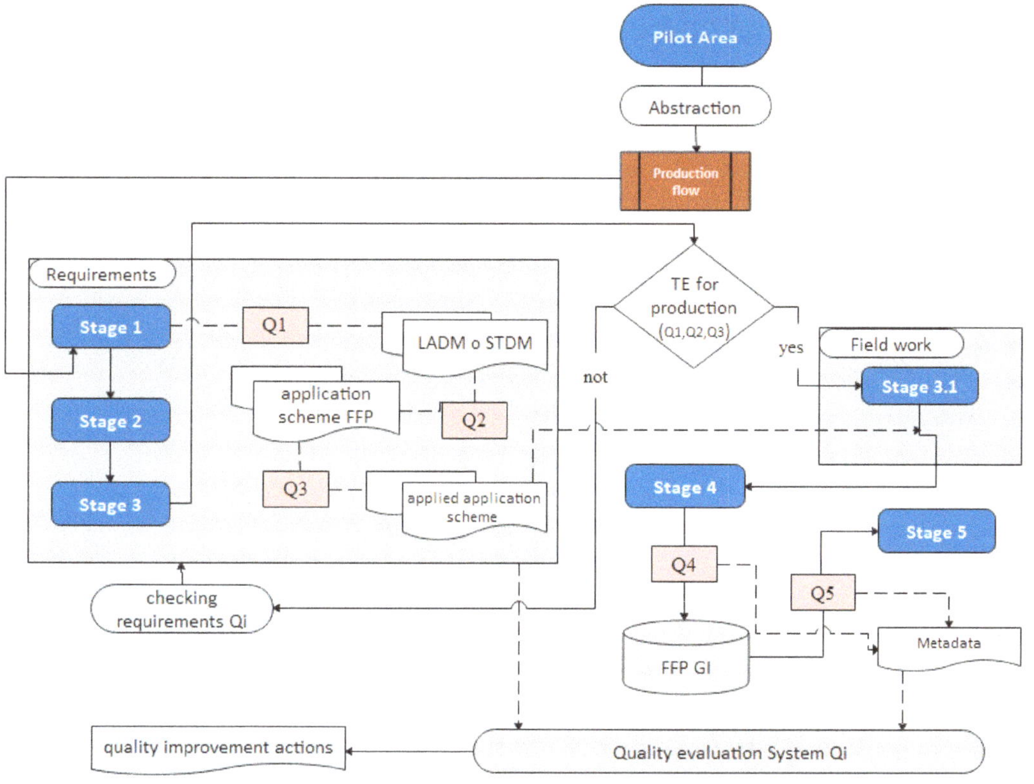

Figure 1. Fit for Purpose (FFP) quality assurance model.

The quality assurance model comprises five types of conformance (Figure 1), namely Q1, Q2, and Q3, associated with the conceptual quality assessment, Q4, associated with the internal product evaluation, and Q5, referring to the valuation of data confronted with those in the terrain. The conformance results in functions Q1–Q5 are registered in the verification system, a conformance report that allows for a clear vision of the product's quality and for improvement actions to be carried out based on that report.

Quality Model

The FFP quality model determines the first activity as defining the quality of the product framed in the product life cycle, which involves five stages, as follows (Figure 2) [6]:

(a) Conceptualization stage: defines the needs and demands of the product and describes the general product features;
(b) Design stage: defines the product itself and includes evaluating its effectiveness, the definition of the quality model and the production methods, and quality control;
(c) Production stage: this activity is based on the Product Specifications (Technical specifications);
(d) Stage of preparation for exploitation: defines the activities before the exploitation of the product itself;
(e) Exploitation stage: defines the activities after the production, and it refers to geographic information or the product itself.

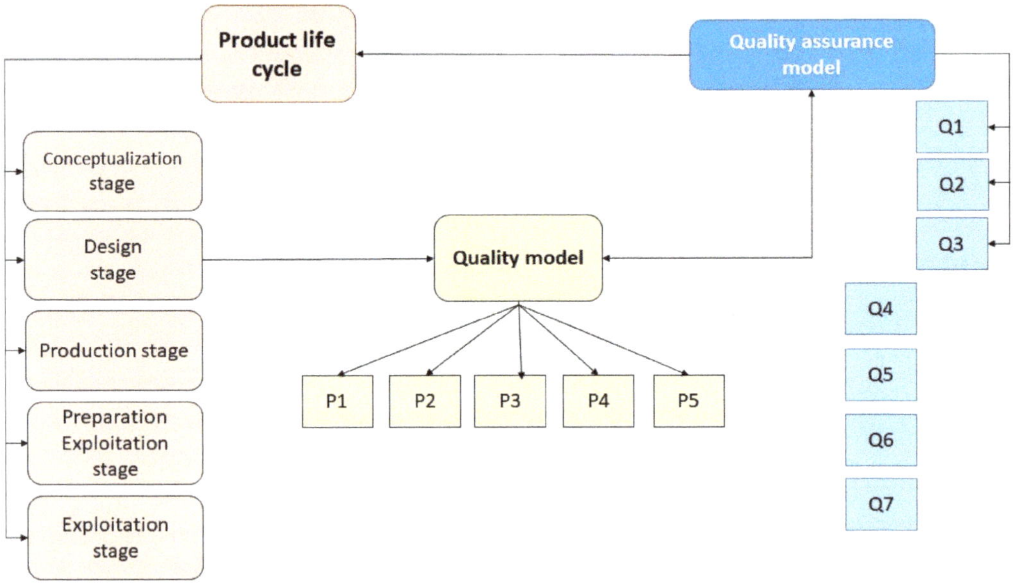

Figure 2. Product life cycle.

In this way, the FFPLA quality model is defined as a model describing the quality of spatial data as a component of the quality assurance model (Figure 2). The conceptual model of the FFP quality model includes three components according to ISO 19157:2013, namely requirements, conformance, and evaluation processes, which address the following items [7]:

P1—Product specification: here, the definition of the specification of the desired product is given such as is specified in the international standard ISO 19131;
P2—Quality objectives: the quality objectives and requirements are identified, addressing the quality elements of the ISO 19157 standard, thus giving complete product specifications;
P3—Production specifications: this item addresses product control, establishing requirements and production guidelines, where established procedures, manuals, and documentation are related to quality;
P4—Quality evaluation, reporting, and monitoring specifications: this item focuses on control methods, routine checks, quality sampling, and aspects to be used to evaluate the quality of the elements of interest. The quality is checked with levels of conformity pre-established according to the ISO 19157 standard;
P5—Quality improvement specifications: this item establishes the need for specific methods and tools for quality improvement in the product and production processes; thus, feedback on the quality of the product in each production process allows for the execution of improvement actions.

In addition, the quality model establishes an action framework for the quantitative information of the dataset from the quality elements; the production through its requirements and technical specifications; the evaluation specification that establishes control and checking methods, sampling methods, the way in which the quality of the product should be informed; and the production stage, articulated by a quality assurance model and a general quality scheme [7].

As a relevant component of the quality model, the "Technical Specifications Profile for the Fit for Purpose LA methodology in Colombia" defines the FFPLA dataset's characteristics. At the same time, it defines appropriate items that allow for the control of elements such as spatial and temporal reference systems, the type of spatial data, the conceptual model of the dataset, and others. Therefore, as part of the action framework, the evaluation of positional accuracy and logical consistency evaluation was carried out considering the FFP quality model and the FFP technical specifications profile [6].

Quality Standards for FFP

For this work, the ISO 19000 standards of geographic information, published by the Spanish AENOR Technical Standardization Committee 148 [8], were adopted by the Colombian government. Therefore, these series of standards were formalized in the year 2020 by the Geographic Institute "Agustin Codazzi", the cadastral and mapping agency of Colombia, to produce cadastral information in the frame of the multipurpose cadastral project. Consequently, for this study, these standards were explored, and the definition of how to apply them to the FFPLA quality assurance model is presented in Figure 3.

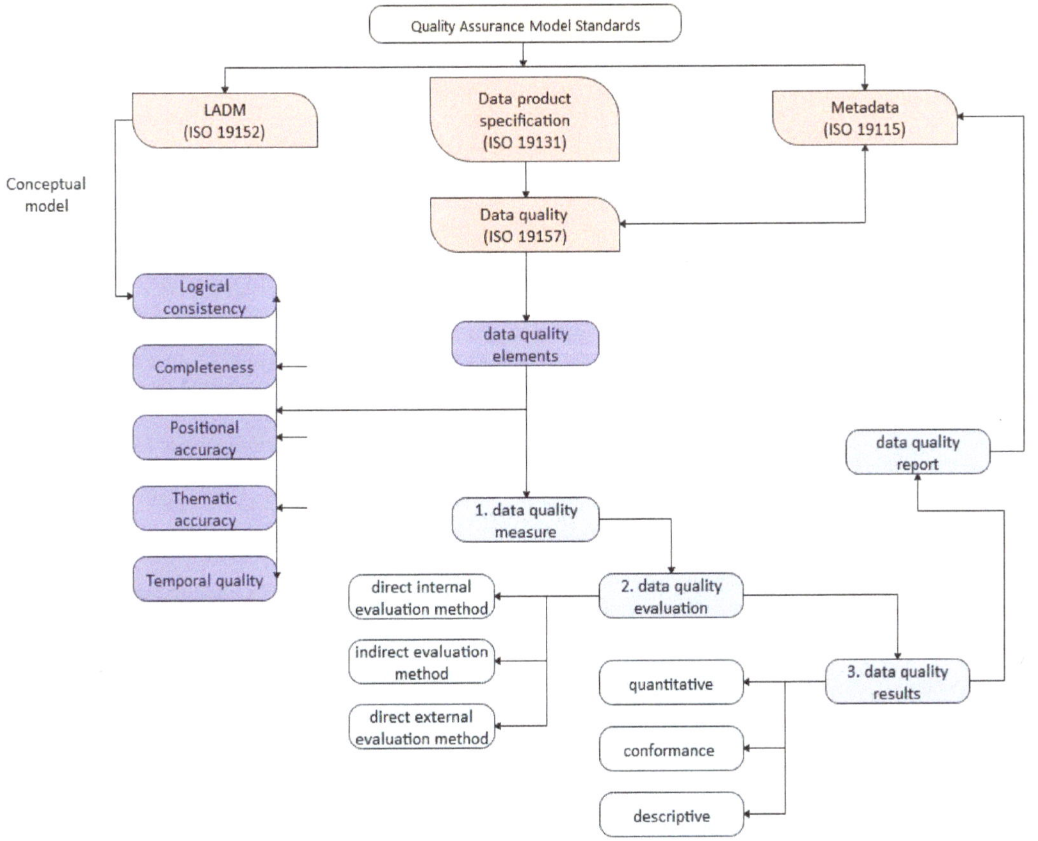

Figure 3. Standards of the FFP quality assurance model.

An analysis of the applicable Colombian regulatory framework for field data collection and geographic information management was also made, taking into account the principles of the Fit for Purpose methodology. In this way, Resolution 471 of 2020 of IGAC [9] establishes the measures, methods, and levels of conformity for each quality element of topographic data, without the concept of the quality assurance model that is open to producers, who have to report the data quality in a document. This resolution makes an adaptation to the quality measures of ISO 19157, such as the Root Mean Square Error

(RMSE), which uses an external direct method for quality evaluation and implies the use of more accurate external sources and a confidence level of 95% [9].

In addition, the authors analyzed the IGAC resolution 643 of 2018 [10], which adopts the technical specifications for planimetric surveying of massive rural property data within the framework of the social ordering of rural property implemented by the National Land Agency (ANT). Therefore, based on this resolution, in the case of rural premises, the positional accuracy must ensure a 95% level of conformity and an RMSE of 0.41 m.

Finally, due to the Colombian context, the FFP methodology focuses on the capture of parcel borders and legal information related to the property. In this work, the authors adopted the specifications for capturing borders given by the ANT, which indicate the application of a direct method using GNSS equipment for small parcels or orthophotos for large parcels in order for the parcels to reflect their real form on the terrain.

Quality Scheme for FFP

The concept of a quality scheme, according to the NTCA_01003, is the materialization and synthesis of a quality model and is a matrix tool with product-related categories and attributes arranged in a simple and orderly manner. The categories are the objects, which are established in columns, and the quality elements to be evaluated are arranged in rows, with their respective measures and sampling methods to be applied according to the nature of the predefined product [7].

The quality scheme is a simple way of organizing and showing the information depicted in the FFP methodology's general quality model applied in Colombia (Figure 4).

The quality scheme in Figure 4 shows the geographic objects defined for the FFPLA approach in the Colombian context, such as spatial unit, anchor point, and reference object. The quality elements to be measured for each object include logical consistency, completeness, positional accuracy, thematic accuracy, and temporal accuracy [11]. The quality scheme also includes the identifiers of the quality measure (ID) [11], the quality measure expressed in rate (T) or indicators as count (Co) and Boolean (B), and the conformity expressed as a percentage (%). The yellow color indicates the sampling criteria objects selected for evaluation, and the peach color means the inspection is conducted for 100% of the data.

For instance, for the Anchor point object in Figure 4, the quality element "absolute positional accuracy" is presented in the following way [5]:

- Quality measure identifier related to the object: ISO 19157 [11];
- Quality measure indicator: expressed as a rate (T);
- Objective's conformance expressed in percentage: 95% conformance.

In the case of the positional accuracy quality element, sampling can be carried out by selecting the geographical objects; however, the selection of the elements based on the area can also be considered. For this work, the population was defined for the sampling—first the object-based selection and then the object area-based selection, according to the specifications given by ANT [12].

A procedure was performed, as shown in Figure 5, to evaluate the quality of the data regarding the positional accuracy quality element [13].

	Logical consistency				Completeness		Positional accuracy	Thematic Accuracy		Temporal quality
	Conceptual	Domain	Format	Topological	Omission	Commission	Absolute	classification correctness	Non-quantitative attribute correctness	Temporal consistency
Spatial_unit	ID13(T)	ID17(T)	ID119(B)	IDXX (Co)	ID7(T)	ID3(T)		ID61(T)	ID67(T)	ID159(b)
objective	100%	100%	False		/	/		/	/	False
GlobalID										
objective										
gps_bearer					ID6(Co)					
objective					/					
landuse		IDE16(co)			ID6(Co)			ID60(Co)		
objective		/			/			/		
landuse_description										
objective										
legal_id									ID65(Co)	
objective									/	
physical_id									ID65(Co)	
objective									/	
spatial_source		IDE16(co)			ID6(Co)			ID60(Co)		
objective		/			/			/		
spatialunit_name					ID6(Co)					
objective					/					
Survey_id										
objective										
Survey_unit		IDE16(co)			ID6(Co)			ID60(Co)		
objective		/			/			/		
unified_id										
objective										
Anchor_point	ID13(T)		ID119(B)	IDXX (Co)	ID7(T)	ID3(T)	ID45/ID47 (T)			ID159(B)
objective	100%		False		/	/	95%			False
GlobalID										
objective										
spatial_unit_id					ID6(Col)					
objective					/					
Reference_object	ID13(T)	ID17(T)	ID119(B)	IDXX (Co)	ID7(T)	ID3(T)		ID61(T)	ID67(T)	ID159(B)
objective	100%	100%	False		/	/		/	/	False
GlobalID										
objective										
Object_name					ID6(Col)				ID6(5Col)	
objective					/				/	
Object_type					ID6(Col)			ID6(5Col)		
objective					/			/		
spatial_unit_id										
objective										

sampling or criteria
100% inspection
(T) = *Rate*
(Co) = *counting*
(B) = *boolean*
IDxx = *Measure identifier in ISO 19157*

Figure 4. Quality scheme.

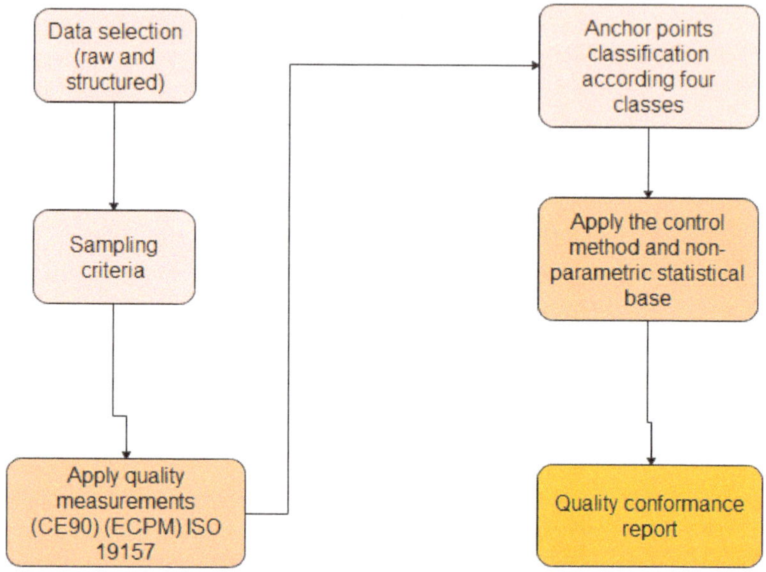

Figure 5. Procedure to evaluate positional accuracy quality element for FFPLA.

3. Results

Once the quality assurance model was determined based on the quality model and the quality scheme for FFP, the quality elements according to the ISO 19157:2013 standard [11] were determined to evaluate the quality of the data collected in the Municipality of Apartadó (Vereda Los Mandarinos) in the year 2018 using the FFP methodology. The results of this work are presented below, and the points related to the parcels are classified as presented in Figure 6 [6]:

(a) Anchor point: a point located in more than two boundaries, or it represents a change in the type of boundary (e.g., the change from river boundary to road boundary);
(b) Vertex point: the point between two anchor points (without a change in owner or type of boundary);
(c) Reference point: this point records the type of object hat borders the boundary; it indicates the element of the landscape that is part of the boundary, such as rivers, roads,, etc.

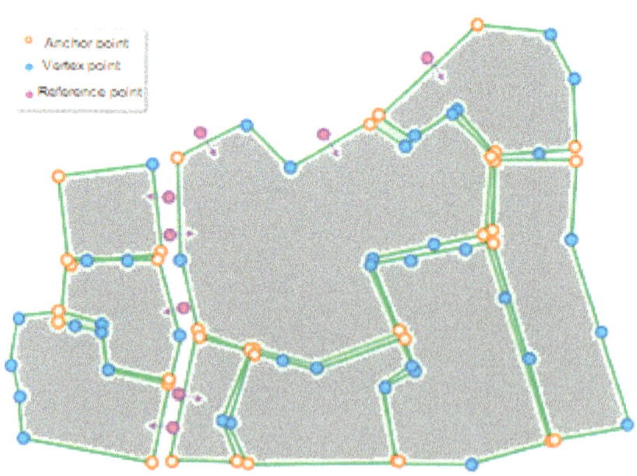

Figure 6. Types of points in the FFP methodology. Source [14].

3.1. *Positional Accuracy Evaluation*

The quality evaluation of the positional accuracy element [11] was performed for data collected in the Vereda Los Mandarinos, using the following product specifications [5]:

- Quality measure: UNE-EN ISO 19157 standard [11];
- Circular error to 95% of significance level (CE95): identifier 45 [11];
- Root Mean Square error: identifier 47 [11].

Quality evaluation method:

The quality evaluation method applied was direct intern, where the population was based on the polygon area, and by criteria sampling. Data acquisition involved fieldwork using a GPS receiver.

Conformance results:

- Result: Boolean, true or false;
- Specification: conformance levels (New classes in Table 1).

Table 1. Classification of anchor points (conformance level).

Class	RMSE-P (m)	Radius (m)	Scale
A	[0–0.30]	[0–0.50]	1:500–1:1000
B	[0.30–0.60]	[0.50–1]	1:1000–1:2000
C	[0.60–1.5]	[1,2]	1:2000–1:5000
D	[1.5–3]	[2,3]	1:5000–1:10,000

The sampling criteria were based on the following elements:

- Anchor points (boundary points) must be collected by fieldwork. The points collected by another source are excluded;
- One anchor point must have two measurements;
- The sampling data must be distributed homogeneously in the dataset.

The anchor points were classified into four classes, as shown in Table 1 [5].

The Root Mean Square Error of planimetry (RMSE-P) corresponds to a measure in meters (m) of additional control that verifies the data's behavior in the evaluation process (Table 1). The radius in meters (m) describes a circle where the true location of a point with 95% probability is located (Table 1). Classes A, B, C, and D establish a quality classification to discriminate different levels of data quality. In this way, at least 90% of the points must have a classification so that the dataset can be considered reliable and conforming (Figure 7).

The result of this evaluation of data quality in Los Mandarinos, regarding the positional accuracy quality element for the total 52 points, was 96%, thus meeting the established conformance levels, and the points were classified within one of the four classes.

3.2. *Logical Consistency Evaluation*

Next, the logical consistency was evaluated in the post-processing phase, which began by loading information into PostgreSQL from the ArcGIS Geodatabase, since the latest version of Collector for ArcGIS can generate information in a PostgreSQL scheme. The post-processing activities are shown in Figure 8.

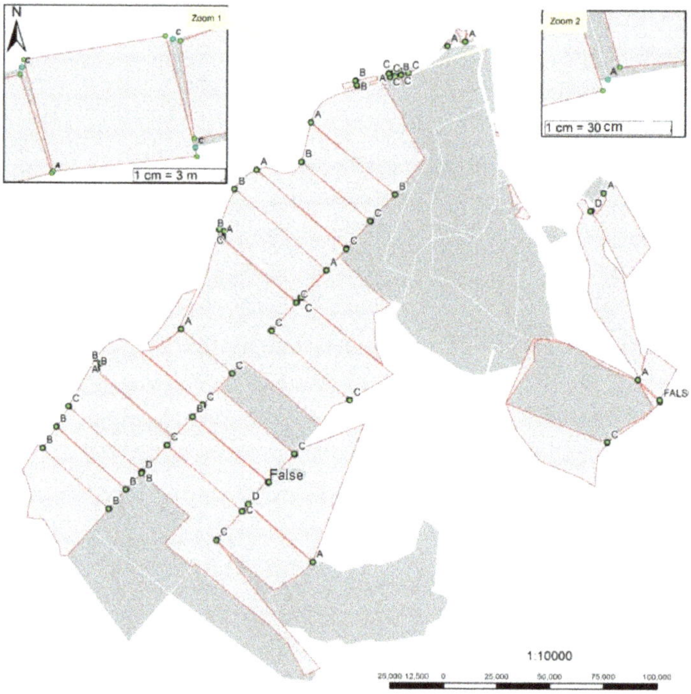

Figure 7. Point distribution in the dataset and classification.

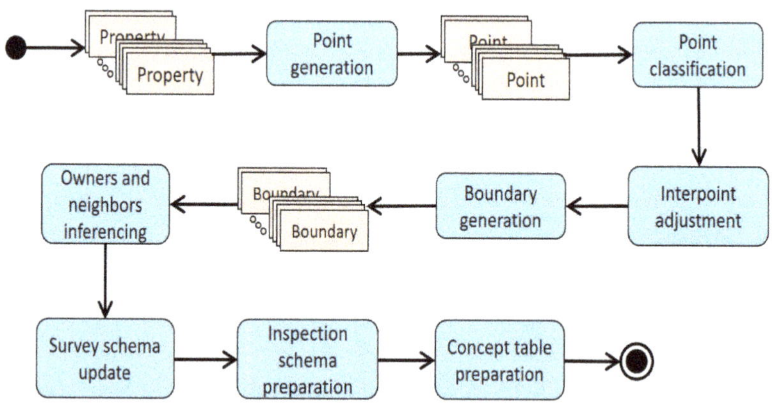

Figure 8. Activity diagram for post-processing stage.

The spatial database was built using the following three schemes with specific purposes (Figure 9) [14]:

(a) Survey: Scheme to store the final information. It contains all the information of the area of study for this case, namely the data of Vereda Los Mandarinos.
(b) Load: Scheme to load the information from external sources. In this schema, the data are cleaned, and the topology consistency verification between neighboring boundaries is carried out.
(c) Inspection: Schema used in the public inspection. Shows and verifies the information from the owner's point of view, including the adjustment between neighbors' bound-

aries. The database is prepared to store all participants' concepts, digital signatures and fingerprints in the public inspections.

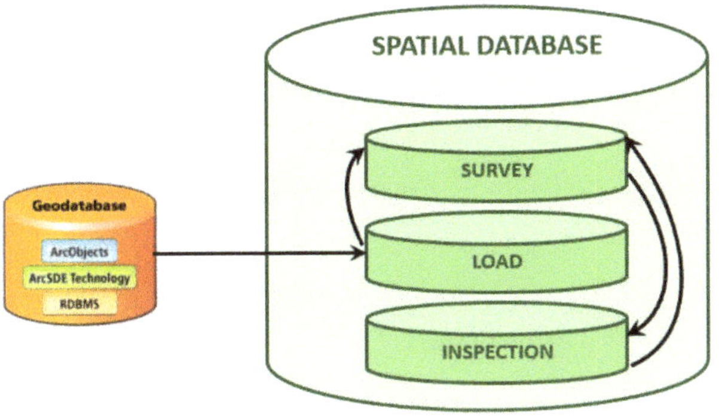

Figure 9. Spatial database schemes.

After loading the information into the database, the spatial information is included in the form of a polygon feature; likewise, the data can also be stored as points and its accuracy. If the points do not exist, it is necessary to generate them, because they will apply the algorithms that adjusts from the polygon representations of the parcel boundaries to real line (boundary)-based topology (starting from the situation as visualized in Figure 6). Figure 10 represents the objects and activities performed in the load schema.

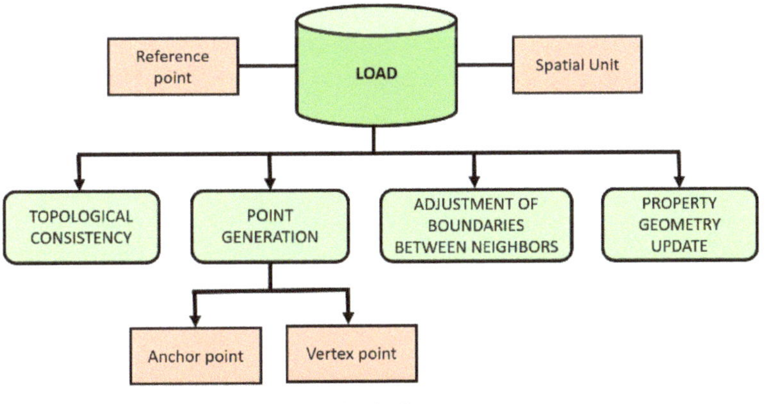

Figure 10. Objects and activities for the load schema.

A distance tolerance between points is defined according to the scale. Those points (from polygon representations of the different parcels) within the tolerance distance will be merged unless a reference point is associated with a boundary parcel that includes the merge points. The adjustment between points is accomplished by the "Weighted Symmetric Adjustment Algorithm" (WSAA), which works with two (Figures 11 and 12), three (Figures 13 and 14), or four points, and the function of which is to find an adjustment point, where the other points converge, using the accuracy of the GPS points to obtain the best locations for the placement of each point.

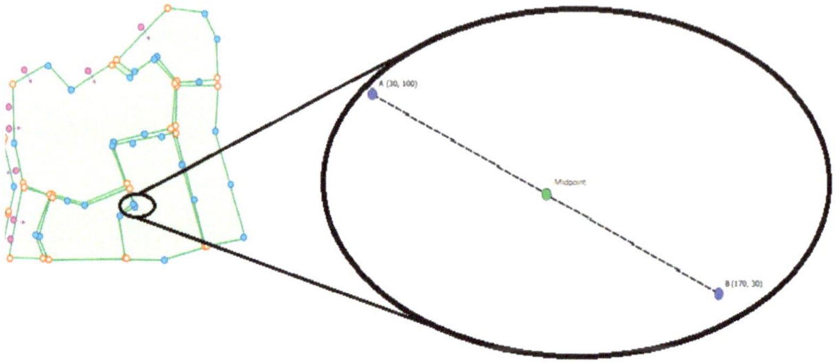

Figure 11. Symmetric point between points A and B.

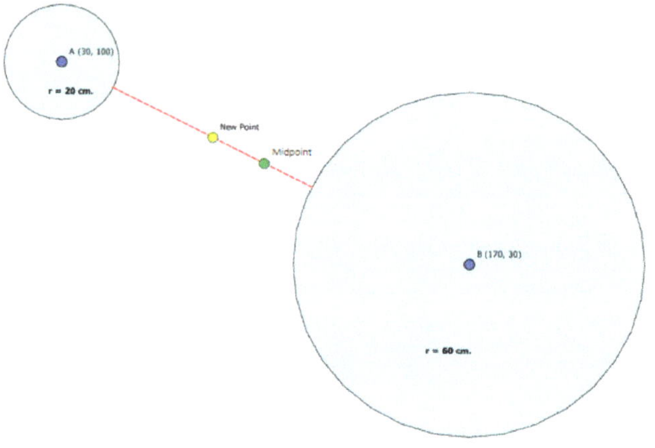

Figure 12. Weighted symmetric point between points A and B.

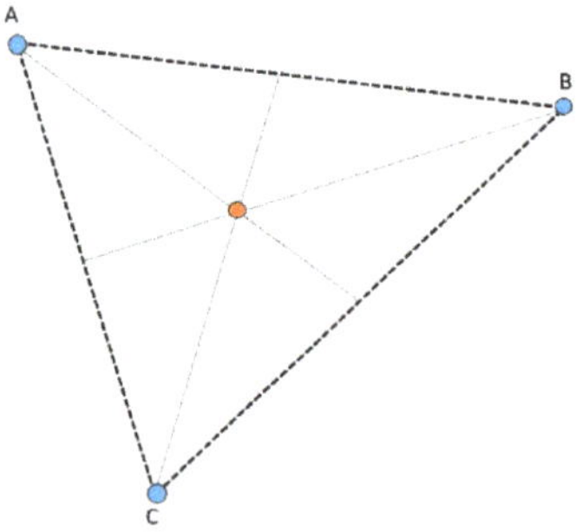

Figure 13. Symmetric point between points A, B, and C.

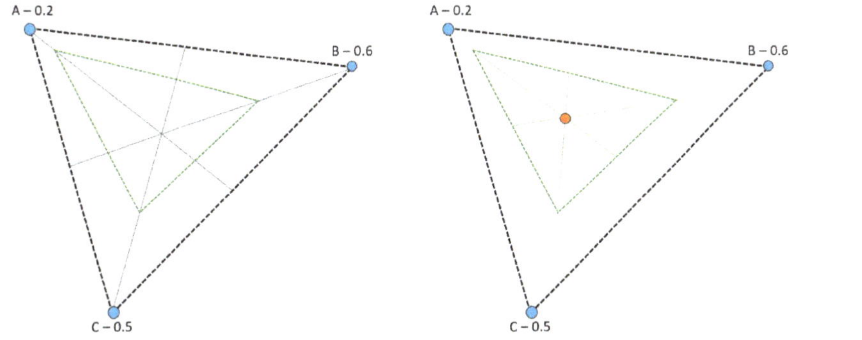

Figure 14. Weighted symmetric point between points A, B, and C.

For example, suppose that two points in two parcel boundaries have been captured with a GPS device (A and B), at different times, but which reference the same point. The distance between points A and B is 156.5 cm. If the precision of the GPS collected points is not captured, the best alternative is the automatic adjustment of the two points in a common position to find the midpoint of the line that joins the two points, assuming that the real point is found somewhere between the two points in the space between them, as shown in Figure 11.

However, as the accuracy values associated with each point are provided, an automatic adjustment can be made that favors the point with the best accuracy. Assuming that the real point must be somewhere between the two points, the two points A and B are identified as the limits of the distances of the accuracy of each point in the direction of the other point. Then, the midpoint of the line formed between the accuracy limits is determined, as shown in Figure 12. In the example, the new point is 19.91 cm nearer to point A with better accuracy.

When there are three points as an input, none of which register the accuracy, the algorithm finds the barycenter of the triangle, as shown in Figure 13.

In the event that the three points in three boundaries register accuracy, a new position is determined for each point as a percentage of the distance, according to the accuracy of the distance of the barycenter of the triangle, assuming that the adjustment error is in the barycenter direction of the triangle formed by the three points. Once the new points are found, a new triangle is created with a new barycenter as an adjustment point (Figure 14).

Figure 15 shows an example with some rural parcels where the Weighted Symmetric Adjustment Algorithm (WSAA) was successfully applied, which was one of the real cases where this algorithm was tested for the total number of parcels (100%) of the study area of Los Mandarinos.

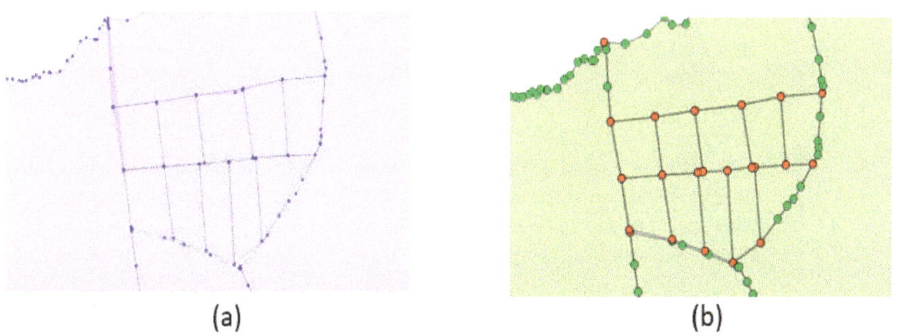

Figure 15. Properties and their points before (**a**), and after (**b**) adjustment.

Once the adjustment is made, the parcels' geometries are updated in the database in the Survey scheme (Figure 16).

Figure 16. Structured dataset of Los Mandarinos.

4. Discussion

The quality assurance model for the "Fit-For-Purpose Land Administration" approach in Colombia is the framework of action of the implementation of the FFPLA methodology, which allows for quality control of the data collected in the field for the purpose of land titling in the country. The model supports the creation of a product quality life cycle and a quality model in the Colombian context, as shown in Figure 2.

In order to apply the quality assurance model, one of the objectives of the project was to measure the quality of FFPLA data. This objective was accomplished because the project showed how well a dataset conformed to its product specification and was able to determine whether FFPLA data were of sufficient quality for the Colombian multipurpose cadaster, specifically in the case of positional accuracy and logical consistency. The results show that the data quality assurance model for the FFPLA approach allows one to evaluate how the dataset meets the criteria of the product specifications defined for FFP data according to the quality scheme shown in Figure 4. In addition, as the quality scheme is a tool for supporting the understanding and application of the quality model, it also allowed us to understand the methods, measures, and the evaluation procedures for the quality elements and attributes of the dataset in order to accomplish the required levels of conformity [10,13].

The technical specification profile for the positional accuracy quality element was an essential issue for the project and permitted us to evaluate the quality of the dataset according to the national standards and ISO 19157. The quality classification permits one to determine whether the classes are heterogeneous; in other words, the anchor points in a parcel may have all the quality classes, but classes C and B are pre-dominant for this evaluation since they are related to scales of representation between 1:1000 and 1:5000 in those areas with low land densities. As a result, the quality evaluation of the positional accuracy for the complete dataset (52 points) was 96% in conformity, a value that is adequate for the FFP perspective.

The weighted symmetric adjustment algorithm (WSAA) allows one to save time in the topology construction, and it processes most cases to adjust the boundaries between neighbors. It can be used with rural and urban properties effectively with the appropriate tolerance values. However, a review of the results is needed to ensure that everything was correctly processed. It is important to highlight that the algorithm was built specifically for this FFP project.

As the FFP quality assurance model is essentially producer driven, it is important to include the level of conformance that permits the usability of the data to be evaluated from the user's point of view. In the same way, it is suggested that the other quality elements, such as completeness, also be evaluated.

One of the problems in measuring the positional accuracy element in the country under study is that Colombia experiences tectonic plate displacements that can reach several centimeters per year, which generate accuracy problems between geographic datasets.

It was, therefore, necessary to adapt the MAGNA-SIRGAS reference system (WKID: 4686) and the reference year to 2018. Therefore, not only should international standards such as ISO 19157 be taken into account, but they must also be adaptable to national standards generated by the geographical and physical conditions of the country.

5. Conclusions and Recommendations

The study results were positive, showing that the FFPLA methodology can be established for the land administration processes in the country, as framed by the Modernization Land Administration project with a highly participatory approach and with a simple user interface [15]. This concerns not only the implementation of processes, but also the quality of the data collected in the field according to the standards used in Colombia for a rural cadastral survey. However, as the FFPLA methodology is still in the process of being implemented, this work is significant because it is the first project carried out in Colombia to apply new FFPLA data quality standards.

Finally, in future work, it is recommended that the FFP quality assurance model be applied in the process of implementation of the FFP methodology to facilitate the comparison and selection of the dataset best suited to application needs and technical requirements in the framework of the multipurpose cadaster and according to variations in the data of the terrain.

Author Contributions: Conceptualization and writing—original draft preparation, all authors; writing—review and editing, L.A.R.; J.M. and A.O. All authors have read and agreed to the published version of the manuscript.

Funding: Funding of the publication costs for this article was kindly provided by the School of Land Administration Studies, University of Twente, in combination with Kadaster International, the Netherlands.

Institutional Review Board Statement: Not applicable.

Informed Consent Statement: Not applicable.

Data Availability Statement: Not applicable.

Acknowledgments: The authors would like to acknowledge the support of Kadaster International, ITC Faculty, University of Twente, and the Universidad Distrital Francisco José de Caldas.

Conflicts of Interest: The authors declare no conflict of interest.

References

1. Consejo Nacional de Política Económica y Social CONPES 3859. *Política Para la Adopción e Implementación de un Catastro Multipropósito Rural—Urbano*; Departamento Nacional de Planeación: Bogotá, Colombia, 2016; p. 4.
2. Molendijk, M.; Santos, T.; Lemmen, C. Land and Peace in Colombia: FFP Methodology for Field Data Collection and Data Handling. Land Governance in an interconnected World. In Proceedings of the Annual World Bank Conference on Land and Poverty, Washington, DC, USA, 19–23 March 2018.

3. Rocha, L.A.; Ortiz, A.E.; Suarez, E. Introducción de metodologías participativas en la modernización de la Administración de Tierras en Colombia acercando la tecnología a la comunidad. *RISTI* **2019**, *E17*, 775–786.
4. Saers, P.; Meijer, C.; Molendijk, M. Integrating Land Data Sets—Towards Inclusive Land Information Systems. Responsible Land Governance: Towards and evidence Based Approach. In Proceedings of the Annual World Bank Conference on Land and Poverty, Washington, DC, USA, 20–24 March 2017.
5. Montoya, J. Diseño del Modelo de Calidad para los Datos Geográficos Generados a Partir de la Metodología "Fit For Purpose". Tesis de Pregrado, Universidad Distrital Francisco José de Caldas, Bogotá, Colombia, 2020.
6. Rocha, L.A.; Montoya, J. *Spatial Data Quality Model for "Fit-For-Purpose" Methodology in Colombia*; FIG Working Week 2020: Amsterdam, The Netherlands, 2020.
7. Comisión Interdepartamental Estadística y Cartográfica. *Modelo de Calidad para la Información Geográfica en Andalucía*; NTCA_01003; Instituto de Estadística y Cartografía de Andalucía: Sevilla, España, 2011.
8. López, F.J.; Rodríguez, P. *Introducción a la Normalización en Información Geográfica: La familia ISO 19100*; Universidad de Jaén: Jaén, España, 2008; ISBN 978-84-612-2075-5.
9. IGAC. *Resolución No. 471 DE 2020. Por Medio de la cual se Establecen las Especificaciones Técnicas Mínimas que Deben Tener los Productos de la Cartografía Básica Oficial de Colombia*; Instituto Geográfico Agustín Codazzi: Bogotá, Colombia, 2020.
10. IGAC. *Resolución 643 DE 2018. Por la cual se Adoptan las Especificaciones Técnicas de Levantamiento Planimétrico para las Actividades de Barrido Predial Masivo y las Especificaciones Técnicas del Levantamiento Topográfico Planimétrico para Casos Puntuales*; Diario Oficial No. 50.611 de 1 de junio de 2018; Instituto Geográfico Agustín Codazzi: Bogotá, Colombia, 2018.
11. ISO 19157:2013, Geographic Information—Data Quality. Available online: https://www.iso.org/obp/ui/#iso:std:iso:19157:ed-1:v1:en (accessed on 16 January 2020).
12. Agencia Nacional de Tierras. *Lineamiento de Exactitud Posicional de los Levantamientos Planimétricos para Procesos Adelantados en el Marco del Barrido Predial Masivo*; Agencia Nacional de Tierras: Bogotá, Colombia, 2018.
13. ICDE. Guía de Implementación Para la Evaluación de la Calidad de Información Geográfica. *Obtenido de Central de Documentos ICDE*. 2017. Available online: http://www.icde.org.co/node/1043 (accessed on 28 January 2021).
14. Morales, J.; Lemmen, C.; de By, R.; Molendijk, M.; Oosterbroek, E.-P.; Ortiz Davila, A.E. On the Design of a Modern and Generic Approach to Land Registration: The Colombia Experience. 2019. Available online: https://doi.org/10.4233/uuid:39072a6f-1ca8-4897-95bc-48f0f422ef38 (accessed on 20 February 2021).
15. ICDE-AT. Proyecto de Modernización de la Administración de Tierras en Colombia. Obtenido de ICDE-AT. Available online: https://www.proadmintierra.info/descargas-2/ide-at/ (accessed on 20 February 2021).

land

MDPI

Article

Do Design Science Research and Design Thinking Processes Improve the 'Fit' of the Fit-For-Purpose Approach to Securing Land Tenure for All in South Africa?

Simon Hull and Jennifer Whittal *

Division of Geomatics, School of Architecture, Planning and Geomatics, University of Cape Town, Private Bag X3, Rondebosch, Cape Town 7701, South Africa; simon.hull@uct.ac.za
* Correspondence: jennifer.whittal@uct.ac.za

Abstract: In South Africa, land tenure security is a challenge for 60% or more of the population who hold interests in land outside of the formal system of registered title. There is a need for the cadastral and land administration systems to be reshaped, and for new land tenure forms to be developed to record all land rights and interests so as to improve land tenure security for all. In this paper, we undertake a reflective retrospective of the processes of land administrative reform in South Africa using a thematic framework that includes fit-for-purpose, design science research, and design thinking processes. Literary sources are coded using the thematic framework to identify potential contributions of foregrounding design science research and design thinking in fit-for-purpose land administration (FFP LA) approaches. Design science research paired with tools of behavioral science add value in understanding the context, problems, needs, and objectives and in communicating the results of critical reflection. The design thinking process has much to offer in capitalizing on the human abilities of empathy, deep understanding, and challenging assumptions, setting the scene for unconstrained creative thinking. Design science research and design thinking within FFP LA may promote innovations in land administration systems reform initiatives that deliver restorative justice in the South African land sector.

Keywords: land rights and tenure; land administration; fit-for-purpose approach; human rights; design science research; design thinking

Citation: Hull, S.; Whittal, J. Do Design Science Research and Design Thinking Processes Improve the 'Fit' of the Fit-For-Purpose Approach to Securing Land Tenure for All in South Africa?. *Land* **2021**, *10*, 484. https://doi.org/10.3390/land10050484

Academic Editor: Stig Enemark

Received: 23 March 2021
Accepted: 29 April 2021
Published: 4 May 2021

1. Introduction

Post-colonial countries may exhibit an awkward combination of *inherent* and *inherited* land administration systems. The *inherent* system is that which was already in existence prior to colonization. It was administered by traditional authorities and structures according to cultural norms and standards. These prioritized the relationships between people and land for purposes of landholding (both individual and communal), grazing, agriculture, and other such land uses [1]. Colonization introduced an *inherited* land administration system (LAS) that is based on the norms and standards of the colonizing nation. This 'formal' LAS was designed to secure property rights for the colonizers and promote a land market [2]. The inherited view of land as a commodity was in opposition to the inherent view of land as an integral part of the community and an essential aspect of the indigenous social network [3]. The inherent LAS was thus largely ignored by the inherited system. However, colonial authorities codified the inherent customary traditions creating official customary law. They also employed traditional authorities with elevated political powers to serve the needs of the colony rather than to continue the system of pre-colonial governance. Thus, the inherent LAS post-colonization is no longer a true reflection of its pre-colonial self and exhibits both inherited and inherent attributes [4].

The resulting duality of the LAS, coupled with intentionally discriminatory colonial and (in South Africa) apartheid policies, has led to land dispossession being a symbol of oppression [5] that cuts to the core of dignity and personhood [1]. This continues to be manifest as land tenure insecurity for those land rights-holders who live outside of the inherited LAS. Estimates suggest that in 2011 this was the reality for 60% or more of the population of South Africa who live in the so-called communal areas (former apartheid 'homelands'), as farm laborers and their families, in informal settlements, and on former mission stations [6]. Although the Constitution of the Republic of South Africa (the Constitution) [7] recognizes these customary and 'informal' land rights, such recognition has not carried through to a national LAS. Instead, South Africa has a fragmented LAS characterized by a legislated and well-functioning 'formal' (inherited) system of land registration and cadaster existing alongside an off-register, informal (amended inherent) system administered by traditional authorities, civil society groups, non-government organizations (NGOs), slumlords, and anyone else with the appropriate acquired or appointed authority. Such fragmentation is detrimental to the realization of the objectives of Section 25 of the Constitution, particularly sub-sections (5)–(9) of the Constitution. These sub-sections obligate the State to promote equitable access to land for her citizens, improve tenure security for those whose tenure is insecure due to colonial and apartheid laws and policies, and allow for restitution of land for those dispossessed of land under the same. According to an advisory panel report on land reform and agriculture constituted by the presidency [8], read with the *Report of the High Level Panel on the Assessment of Key legislation and the Acceleration of Fundamental Change* [9], this fragmented LAS is a major contributor to land tenure insecurity in South Africa.

Thus, there is a need for the South African land administration and cadastral systems to be reshaped to allow for the recognition and recordation of rights and interests in land in the country and the improvement of tenure security for all. However, the manner of this reshaping must consider the pressing need for restorative justice, not only related to the outcomes of reform, but also in respect of the process through which it is conducted. Such reform should also improve economic development through improved investment opportunities (for local and international investors) and clear administrative procedures. South Africa's National Development Plan [10] has broad and bold aims to eliminate poverty and reduce inequality by 2030. These 'Vision 2030′ objectives are linked to transformation and development and have relevance to a FFP LA system design accompanied by a range of strategies and programs—some formal and others that have not been passed through democratic processes of governance [1].

The goal of increasing recognition and recordation of rights and interests in land and of delivering security of tenure for all is in keeping with Aspirations 1, 3, 4, 6, and 7 of the African Union's Agenda 2063 and associated calls to action [11], as well as several of the Sustainable Development Goals (most notably goals 1, 2, 5, 11, and 15) [12], summarized in Table 1. The FFP LA approach is promoted as one way of addressing this need [13].

Table 1. Aspirations of the African Union and SDGs that align with FFP LA.

African Union's Agenda 2063	United Nations Sustainable Development Goals
1: A prosperous Africa based on inclusive growth and sustainable development	1: End poverty in all its forms everywhere
3: An Africa of good governance, democracy, respect for human rights, justice, and the rule of law	2: End hunger, achieve food security and improved nutrition, and promote sustainable agriculture
4: A peaceful and secure Africa	5: Achieve gender equality and empower all women and girls
6: An Africa whose development is people-driven	11: Make cities and human settlements inclusive, safe, resilient, and sustainable
7: Africa as a strong, united, and influential global player and partner	15: Protect, restore, and promote sustainable use of terrestrial ecosystems.

FFP LA is intended to include processes that lead to a deep understanding of problem situations and needs, and to design solutions to meet these. FFP LA seeks to "understand the social, cultural, legal and institutional dynamics of their own Communities" [13], but achieving this at all scales and for all role-players, particularly in complex contexts, is challenging. The approach focusing on the spatial, legal, and institutional frameworks of FFP LA may lead to false assumptions of hegemony and fail to reveal contest between needs within an area or state. Behavioral science and human-centered approaches have different points of entry and focus. Foregrounding their strengths within the FFP LA approach may be useful to address South African land challenges that endure even after more than two decades of change—a deep understanding and creative approaches are required. The consequent aim of this paper is to identify contributions of design thinking and design science research paired with behavioral science to enhance the appropriateness of the FFP process in addressing the shortcomings of land administration and improving tenure security for all South Africans.

It may be noted that the authors do not separate 'fit' from 'purpose'. Purpose may stand alone, and the purpose of LAS reform is discernible from the reviewed literature: in South Africa, the general purpose is to address land tenure insecurity. This problem situation demands creative solutions developed from a deep understanding of complexity in context. The method/approach adopted should fit this purpose. In this paper, we argue that the method/approach can benefit from foregrounding design science research (paired with behavioral science) and design thinking into an FFP approach. This would lead to a better 'fit' of the method/approach to the purpose.

This paper adopts a critical realist ontology as per Whittal [14] and Carlsson [15]. This overcomes the toolbox approaches of pragmatism and experimentation associated with positivist ontologies, and the purely social/behavioral approaches associated with constructivism. Carlsson [15] critiques ad-hoc pluralist approaches that fail in assessing causes for success and failure, and the attempts of others to mix the approaches of pragmatism, experimentation, and social/behavioral science without a defensible ontological position. Critical realism overcomes the drawbacks of some common approaches and even of the ad-hoc combinations of these. It also supports the use of a suitable suite of mixed methodologies. Critical realism underpins this investigation and should be considered as a suitable theoretical approach for interventions aimed at improving land administration systems that are 'fit-for-purpose'.

2. Theoretical Framework

2.1. Overview

The FFP approach has been developed with extensive consultation and testing. It is not the aim of this paper to critique the FFP approach directly, but to explore approaches that may offer improvements in designing interventions for challenging situations, especially for the South African context. The findings should have relevance for similar post-colonial states. Design science research partnered with behavioral science (DSR-BS) and design thinking (DT) are identified as potentially complimentary approaches to the FFP. DSR-BS is chosen since it is highlighted by Çağdaş and Stubkjær [16] as a suitable over-arching approach from their detailed analysis of literature on cadastral research. The DT approach is included since it is starting to receive traction in organizational and policy change where its human-centered approach to innovation is valued [17,18].

The research seeks to reflect on land administration systems (LAS) reform in South Africa in a retrospective analysis of relevant literature. The thematic framework of the DSR-BS, FFP, and DT approaches is used to assess whether the FFP LA approach can be improved by the inclusion of DT and DSR-BS approaches in guiding a process of LAS reform. Improvements to the approach adopted in reform processes may increase the likelihood of meeting identified outcomes (success), delivering solutions that last (sustainability), and that meet the needs of all stakeholders—the State and its organizations, and the land rights-holders, both individually and collectively (significance) [19].

2.2. Fit for Purpose

Significance is an important factor for ensuring the *success* and *sustainability* of LAS [19]. This means that LAS should use tools and procedures and deliver outcomes that are relevant to the end-users: the citizens and communities served by the system. A lack of significance may lead to the proposed system being abandoned by the would-be beneficiaries and a reversion to traditional, indigenous, extra-legal ways of doing things [20]. The issue of significance is partly addressed through a LAS designed along FFP guidelines.

FFP LAS "should be designed for the purpose of managing current land issues within a specific country or region" using an approach that is participatory, inclusive, flexible, and focused on citizens' needs [21]. Twelve principles grouped into three frameworks provide the foundation of the approach [13] (see Table 2). The spatial framework is concerned with how land is occupied and used. The legal (and regulatory) framework is necessary to support the adoption and implementation of the FFP approach. The institutional framework is required for effective management and administration of land rights and resources, and the delivery of accessible and inclusive services.

Table 2. The key principles of a fit-for-purpose approach to land administration [13].

Spatial Framework	Legal Framework	Institutional Framework
Visible (general) boundaries	Flexible, administrative	Good land governance
Aerial imagery	Continuum of tenure	Integration
Accuracy for purpose	Flexible recordation	Flexible ICT approach
Updating, upgrading, ongoing improvement	Gender equity	Land information: transparent, affordable, accessible

Implementation of the FFP approach to land administration is proposed to follow several steps [22], beginning with an analysis of the country context and ending with a benefits analysis (see Figure 1). The country context analysis is a crucially important first step because interventions should be designed to fit the local context [23,24]. Thus, the FFP approach is noted to be a set of *guidelines to be tailored for individual country contexts*, not a set of principles for guaranteed success [25]. An analysis of the existing spatial, legal, and institutional frameworks provides a baseline assessment for the gap analysis: what is the existing situation, what is the desired situation, and what needs to change to achieve the desired situation [26]? Using the FFP principles and frameworks listed in Table 2 as a guide, the country-specific strategy can then be implemented. Capacity development is a core concern and implementation should follow an incremental approach using intermediate goals and objectives. Detailed instruction manuals should be developed to provide guidance for all stakeholders to ensure that a consistent, cohesive approach is followed. Finally, an analysis of anticipated economic, environmental, and social benefits should include a cost comparison to indicate the anticipated benefits of adopting a FFP approach to land administration and hence garner political support for the process.

Taking the cautions raised by Barry [25] into consideration, these implementation steps should not be considered linear because any intervention changes the context. An analysis of the new country context should thus follow the benefits analysis cyclically to make sure that the newly designed and implemented LAS is 'fit' for the changed context (see Figure 1).

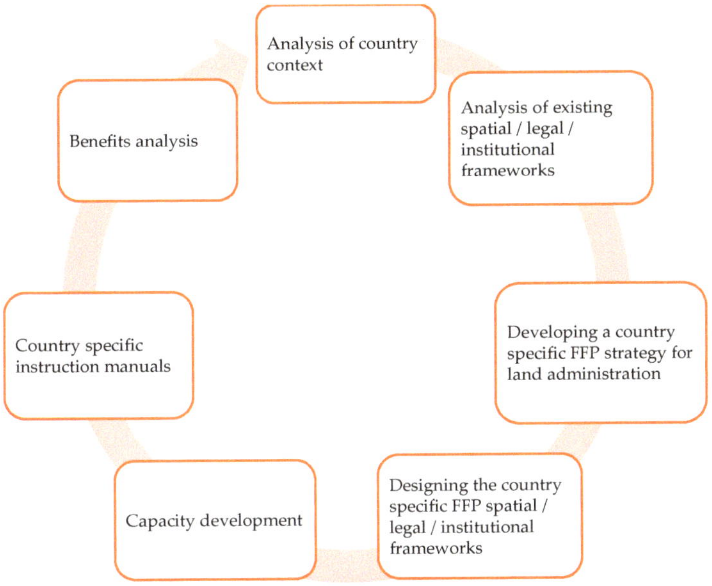

Figure 1. Country-level FFP implementation approach [22]

2.3. *Design Science Research Paradigm*

The design science research paradigm is becoming mainstream in Information System (IS) research. Most modern organizational management theories rely on a theory of design, focusing on creating or reforming the organization rather than an artefact or service [18]. These include systems thinking approaches to organizational change [18] as a necessary application of behavioral science to ascertain the problem situation in context prior to any intervention. This methodology has been followed in many cadastral research projects [16]. The design science research paradigm draws on existing knowledge and theories in the design of an artefact to meet the identified need in the form of an IS or to reform an existing IS. It considers both the design of the process of reform as well as the design of the product or artefact that results [27]. Thus, assessments of performance management should include assessment of both the process and of the product/artefact.

2.3.1. The Alignment of the Design Science Research Paradigm with Theories and Methods Underpinning LAS Reform

Çağdaş and Stubkjær [16] promote the cadastral design science research paradigm consisting of socio-technical systems theory for the design of a Land Tenure Information System (LTIS) artefact or intervention, possibly augmented by Searle's theory of social reality (STS) [28]. STS includes technical elements that do not display intent, stakeholders (natural persons, collectives/groups, legal persons, and organizations) that do possess intentionality, and social elements such as laws, norms, and practices. Searle's theory separates physical reality (tangible) from institutional reality (which is real but not necessarily tangible). Institutional reality in the cadastral domain includes the rules of land rights and tenure codified into cadastral statutory law, African customary laws of landholding and administration, and legitimate, extra-legal but uncodified rules that may exist in informal urban social tenure contexts, as examples.

Additionally, within the design science research paradigm, methods from behavioral science, such as soft systems methodology (SSM) [29], are promoted by Çağdaş and Stubkjær [16] to understand the context in which the LTIS is to operate. SSM has been (and continues to be) used to good effect in a number of case studies conducted within the research group of the authors, such as in Whittal [14] and Mabesa [30]. More recently,

Augustinus has highlighted the explanatory power of SSM [31]. The Two Streams Analysis of SSM includes Gap Analysis, in which the difference between the current state and the desired state is identified [29]. Change management processes are necessary to move from the present state to the desired state [14]. Social Network Analysis (SNA) [32] adds the sensitivity to transactions between stakeholders. Power differentials, differing and even competing goals, and other aspects of relational complexity may be appreciated and modeled using SNA. New institutional economics (NIE) [33] focuses on institutions (intangible rules, transactions, and laws) and organizations relating to the cadaster (such as the Surveyor-General's and Deeds Offices, and professional registration bodies). NIE is promoted as underpinning the evaluation of efficiency and adoption of the artefact [16].

The pairing of design science research processes and behavioral science (DSR-BS) adds important sensitivity to context—design science research on its own is inadequate.

2.3.2. Relevance of DSR-BS Processes to LAS Reform

Design science research is conducted according to the following six steps [16]:

1. Identify the problem—a process underscored by behavioral science.
2. Define objectives for a solution.
3. Design the artefact—this can be a construct, a model, a method, or an implementation [34].
4. Demonstrate the use of the artefact to solve the problems identified in (1).
5. Evaluation of the artefact.
6. Communication of findings.

As mentioned, for design science research based on a critical realist ontology, the socio-technical system and process of change, not only the artefact, would be evaluated and results communicated in steps 5 and 6 [27]. This compliments the FFP process. Additionally, the pairing of behavioral science methodologies with the design science research process (DSR-BS) adds sensitivity to context and hence value to the FFP process. However, focus on contextual analysis in FFP is at the spatial/legal/institutional level—the DSR-BS approach does not refocus the analysis to the individual, family, kinship group, or customary area, which may be necessary. Additionally, the DSR-BS does not appear to add value in the design phase.

2.4. *Design Thinking Process Model*

Design thinking (DT) is at its heart a human-centered approach that may help to solve complex problems that will not submit to technical-only solutions [35]. DT is used in policy reform and service design and follows developments in organizational theory (reinventing government, New Public Management (NPM), entrepreneurial state) [17]. There is an assumption that innovation goes hand in hand with disruption rather than through gradual processes of incremental change as are preferred in bureaucracies (though it is noted that this assumption can be challenged) [17]. What DT brings to the change management field is "a framework for more participatory and cross-disciplinary approaches to social problems" [17].

Public Sector Innovation Labs (PSI labs) adopt DT—an inclusive and experimental approach to problem solving that provides a balance to rational positivism with the inclusion of empathy, curiosity, emotion, and intuition. With respect to policy intervention case studies, various steps are identified by McGann [17]. However, these are not well-aligned to pure DT, relying instead on the instrumental rationality of positivist framings of 'evidence-based policymaking' [17].

Of important interest to FFP approaches to land administration reform, DT is complementary as it identifies that "Participatory, user-centred [*sic.*] approaches may excel in producing ethnographically informed insights and in collaboratively generating ideas that have 'buy in' from stakeholders." [17] Furthermore, DT promotes "enhancing the lives of individuals" [18]. This aim aligns with the aspect of Significance highlighted by Hull and Whittal [19,36]. In addressing the challenges of FFP LA in complex situations, DT may offer an interesting approach to "deliver implementable solutions to problems that are structurally complex and necessitate interconnected solutions" [17]. A core premise of DT is that the world is complex.

The usual approach in design thinking follows these steps:

1. Empathize—research the user's needs.
2. Define—state your user's needs and problems.
3. Ideate—challenge assumptions and create ideas.
4. Prototype—start to create solutions.
5. Test—try out your solutions.

Some of the steps are familiar to well-known change processes (steps 1, 2, 4, and 5). However, the framing of step 1 as 'empathize' firmly positions the approach as human-centered and encourages the researcher to see through the eyes of the other. Step 3 is novel—this step encourages participants to think differently, expand their understanding, and be innovative in 'ideating'.

In contrast to DSR-BS, DT has at its essence imagination, cognitive processes, the "spirit of creativity and value" [18] within all elements and units of an organization, and creative inquiry (including synthesis—informed intuition) in the process of problem solving. Creative inquiry includes four 'moments' [18]:

1. Invention: creation of new ideas—breaking new ground.
2. Judgement: this is based on the criteria of desirability—does the invention meet the needs (significant for the users), is it feasible (suitable in the context), is it viable (sustainable over time)?
3. Connection and development: based on the criteria of usefulness (does it perform the task?), useableness (is it compatible with a human user?), desirability (does it deliver emotional satisfaction?)
4. Integration and evaluation: the worthiness of the solution—should it be implemented (for stakeholders, especially users, society, the state)?

Buchanan highlights creative inquiry as the most essential feature of DT, but DT does not end with concepts and ideas—embedded in DT is making and doing [18].

Most importantly for LAS reform,

> "The principle of design ... is grounded in the quality of experience for all of those served by the organization. This includes the individuals who directly use the products and services of the organization, but it also includes those who are affected by the internal and external operations of the organization and by those in society at large who are ultimately affected by the vision and strategies of the organization". [18]

Buchanan explores the meaning of "experience". It can refer to an individual's perceptions and sensations, but could also refer to an individual's relationship with the environment (consisting of "objects and activities, signs and symbols" [18]). This relationship is initiated by human intent and human selection of certain aspects of the environment with which to engage. Human engagement creates meanings for that individual. Ideally, DT enhances the unity between humans and their environment—it removes obstacles to meaningful human experiences. These obstacles can be physical/practical, ones of intellectual understanding, and emotional engagement. Furthermore, Buchanan reflects on Asmal, in reference to the Constitution, that:

> "Human-centered design is fundamentally an affirmation of human dignity. It is an ongoing search for what can be done to support and strengthen the dignity of human beings as they act out their lives in variety social, economic, political, and cultural circumstances ... the quality of design is distinguished not merely by technical skill of execution or by aesthetic vision but by the moral and intellectual purpose toward which technical and artistic skill is directed.". [35]

A truly FFP LAS will strive to deliver this type of unity between all stakeholders (internal and external) and the system in order to deliver on sustainability, success, and significance [19,36], but perhaps most importantly in South Africa, to deliver on human dignity for all who suffered extreme loss of dignity in the colonial and Apartheid years. DT may assist in this process.

2.5. Comparison

In comparison, these methodologies all begin with understanding the context. Design thinking stands apart in its human-centered approach to this process. It highlights empathizing with the stakeholders, involving engagement with their feelings. Feelings are necessarily related to the stakeholders' experiences. The FFP process is specific in its analysis of "country context" which could include aspects such as history, development, legal pluralism, and culture—both organisational and societal. The design science research paradigm is importantly paired with behavioural research methodologies for this phase—on its own it would be insensitive to context. It is complementary to the FFP process.

Each of the methods includes a stage of identifying the problem. All methods include evaluation/analysis of the results. The DSR-BS methodology requires the design and development of an intervention and a demonstration of the effectiveness thereof. The DT approach has a significant addition of the ideate phase, while its human-centered focus throughout the process is significantly different from the other two.

The thematic framework thus derived appears as Table 3 and forms the basis for the analysis.

Table 3. Thematic framework: the DSR-BS, FFP, and DT approaches.

(Behavioral and) Design Science Research (DSR-BS)	Fit-For-Purpose (FFP)	Design Thinking (DT)
(Describe the situation/context)	Analysis of country context. Analysis of existing spatial/legal/institutional frameworks	Empathize—needs
Problem ID and motivation	(incorporated in analysis above)	Define—needs and problems
Definition of objectives	Developing a country-specific FFP strategy for LA	
Design and development	Designing the country-specific FFP spatial/legal/institutional frameworks	Ideate—challenge assumptions and create ideas—*the invention moment* Prototype—solution
	Capacity development	
Demonstration		Test—*the judgement moment*
Evaluation	Economic benefits analysis	*the connection and development moment* *the integration and evaluation moment*
Communication	Country-specific instruction manuals	

3. Materials and Methods

3.1. Materials

This paper draws on a range of secondary data sources including papers, reports, policies, critiques, books, media articles, and others to identify challenges relating to land reform processes and delivery in South Africa since 1994: the advent of democratic government. These sources reflect on land reform policy, general nationwide implementation, and specific land reform interventions that cover a variety of contexts such as urban, peri-urban, rural, and customary communal land reform to improve land tenure security. International and sub-Saharan literature, where identified as relevant to the use of these approaches in the South African context, is also included.

The strategy used to identify the sources was as follows:

1. Using various combinations of keywords (FFP LA, FFP, fit-for-purpose, land administration, land reform, South Africa, Africa), online search engines (Google Scholar, Google Books, Google, ResearchGate, International Federation of Surveyors (FIG), University of Cape Town (UCT) library databases including EBSCOhost, Elsevier, Emerald, HeinOnline, Springer, Taylor and Francis, Thomson Reuters) were inter-

rogated for peer-reviewed journal articles, doctoral theses, conference proceedings, books, policy documents, and technical reports.

2. Literature that included FFP case studies and high-level critiques of the FFP approach was considered. Books and chapters that investigate, explain, or critique land reform processes in sub-Saharan Africa, and particularly South Africa, are included. (Many of these are known to the authors or would have appeared on ResearchGate and other searched sites as well as UCT libraries.) National land policies and high-level country analyses provide rich data—these are well-known to practitioners and researchers in South Africa.
3. Literature focusing on technical interventions was excluded.

A list of the 37 sources and coding data is presented in Appendix A. Saturation sampling logic was used—the sources are not intended to be inclusive, but representative. Sufficient sources were included such that additional sources are not likely to cause data divergence or change the research outcomes.

3.2. Coding and Extraction Process

Guba [37] suggests that researchers should establish an audit trail that allows someone else to examine the process of data collection and analysis. To this end, we made use of computer-assisted qualitative data analysis software (CAQDAS) called Nvivo. CAQDAS is useful for making sense of dense, detailed qualitative data in a variety of different formats: textual documents, audio-visual recordings, and pictures [38–40]. Categorization of the source documents and coding were undertaken using this software. This allows for transparency of data analysis, improves the credibility of the findings, and makes it possible for others to replicate the research (dependability).

The source texts were imported into NVivo version 12 and categorized as non-South African and South African. Each source was then read through, coding text that relates to the thematic framework in Table 3 as illustrated by the concept map in Figure 2. The coding process involves selecting text in NVivo and associating that text with one or more elements in the thematic framework—these are called codes. Each theme and its associated coded text were then extracted using the export functions of NVivo. The result is a set of documents, each one containing the selected source texts that relate to that code.

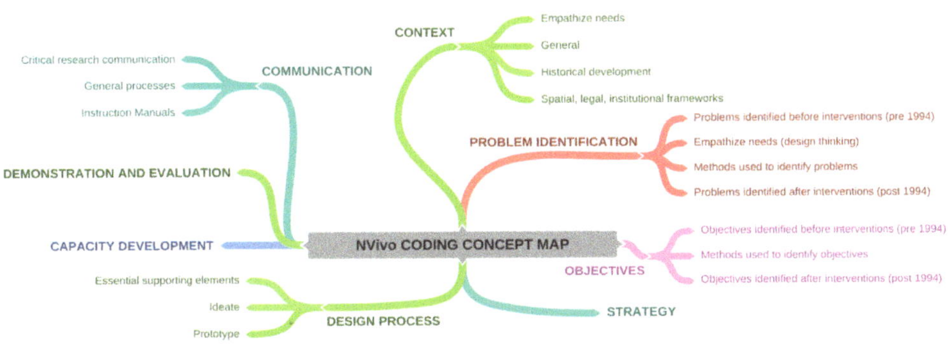

Figure 2. Concept map of coding in NVivo.

4. Results

The text extracts relevant to each code and subcode were analyzed. This involved the researchers synthesizing this information using the thematic codes illustrated in Figure 2 The issues identified are not as relevant to this paper as whether the DSR-BS, FFP, and DT processes would be sensitive to them and could direct appropriate responses. The full suite of themed issues resulting from the coding would be valuable data for another study.

4.1. The Process of Understanding the Context

The assessment of context is tackled differently in the various approaches as shown in Table 4.

Table 4. Extract of context aspects.

(Behavioral and) Design Science Research (DSR-BS)	Fit-For-Purpose (FFP)	Design Thinking (DT)
(Describe the situation/context)	Analysis of country context. Analysis of existing spatial/legal/institutional frameworks	Empathize—needs

DSR-BS relies on a partnership with behavioural science tools that focus on description since DSR-BS does not explicitly include this aspect. FFP focuses on analysis and in particular on the spatial/legal/institutional frameworks, while DT focuses on empathizing with the needs. Embedded in empathy is understanding through the eyes of the stakeholders, in particular, the beneficiaries of land tenure projects. For this reason, the DT "empathize needs" is placed at this level of the intervention—it is impossible to empathize without an understanding of the context and situation—so the process of understanding the context is implied. Similarly, analysis of the context and frameworks relies on knowledge of these—usually in some kind of descriptive form. The three approaches are thus seen as complementary—FFP and DT relies on DSR-BS while DT relies on FFP as well.

There are few if any land tenure interventions that do not spend a good deal of effort on describing the history of the land issues in a country—South African texts show no exception. The essential aspect of understanding land institutions and their functioning in context is also a given. In many cases, the increasingly popular systems thinking approaches are used to describe and assess frameworks, for example legislative, technical, and political systems; e.g., [13,29,41]. The Human-Rights-Based Approach is endorsed as promoting integration between these systems while also promoting the acknowledgement and fulfillment of the South African State's vertical obligations [42] (i.e., the obligations of the state towards their citizens). Unfortunately, a Human-Rights-Based Approach is seldom or only very briefly mentioned in State policy (e.g., in the Communal Land Tenure Policy [43]) with a shift away from this approach to one of indirect rule used in past oppressive regimes [26].

A combined approach involving description and analysis of the context, while empathizing to include sensitivity to stakeholder needs, would be more useful than any of the approaches alone. This would help in complex cases, where there are unhelpful power relations, and to understand the impact of new technologies in the context [25,26,41,42]. It would include identification of the drivers of change in context—this is an important aspect that influences design later on in the process [14,26]. Hull and Whittal [42] highlight that conceptions of human rights and land rights are culturally nuanced—land rights should be viewed through a socio-cultural lens in the understanding of context. Seeing through a socio-cultural lens different from your own is the essence of empathy, endorsing the DT approach.

Pushing this step further to empathize with the needs of stakeholders also has the potential to:

- build cooperation and partnership [41];
- understand the social aspects of land value (land as a sense of place incorporating both past and present people [1,44,45]);
- explore perceptions of land tenure security as opposed to legal/functional land tenure security [46]. An example is the perceived tenure security through the physical holding of title or occupation documents [26];
- explore the spiritual nature of land, which is tied up with individual and collective identity [9];
- understand different conceptions of the subjects of land holding, such as the strongly expressed view that people belong to the land as well as that land belongs to peo-

ple [26], and multigenerational views of landholding that include subjects that are deceased or not yet born [47]; and
- understand customary laws and institutions and community perceptions of their functioning [26].

In the South African context, appreciating such aspects is key to restoring dignity through delivering secure land tenure to all. Furthermore, empathizing facilitates an understanding of land as an essential aspect of a way of life underscored by ubuntu and human solidarity—prioritized in South African land policy [44,48].

4.2. The Problem Identification Process

Assessing the problems prior to any intervention to improve land tenure security is a widely accepted and essential element of development programming. Without adequate understanding of the problems, a solution cannot be designed or will be poorly designed [25]. Table 5 gives the problem identification aspects of the various approaches.

Table 5. Extract of problem identification aspects.

(Behavioral and) Design Science Research (DSR-BS)	Fit-For-Purpose (FFP)	Design Thinking (DT)
Problem ID and motivation	(incorporated in analysis?)	Define—needs and problems

The DSR-BS approach specifies problem identification to motivate change. With the FFP approach, this appears to be implied by the analysis of the context discussed in the previous section. The DT approach specifies the definition of needs and problems. These will be influenced by the rich data resulting from empathizing in the prior stage. The aim of this section is to assess the process of problem identification rather than to reflect the problems, needs, and motivations so identified.

4.2.1. Use of a Framework

In some cases, an established framework is used to assess problems. Bennet and Alemie [49] use the framework of Zimmerman—their paper highlights a number of issues, some of which have relevance to the overall process discussed here: capacity development and institutional support, communication and public–private collaboration, as well as use of standards in assessment and monitoring and evaluation processes. Enemark and McLaren [22] suggest using the Land Governance Assessment Framework (LGAF) developed by the World Bank [50]. Zein assesses business processes using SWOT analysis and assessing the legislation [51]. Flores et al., in designing a FFP response, relied on an extensive literature review and grounded theory case study approach [52]. They propose the FGAF–FFP Governance Assessment Framework, which views the seven FFP principles as governance elements. These are assessed using their Governance Assessment Tool (GAT). Hull proposes a framework developed to deliver land administration systems that are significant, successful, and sustainable [26]. This framework incorporates many aspects of others but is assessed for application in customary land rights contexts. Hull also reviews the theories underpinning land administration reform and places these on a continuum from conservative to adaptation and replacement theories [19]. He contends that "the theory informing development [often] goes unspoken and unnoticed. Conscious decisions at the theory level are important, especially when seeking to undertake cadastral systems development in contexts differing from well-understood western norms, because the value and meaning of land to land rights-holders is context-specific" [26]. Some frameworks will be more sensitive to certain types of problems than others—the choice of framework is important.

4.2.2. Reading and Listening

With regard to methods used to identify problems in South Africa, there have been some very extensive summits, investigations, and reports. The 2005 Land Summit engaged

1500 stakeholders and academics. The investigation that resulted in the High Level Panel report of 2017 involved multiple stakeholder engagement and public hearings across the country supported by two national colloquia on land reform [9]. Submissions from stakeholders and academics as well as working groups, consultative round tables, and commissioned reports were used to identify issues relating to a task list of identified themes. The 2019 Final Panel Report of the Presidential Advisory Panel on Land Reform (the Advisory Panel Report) underwent a diagnostic process that began with existing information and included a detailed consultative process [8].

Empathy for problems and stakeholder needs is an aspect of DT. The White Paper on South African Land Policy of 1997 identified deep resentment over layers of ongoing dispossession of land rights and interests [44]. The 2005 Land Summit suggested a Restitution Truth and Reconciliation Committee at which people's experiences of dispossession could be heard. The aim would have been to bring healing and to restore relationships, but this suggestion early in the process of land reform was never enacted. Healing the wounds of the past is also identified in the National Development Plan of 2012 [10]. However, communal area residents have little voice [5]. The approach reflected by Abubakari et al. [45] also borders on empathy with detailed investigations into the effects of codification, and who was affected. The High Level Panel report of 2017 as well as the 2019 Advisory Panel Report paid particular attention to hearing first-hand the lived experiences of people who live in South Africa [8,9]. The Advisory Panel Report highlights that "rural and urban spaces define people's identities, social standing and the participation in the mainstream economy". This reveals that empathy for how problems affect stakeholders is an important aspect of the process, further endorsing the DT approach. Abubakari et al. [45] highlight the importance of identifying problems with sensitivity to geographical and actor variations in the practices and norms, as well as including socio-political factors and sensitivity to the effects of codification. This counters the drive to find solutions that can be delivered at scale and promotes a more nuanced approach that could also benefit from DT at the design phase since it challenges assumptions and promotes creative thinking.

4.2.3. Sensitivity to the Problems

The problems identified in South Africa with respect to land delivery have been extensively reviewed [1,6,8,9,19,42,44,48,53–58]. With comprehensive reports forming part of the review, saturation sampling was reached—the list of problems is comprehensive and spans those identified in the early stages prior to the implementation of land reform, as well as those identified reflectively. Similar problems have been identified in applying a FFP approach internationally as reflected in the FFP literature and selected case studies [13,21,22,45,49,51,52,59,60]. However, the intention of this paper is not to summarize the problems but to assess the combined thematic framework.

The methods of DSR-BS (see Section 2.3) are likely to be sensitive to problems that would not be revealed by the employment of exclusively non-social science methods. Barry highlights some of these—competing goals, the influence of power, corruption, and contrasting organizational cultures [25]. Abubakari et al. identify political, economic, and sociocultural contingencies to which the FFP approach is not particularly sensitve [45], although the FFP approach's use of spatial, legal, and institutional components reveals some level of systems thinking. Sensitivity to gender issues with respect to land is also important in South Africa [8,44]. Systems thinking could reveal South African insitutional issues such as:

- lack of capacity [10,44,58];
- inappropriate policy, weak and bureaucratic insitutions, and funding issues [10,57];
- dualistic and inequitable (especiailly in customary areas) insitutional frameworks relating to both land governance and management that are compounded by the silo culture and confusion of mandates in South African land institutions [5,8,10,58];
- disfunctional political/parliamentary systems [9,10];

- inadequate legal structures and processes that are not transparent and cannot accommodate opposing views [8,10];
- inconsistencies, contraditions, and irrationality in the existing laws, even to the extent of being unconstitutional, that bedevils land development and management, especially in relation to communal land, traditional land, and mineral and petroleum resources development [8–10]; and
- social strategies for gaining access to land and securing tenure (often multiple and overlapping holders and rights), especially within informal and customary settings [61].

The DT approach that emphasizes human engagement is clearly aligned with the High Level Panel Report and Advisory Panel Report processes. The meaning of the process for individuals and collectives can be enhanced through the shared understanding of problems and emotional engagement. Thus, the unity between individuals, collectives, and their environments may be improved. The usual aspect of public and stakeholder participation is executed at a deeper level using social science methods with the aim of building a deep understanding and empathy. Since the identification of problems is undertaken at the start of the process, deep engagement at this stage sets the scene for success. Although participation is a FFP element, the DSR-BS and DT approaches offer value to augment the FFP approach at this stage of problem identification.

4.3. *The Process of Identifying Objectives and A Strategy*

Table 6 shows the identification of objectives, a strategy, and goals in the various approaches.

Table 6. Extract of the aspects of the identification of objectives, a strategy, and goals.

(Behavioral and) Design Science Research (DSR-BS)	Fit-For-Purpose (FFP)	Design Thinking (DT)
Definition of objectives	Developing a country-specific FFP strategy for LA	

Although the DT approach does not specify the identification of goals, objectives, or a strategy, the DSR-BS approach focuses on defining the objectives of an intervention and the FFP approach focuses on developing a strategy. 'Strategy' is the approach used to implement change in order to achieve each goal. The interim steps are the 'objectives'—these form part of the overall strategy. Thus, the DSR-BS and FFP approaches are not substantively different although the terminology is not identical.

Hull, Kingwill, and Fokane [62] highlight the systemic nature of land administration, as illustrated in Figure 3. At the policy level, the vision is set. At the management level, strategies for achieving the vision are laid out. At the administration level, actionable objectives are carried out to realize the strategies and hence the vision. Such an articulation highlights the synergy between the DSR-BS and FFP approaches and their appropriateness for land administration studies.

Figure 3. Hierarchy and functions of elements of Land Administration Systems [62].

4.3.1. Underlying Assumptions and How They Influence Goal, Objective, and Strategy Formulation

South African land reform to date is criticized for failing to interrogate assumptions about land access, land rights and tenure, and land administration. Strategy is linked to the type of intervention and whether it follows a conservative, adaptation, incremental, replacement, or systematic titling approach [26]. A bias in favor of replacement theory is based on the belief that a formal land administration system, aligned with the civil law system, is superior to other options aligned with a hybrid civil, common, and African customary law system [26]. The capital value of land to landholders as espoused by de Soto underpins this belief but is strongly criticized for the South African context [61]. Strategies to deliver land rights in customary communal areas are based on contested assumptions about the processes of land administration employed by traditional authorities and the rights in land and land-based resources of individuals and families in these areas [26]. False assumptions about the hegemony of groups such as customary communal and informal settlement residents, the poor [61], and women may also result in misdirected goals, objectives, and strategies for South African land administration reform [8].

4.3.2. Methods Used to Identify and Report on the Goals, Objectives, and Strategy

In much of the literature both internationally and relative to South Africa, the methods used to identify goals, objectives, and a strategy are not explicitly identified. Authors in research institutions may be specific, such as identifying spatial and social science methodologies [46], while others, for example Balas et al. [59], identify frameworks such as those of Williamson et al. [2] and Lewis [63]. Many frameworks used to identify the context and problems (including the FFP approach, Voluntary Guidelines on the Responsible Governance of Tenure—the VGGTs) are also used indirectly to identify goals and objectives of a change process. The literature appears thin regarding information on strategy, except where change management processes (The FFP approach is also described as a change management process [22]) are explicitly identified, e.g., [13], where there is an extensive process such as national land summits [50,53], detailed investigative reports [8,9,58], or in new land policy, such as South Africa's White and Green papers [44,48] and Rwanda's policy and legal reform, in which clear goals and objectives are identified. Strategy is then linked to state-led programs such as the South African Reconstruction and Development

Programme and the current National Development Plan with its Comprehensive Rural Development Plan. The National Development Plan resulted from a commission of experts that consulted widely [10].

Acknowledging that the IT/IS aspects are important for any LAS (noting that land information forms the foundation for the LAS and feeds into every level as illustrated in Figure 3), the IT strategy, specifically for capacity development, is sometimes highlighted [22]. The FFP approach specifies formal acceptance of the strategy by senior civil servants and politicians [22]. Apart from embedded strategic components, the FFP approach is also referred to as a strategy [22], but the FFP as a strategic tool is critiqued by Barry in favor of the rational comprehensive strategic planning process [25].

Barry [25] highlights that once a strategy, or pathway, is formulated, there is an important process of stakeholder participation that should be followed prior to implementation. This is a focus of Moreri et al. [60], who highlight that a strategy for public participation is needed. Hull and Whittal promote a Human-Rights-Based Approach that views participants in the process as active rights-holders rather than passive beneficiaries [42]. Top-down and bottom-up processes of engagement and formulating goals, objectives, and strategy should be considered [60]—the duties of the state are paired with citizen and community rights [42]. On the same note, the strategy of FFP is critiqued by Barry [25] for a missing link between community-level problems and national policy. As can be seen in South Africa, policy exists and is in the process of being redrafted. The deep and inclusive analyses, such as expressed in comprehensive reports in South Africa over the last 20 years, go a long way to linking problems to policy amendments, highlighting goals, formulating measurable objectives, and defining a comprehensive integrated strategy for implementation that is approved by the government departments tasked with implementation. These comprehensive reports have benefitted from years of behavioral science research in South African communities to generate a deep understanding.

4.3.3. The Role of Politics in Land Administration Reform Strategy

The strategies employed in South Africa are categorized by Cousins [5] according to successive presidential terms associated with general policy shifts: the Mandela years (1991–1999), the Mbeki era (1999–2000), and the Zuma period (2000–2016). This reflects that the strategic direction is strongly linked to the politics of the day. This is not surprising given the importance of land in political rhetoric, which can often distract the strategy and blur goals and objectives [5].

4.3.4. Commentary on the Process

The gap in this process for the DT approach (see Table 6) necessitates a jump from knowledge of the problems and needs to developing solutions. This seems counter to mainstream change management theory such as Kotter' eight stages [64] and the nudge theory [65] that rely on a vision for the development or reform process. Bringing the strength of DT's creative thought into this stage in some manner may assist stakeholders with deep knowledge and empathy for problems and needs to formulate suitable high-level goals and objectives. This would provide some direction for the next state in DT while maintaining a space for creative and free thought in the design phase.

Linear models in which one stage follows the other are critiqued by Whittal [14] and Barry [25] in favor of a spiral model of continual reflective engagement with prior stages, acknowledging that over time, goals, objectives, and the strategy must change to meet changing circumstances. The process of change is never considered to be over—once objectives have been achieved and goals have been met, the context is likely to change, necessitating ongoing adaption. A slow, methodical, and reflective process is advocated rather than a 'big bang' rapid approach to change. Ongoing assessment as to whether an intervention is realizing its intended objectives is required [14,25]. Furthermore, cultural change is highlighted as Kotter's eighth stage [64] and is endorsed by Whittal [14] and Barry [25], not only within land administration institutions, but regarding societal changes in relationship to land through LAS. The FFP approach is intended to be repetitive as illustrated in Figure 1, acknowledging that FFP initiatives are socially constructive—this is appropriate for the South African context.

4.4. The Design Process

The design phases of the various approaches are summarized in Table 7. The DSR-BS and FFP approaches are well-aligned with respect to design, with the FFP approach adding some domain-specific detail regarding the spatial, legal, and institutional frameworks. The FFP design phase is thus directly informed by an understanding of these frameworks from the context assessment stage.

Table 7. Extract of the design process aspects.

(Behavioral and) Design Science Research (DSR-BS)	Fit-For-Purpose (FFP)	Design Thinking (DT)
Design and development	Designing the country-specific FFP spatial/legal/institutional frameworks	Ideate—challenge assumptions and create ideas Prototype—solution

The FFP design criteria are well-expressed in the key FFP texts [13,21] and summarized in Section 2.2. There are an additional 10 special design elements for pro-poor land rights recordation [13]. The components of the FFP approach should be weighted up in view of the identified needs and essential elements for the South African context given in Sections 4.4.2 and 4.4.3.

DT offers a fresh approach to problem solving, bringing together human capacity to think of new ideas and possibilities, to be innovative and creative, to encourage the discipline and practice of asking and answering questions, to broaden our cognitive processes to understand, create meaning, synthesize, and make decisions, and making the product in practice. This is described as 'pluralism of design thinking'.

As highlighted in Section 2.4, Hull's framework of Success, Significance, and Sustainability [26] is aligned exactly to the judgment 'moment'—one of the four key 'moments' of design thinking. Design thinking is human-centered and affirms human dignity. In reflecting on the importance of this aspect in South Africa's land history in Section 2.4, DT appears to have a great deal to offer as an addition to the DSR-BS and FFP approaches. The FFP approach is stated to be a human-rights approach [21] and is also intended to be human-centered—the focus of DT is compatible and complementary in the design phase.

4.4.1. International Innovative Design

New approaches are promoted in the FFP strategy [21] although the principles appear prescriptive [13] and may limit design thinking. Innovative thinkers now reject the binaries of informality versus formality, communal versus individual, traditional, customary, and neo-customary versus modern Western forms of rights and tenure as these are identified as possibly unhelpful and limiting in understanding and hence in design. Innovative ideas such as spaghetti boundaries, point cadasters [21], rights and interests recordation (rather than registration—an example is a pro-poor land recordation tool [46]), and flexible land tenure systems show evidence of DT. Spaghetti boundaries have not gained much traction, probably failing on one or more of the four DT 'moments' (see Section 2.4). However, point cadasters are promoted as a first step in recording [21], while flexible land tenure systems are now legislated and piloted in Namibia along with the innovative concept of para-professionals. Multidirectional flexibility in landholding type is further evidence of the need for innovative thinking [66].

Conceptual models and domain models such as the new continuum of land rights [21,67] and the Social Tenure Domain Model [68] also reflect innovative thinking. The UN Habitat Global Land Tools Network (GLTN) has for many years supported development of new FFP tools. Its4land prioritizes innovative tool design, revealing some DT elements such as: transdisciplinary work, gender-sensitive analysis, technical innovation for community mapping, localized domain model development, comprehensive stakeholder participation, design focused on end-user interests and actor preferences, as well as generating innovation capacity and knowledge sharing [41]. Technological innovations are often implemented in pilot test projects; some include social methods of data collection, which also enhance participation [21,69].

Fisher and Whittal [1] highlight future areas of innovation in property definition. These may include fuzzy areas, time-varying areas, multiple overlapping rights and interests, family titles, multidirectional flexibility and conceptions that can reflect mobility of rights holders. Legal freehold space units (or property objects), a three-dimensional cadaster with volumetric representations and registry procedures for titling are also discussed [1]—some of these concepts are in the process of development for dense first-world cities [70]. These concepts have application in the offshore environment as well as for underground and above ground space. The recordation of land-related debts and claims not based on contract, but related to land, such as claims against the State, licenses, permits, and quotas are additional innovative aspects of land rights and interests that are highlighted as possible additions in an inclusive land administration system to improve land tenure security [1]. The suitability of unique parcel identifiers is questioned while spatial location is argued as accessible and thus highly usable for all stakeholders. A mind-shift away from the parcel identifier as the basic unit of the cadaster would be required. The shifting nature of South African Constitutional property law, which now embraces contextual and non-hierarchical thinking, sets the platform for DT from the top down [1].

4.4.2. South Africa's Need for Innovative Design

Case studies and reports on South Africa's land reform process between 1997 and 2021 [1,6,8,9,26,42,44,48,54,56,61,67,71] include repeated calls and suggestions for innovative approaches ("a radical and rapid break from the past" [58]) such as:

- interrogating the legal and social system of landholding;
- design of new ways of recording rights and interests in land and land-based resources (particularly in customary communal and other complex settings, possibly family titles, locally nuanced, that consider new developments such as blockchain);
- design new types of proprietary land unit that may
    - challenge the parcel as the basic unit of the cadaster;
    - include boundaries that are fluid (shifting over the short or longer term), fuzzy (imprecisely defined in space), and adaptable (changing in nature);
    - represent the third spatial dimension;
    - define spatial rights that may be of variable nature, nested, overlapping, and time-varying;
    - relate to rights holders that are individual, family, and kinship (including multigenerational landholding: living, dead, and unborn), and other collectives based on voluntary affiliation (preferred over tribal affiliation);
- overcoming institutional and process issues of delivery and controlling and promoting effective land use;
- passing new laws through interdepartmental work, and modifying and integrating silos in legal and land administration systems;
- adopting a social systems approach to solutions; dealing with complexity; understanding humankind–land relationships through an African lens (or lenses);
- promoting a continuum of land rights in practice;
- promoting new technical (for example IS and surveying) tools;
- promoting gender equity; promoting stakeholder participation;
- dealing with land acquisition by the State for land reform purposes;
- designing dispute and conflict resolution mechanisms; and
- managing differential power relationships and corruption.

4.4.3. Essential Elements Identified for South Africa

The following essential elements are identified from sources (only some of a plethora of sources are referenced) relating to South Africa as key aspects of the land administration design and reform process to improve land tenure security; these essential elements are echoed in international sources:

- a rural development focus and tangible and sustained support [54,72];
- a capable State [8,9,42,54] that meets its obligations [42];
- integrated state institutions [54], integrated traditional administration/leadership, developmental state [9], functional parliament [9];
- good governance [14,56], curbing corruption [8,9], accessible land and land tenure [44], the cost of these [56];
- public participation, especially State-enabled and citizen-empowered [5,8,42,56] and with private sector partnerships [8];
- strong leadership by the State [5,56] and local project leaders [73]; managing risks [8];
- strong policy and objectives, political will and commitment, resourcing [8,9]; shared vision [8];
- communication, especially when innovative policies, laws, and approaches break from the past [5,8];
- reskilling land development and administration professionals [5]; and
- pre-empting, preventing, and resolving disputes [5].

In the South African context, the essential elements and need for innovative design overlap with aspects of the FFP approach. However, in order to retain the creativity of the ideate process of DT, the tools and methods favored in the FFP approach (see Section 2.2) should only be considered as options late in the design process and should not be prescriptive (noting that FFP is a guide and is not intended to prescribe).

4.5. The Capacity Development Process

Capacity development is highlighted in almost all literature dealing with land reform in the developing world. It is interesting to note that it is not an explicit stage in either DSR-BS or DT (Table 8). It is identified as a need at the innovative design stage and is an essential element for implementation.

Table 8. Extract of the capacity development process aspects.

(Behavioral and) Design Science Research (DSR-BS)	Fit-For-Purpose (FFP)	Design Thinking (DT)
	Capacity development	

The need for capacity development is not restricted to technological interventions such as the IT/IS strategy although this is often highlighted [22]. People working within the land administration sector also need knowledge and skills development—this extends to capacity building at societal and organizational (governmental, private sector, community) levels [21]. Similarly, capacity development is highlighted as important for the would-be beneficiaries of land reform programs. Some evidence suggests that the concept of land rights is still poorly understood by customary land rights-holders [74]. They need training and resources to claim their land rights and support use of the new systems (technological, legislative, administrative) that have been developed. There is often a gulf between beneficiaries' capacity and the plans of developers [5]. This is a noted shortcoming in South Africa regarding Communal Property Associations [75] and traditional authorities [76]. Hence, while land rights might be recognized and protected in law, administrative incapacity means that people are unable to realize them in practice [77].

Capacity development is identified as a process, not an event, in the FFP literature [21]. Public–private partnerships may assist in the process, which can benefit from a well-formulated incremental strategy [13,22,61,78] with performance measurement [13,22,41].

Balas et al. [59] identify issues such as land law, gender, equity, land administration, technical spatial data collection, as well as monitoring and control as core areas of capacity development. Training should also be focused on the importance of a harmonized methodology [59]. In addition to the community-level support highlighted above, specific South African needs for capacity development are identified: provision of local services [44]; assistance with agricultural processes and infrastructure [44,57]; financial services to beneficiaries [44,57]; understanding of authority and responsibility to act; and empowerment [42]. Implementation of policies and laws has been restricted by insufficient institutional capacity. This limitation was noted in both the South African policy White [44] and Green [48] Papers (published in 1997 and 2011, respectively), yet since then there appear to have been no plans to address these constraints and it seems as if the situation is worsening [5,79]. At national, provincial, and local levels, the South African government appears to lack the capacity for proper land administration [75]. Cousins & Hall [80] and Cousins [5] noted that State capacity for implementation of land reform law and policy is inadequate. The High Level Panel [9] found that many policies and laws were sound, but there are serious concerns around their implementation and enforcement.

From the foregoing, and drawing on the South African experience, it is clear that capacity development is an important stage in the FFP process that would possibly be omitted if the DSR-BS or DT processes were used without the benefit of the FFP approach.

4.6. The Piloting/Demonstration and Evaluation Processs

The piloting/demonstration and evaluation stages of the approaches are highly similar and hence grouped together as shown in Table 9.

Table 9. Extract of the piloting/demonstration and evaluation aspects.

(Behavioral and) Design Science Research (DSR-BS)	Fit-For-Purpose (FFP)	Design Thinking (DT)
Demonstration		Test—*the judgement moment*
Evaluation	Economic benefits analysis	*the connection and development moment* *the integration and evaluation moment*

Only the DSR-BS approach has a separate demonstration and evaluation process, while in practice these processes are usually undertaken as pilot studies that are then evaluated. This stage is key to uncovering aspects in the design that are not FFP, or fit for *local* purpose, and modifying the design accordingly. In the FFP approach, the evaluation appears to be focused on economic benefits analysis, but in the FFP literature this is extended to include economic, environmental, and social benefits. In DT, the *judgement moment* assesses desirability—whether the intervention meets the needs (significant for the users), feasibility (suitable in the context), and viability (sustainable over time). The DT *connection and development moment* assesses usefulness (does it perform the task), useableness (is compatible with human users), and desirability (does it deliver emotional satisfaction), while the *integration and evaluation moment* assess the worthiness of the solution for implementation considering all stakeholders, especially users, society, and the state.

4.6.1. Pilot Studies

The literature reveals that pilot studies are used to test innovative technical solutions; these may include noting of boundary disputes for further processing [21,52,81]. The 2019 Advisory Panel Report advocates for the testing of new approaches, highlighting various challenging areas of land tenure delivery in South Africa [8].

4.6.2. Performance Measurement

Change management is incomplete without measuring the performance of the new design. The land administration reform literature is replete with examples of performance measurement—called quality assurance, auditing, meeting of goals, etc. Broadly speaking, "unlocking social and economic benefits" is aligned with successful reform [13]; however, metrics for this are necessary. The World Bank's "Doing Business" reports, the Corruption Perception Index of Transparency International, LGAF, and FGAF&GAT are suggested [13,22,52]. Whittal promotes the 7Es framework for performance measurement, including efficiency, effectiveness, elegance (acceptability), empowerment, emancipation, exception (inclusivity), and emotion [14]. The 7Es framework aligns well with the human-centered DT and could partner well in a process that accommodates DT.

Benchmarking is another possibility [49]. Balas et al. [59] suggest that key performance indicators should be designed for each stage. A well-designed monitoring and evaluation framework is necessary to provide feedback for improvements [14,22]. This promotes a culture of self-critique and self-motivated improvement in land administration organizations [22].

Evaluation and reporting on land reform outcomes include the appropriateness of technology, number of parcels registered, and the cost to do so. Quality assurance is particularly important with data collection using volunteered geographic information (VGI) [60]. There is a caution to not only use numeric indicators; rather, transformation (emancipation) indicators should be used such as those related to poverty, unemployment, and inequality, particularly related to land access [26,44]. Indicators of efficiency, effectiveness, acceptability, security, accessibility, timeliness, affordability, attainability, sustainability, upgradeability, flexibility, inclusiveness, transparency, clarity (especially in governance roles), simplicity, correctness/reliability [14,21], user-friendliness [49,59,81], and resilience (to disasters) are suggested.

Specific issues for South Africa are monitoring of land reform over time [44], monitoring the effectiveness of governance (especially the lack of Parliamentary accountability [9,10]), and the lack of traditional authority accountability [9]. There is a lack of institutions to execute the evaluation process [8], while there is also a need to develop and use 'outcomes indicators' that are appropriate to the local context. Hull and Whittal highlight the responsibility of the State to deliver on fundamental Economic, Social, and Cultural Rights [42]. Evaluation of the process of reform and the medium and long-term impacts is important in any process but is especially relevant in South Africa considering the need for dignity and restorative justice [14]. Processes of active participation, listening to suppressed viewpoints [26], consideration of emotion [14], and empowerment/capacity development (a FFP stage), should be evaluated and reported. It is suggested that the evaluation process should be undertaken independent of South African State institutions to avoid bias [26] and even as a political necessity [14].

4.6.3. Commentary on the Process

The demonstration and evaluation included in the DSR-BS and FFP approaches are clearly imperative to understand whether the proposals meet the needs. Although these can include a human-centered approach, this is specified in the DT approach. It highlights sensitivity in line with the needs of all stakeholders and includes assessing desirability and emotional aspects. As discussed above, these are important in the post-colonial and post-apartheid context in South Africa.

4.7. The Communication Process

Change management theory promotes communication in line with meaningful public participation at all stages of an intervention. The placement of communication as an end stage belies the importance of this activity throughout a change process. Table 10 shows the communication aspects in the three processes.

Table 10. Extract of the communication process aspects.

(Behavioral and) Design Science Research (DSR-BS)	Fit-For-Purpose (FFP)	Design Thinking (DT)
Communication	Country-specific instruction manuals	

DSR-BS focuses strongly on communicating findings, which is key to any research endeavor. FFP and DT are processes designed for project implementation domains rather than research domains and so communication of findings is not explicitly advanced. The FFP approach focuses on preparing manuals, outreach materials, and capacity building materials [59]. The latter may include gender equity, land law, conflict resolution, and other aspects of training required in order to ensure consistent implementation [22]. VGI data collection methods require special attention to communication [60].

Participants in communications include government officials, politicians and policymakers, and beneficiaries [59]. Where open-source tools are developed [41], there are accepted methods of communicating these beyond the project stakeholders. In land restitution in South Africa, communication between claimants and the Land Claims Commission has been a concern [57].

Communication is an important stage of the process of change and in this respect the DT process is deficient. The DSR-BS and FFP processes adequately cover this stage with complimentary emphases.

5. Conclusions

The analysis of the sample of literature against the thematic framework of the DSR-BS, FFP, and DT approaches reveals that the land reform program in South Africa could benefit substantially by mainstreaming elements of both the DSR-BS and the DT approaches within the FFP approach. This finding may be generalized (naturalistic generalization) to implementation of FFP LA in other similar contexts. We thus conclude below on the implications of this work for theory and for the South African land question.

The DSR-BS approach is suited to provide rich descriptions of and insights into the context and in identifying problems and needs. It also offers strength in the communication of a critical reflective review of change processes. The use of methods from the behavioral sciences is reflected in land administration reform literature, but such methods are not prominent in the FFP guidelines. The DT process is a human-centered approach that stresses empathy when assessing context, problems, and needs. Although the FFP is intended to focus on human needs and incorporate creative design, human-centered creative design thinking is not mainstreamed in the process and can easily be ignored, leading to mechanistic implementations.

The core strengths of design science research and design thinking are shown to be complementary to the FFP approach. Foregrounding these in FFP LA implementations will provide a focal point at the start of the process. Viewing the problem context (the first stage in assessing the purpose) through a local socio-cultural lens different from that of the implementors is the essence of empathy, which is fundamental to the DT approach (Sections 4.1 and 4.2). At the design phase, the creativity of the ideate process of DT should precede the consideration of the tools and methods of implementation favored in the FFP approach. The DT field of research and practice is relatively new—case studies explicitly implementing DT in land administration reform will most likely follow.

With the land question in South Africa still center stage after 27 years of democracy, it is very important to be boldly creative—approaches with different emphases may help unlock solutions. In dealing with the current complexities in South African land administration, design thinking may facilitate a deep understanding, challenge assumptions, and set the scene for unconstrained creative thinking. Human cognitive abilities such as empathy, questioning, judgement, creativity, and creating meaning have the potential to promote the design of interventions that are responsive to personal and social aspects in addition to material/technical/procedural aspects (institutional, legal, and spatial frameworks) of LAS, as identified in Section 4.4.2. In seeking to restore dignity and deliver restorative justice in the South African land sector (considering its particular land history), and to deliver sustainability, success, and significance for all stakeholders, DT processes may help to dismantle barriers in thinking and lead to the unity so desperately needed. If an emphasis on a deep understanding and design thinking may help South Africa achieve its land reform aims, there is no doubt that the process could benefit other contexts with complex and intractable land administration challenges such as are found in other sub-Saharan African and post-colonial countries.

Author Contributions: This paper resulted from collaboration in the following manner: conceptualization, J.W. and S.H.; methodology, J.W.; formal analysis, J.W.; investigation, J.W.; resources, S.H. and J.W.; writing—original draft preparation, J.W.; writing—review and editing, S.H.; visualization, S.H. and J.W.; funding acquisition, S.H. and J.W. All authors have read and agreed to the published version of the manuscript.

Funding: Funding of the publication costs for this article has kindly been provided by the School of Land Administration Studies, University of Twente, in combination with Kadaster International, The Netherlands.

Informed Consent Statement: Not applicable.

Conflicts of Interest: The authors declare no conflict of interest. The funders had no role in the design of the study; in the collection, analyses, or interpretation of data; in the writing of the manuscript; or in the decision to publish the results.

Appendix A

Table A1. Sources Coded in NVivo.

Source (Reference-Date-Ordered)	Number of Codes Used	Coded References
Republic of South Africa White paper on South African land policy; March 1991.; Government Printer, Pretoria, 1997; ISBN 0514308869	42	572
Hall, R.; Mbilinyi, M.; Rusimbi, M.; Omeje, K.; Plaut, M.; Gonzales, D. Briefings. *Rev. Afr. Polit. Econ.* 2005, 32, 621–651, doi:10.1080/03056240500467138	42	165
Cousins, B.; Cousins, T.; Hornby, D.; Kingwill, R.; Royston, L.; Smit, W. Will formalising property rights reduce poverty in South Africa's 'second economy'? Questioning the Mythologies of Hernando de Soto. *PLAAS Policy Br No. 18*. Programme for Land and Agrarian Studies (PLAAS), University of the Western Cape. **2005**, pp. 1–6.	16	51
Whittal, J. Fiscal Cadastral Systems Reform A Case Study of the General Valuation Project 2000 in the City of Cape Town, Ph.D. Thesis, University of Calgary, Canada, 2008	13	17
Republic of South Africa *Green Paper on Land Reform*; Government Printer, Pretoria, 2011	25	103
McLaren, R. Crowdsourcing support of land administration—A partnership approach. International Federation of Surveyors, Article of the month: December 2011.	12	41
Hall, R. *Land Reform Policy Discussion Document*, South African History Online, 2012.	35	436
Association for Rural Advancement (AFRA), Traditional Courts Bill uses apartheid laws to subjugate communities. Cape Times, 21 September 2012, 11	9	9
National Planning Commission National Development Plan 2030 Our Future-make it work; National Planning Commission: The Presidency; Pretoria, South Africa, 2012; ISBN 9780621411805	24	95
Nkwinti, G. Land tenure summit key summit thematic areas, National Land Tenure Summit, Presentation in Johannesburg, 4–6 September 2014.	16	46
Archary, L. Strengthening relative rights of people working the land, National Land Tenure Summit, Presentation in Johannesburg, 4–6 September 2014.	17	40
Whittal, J. A New Conceptual Model for the Continuum of Land Rights. *South African J. Geomatics* 2014, 3, 13–32	17	43
Enemark, S.; Bell, K.C.; Lemmen, C.; McLaren, R. *Fit-For-Purpose Land Administration*; FIG & World Bank: Denmark, 2014; ISBN 9788792853103.	34	401
COGTA *Traditional and Khoi-San Leadership Bill*; Minister of Cooperative Governance and Traditional Affairs: Cape Town, 2015	9	19
Enemark, S.; McLaren, R.; Lemmen, C. *Fit-For-Purpose Land Administration Guiding Principles*; United Nations Habitat Global Land Tools Network: Nairobi, Kenya, 2015	27	349

Table A1. *Cont.*

Source (Reference-Date-Ordered)	Number of Codes Used	Coded References
Bennett, R.M.; Alemie, B.K. Fit-for-purpose land administration: Lessons from urban and rural Ethiopia. *Surv. Rev.* 2016, *48*, 11–20	41	219
Hull, S.; Whittal, J. Towards a framework for assessing the impact of cadastral development on land rights-holders. *FIG Work. Week 2016 Recover. from Disaster*, Christchurch, New Zealand, 2–6 May 2016; pp. 1–21.	17	102
Zein, T. Fit-For-Purpose Land Administration: an implementation model for cadastre and land administration systems. *L. Poverty Conf. 2016 Scaling up Responsible L. Gov.* 2016	14	35
Cousins, B. Land reform in South Africa is sinking. Can it be saved? A provocation commissioned by the Nelson Mandela Foundation DST/NRF Research Chair in Poverty, Land and Agrarian Studies, University of the Western Cape	16	76
Hornby, D.; Royston, L.; Kingwill, R.; Cousins, B. Introduction: Tenure practices, concepts and theories in South Africa. In *Untitled: Securing Land Tenure in Urban and Rural South Africa*; Hornby, D., Kingwill, R., Royston, L., Cousins, B., Eds.; University of KwaZulu-Natal Press: Pietermaritzburg, 2017; pp. 1–43	16	96
Asiama, K.; Bennett, R.; Zevenbergen, J. Participatory land administration on customary lands: A practical VGI experiment in Nanton, Ghana. *ISPRS Int. J. Geo-Information* 2017, *6*	22	75
Balas, M.; Murta, J.; Matlava, L.; Marques, M.R.; Joaquim, S.P.; Carrilho, J.; Lemmen, C. A Fit-For-Purpose Land Cadastre in Mozambique. *2017 World Bank Conf. L. Poverty-Washingt. DC, March 20–24*, 2017, 26	45	297
Enemark, S.; Mclaren, R. Fit-for-Purpose Land Administration: Developing Country Specific Strategies for Implementation. *2017 World Bank Conf. L. Poverty* 2017, 1–18	43	137
Report of the High Level Panel on the Assessment of Key Legislation and the Acceleration of Fundamental Change	43	213
Koeva, M.; Bennett, R.; Gerke, M.; Crommelinck, S.; Stöcker, C.; Crompvoets, J.; Ho, S.; Schwering, A.; Chipofya, M.; Schultz, C.; et al. Towards innovative geospatial tools for fit-for-purpose land rights mapping. *Int. Arch. Photogramm. Remote Sens. Spat. Inf. Sci.-ISPRS Geospatial Week.* 2017, 42, 37–43, doi:10.5194/isprs-archives-XLII-2-W7-37-2017	50	289
Barry, M. Fit-for-purpose land administration–Administration that suits local circumstances or management bumper sticker? *Surv. Rev.* 2018, 50, 383–385, doi:10.1080/00396265.2018.1501130	28	73
Balas, M.; Joaquim, S.; Carvalho, J.A.; Murta, J.; Carrilho, J. SiGIT Land Information System and the Challenges Imposed by the Fit For Purpose Approach to Land Administration, *Proc. FIG Congress*, Istanbul, Turkey, May 6–11, 2018	16	34
Moreri, K.; Fairbairn, D.; James, P. Issues in developing a fit for purpose system for incorporating VGI in land administration in Botswana. *Land Use Policy* 2018, 77, 402–411, doi:10.1016/j.landusepol.2018.05.063	42	182
van Asperen, P.; Hendriks, B.; Zevenbergen, J. Scaling up Pro-poor Land Recordation: Findings and Consequences of three peri-urban cases from Sub-Saharan Africa. *African J. L. Policy Geospatial Sci.* 2019, 2, 13–39	11	34
Hull, S. A Framework for Guiding Cadastral Systems Development in Customary Land Rights Contexts, Ph.D. Thesis, University of Cape Town, Cape Town, South Africa, 2019–Chapter 9	26	339
Mahlati, V. *Final Report of the Presidential Advisory Panel on Land Reform and Agriculture*; Presidential Advisory Panel on Land Reform and Agriculture: Pretoria, South Africa, 2019	46	534
Fisher, R.; Whittal, J. *Cadastre: Principles and Practice*; Roger Fisher, Jennifer Whittal, and the South African Geomatics Institute: Cape Town, 2020; ISBN 978-0-620-82878-9	40	539
Flores, C.C.; Tan, E.; Buntinx, I.; Crompvoets, J.; Stöcker, C.; Zevenbergen, J. Governance assessment of the UAVs implementation in Rwanda under the fit-for-purpose land administration approach. *Land Use Policy* 2020, 99, 104725, doi:10.1016/j.landusepol.2020.104725	36	127
Abubakari, Z.; Richter, C.; Zevenbergen, J. Evaluating some major assumptions in land registration: Insights from Ghana's context of land tenure and registration. *Land* 2020, 9, doi:10.3390/LAND9090281	24	99
Hull, S.A.; Whittal, J. Achieving Success and Sustainability Through Significance: a Cross-Case Analysis of Cadastral Systems Development. In *Proc. of the FIG Working Week 2020: Smart Surveyors for land and water management*; Amsterdam, The Netherlands, 2020; pp. 10–14	2	2
Williams-Wynn, C.; Applying the Fit-For-Purpose Land Administration Concept to South Africa: Will it Work? In *Proc. of the FIG Working Week 2020: Smart Surveyors for land and water management*; Amsterdam, The Netherlands, 2020; pp. 1–15	17	39
Hull, S.; Whittal, J. Human rights and land in Africa: highlighting the need for democratic land governance. In *Human Rights Matters*; Corrigan, T., Ed.; IntechOpen, 2021 ISBN 978-1-83968-874-4	43	155

References

1. Fisher, R.; Whittal, J. *Cadastre: Principles and Practice*; Fisher, R., Whittal, J., Eds.; The South African Geomatics Institute: Cape Town, South Africa, 2020; ISBN 978-0-620-82878-9.
2. Williamson, I.; Enemark, S.; Wallace, J.; Rajabifard, A. *Land Administration for Sustainable Development*; Esri Press: Redlands, CA, USA, 2010; ISBN 9781589480414.
3. Nkwae, B. Conceptual Framework for Modelling and Analysing Periurban Land Problems in Southern Africa. Ph.D. Thesis, University of New Brunswick, Fredericton, NB, Canada, 2006.
4. Hull, S.; Sehume, T.; Sibiya, S.; Sothafile, L.; Whittal, J. Land allocation, boundary demarcation and tenure security in tribal areas of South Africa. *S. Afr. J. Geomat.* **2016**, *5*, 68–81. [CrossRef]
5. Cousins, B. Land Reform in South Africa is Sinking. Can it be Saved? A Provocation Commissioned by the Nelson Mandela Foundation DST/NRF Research Chair in Poverty, Land and Agrarian Studies, University of the Western Cape 1. Available online: https://www.nelsonmandela.org/uploads/files/Land__law_and_leadership_-_paper_2.pdf (accessed on 16 March 2021).

6. Hornby, D.; Royston, L.; Kingwill, R.; Cousins, B. Introduction: Tenure practices, concepts and theories in South Africa. In *Untitled: Securing Land Tenure in Urban and Rural South Africa*; Hornby, D., Kingwill, R., Royston, L., Cousins, B., Eds.; University of KwaZulu-Natal Press: Pietermaritzburg, South Africa, 2017; pp. 1–43.
7. Constitution of the Republic of South Africa. 1996. Available online: https://www.gov.za/documents/constitution/constitution-republic-south-africa-1996-1 (accessed on 23 March 2021).
8. Mahlati, V. *Final Report of the Presidential Advisory Panel on Land Reform and Agriculture*; Presidential Advisory Panel on Land Reform and Agriculture: Pretoria, South Africa, 2019.
9. Report of the High Level Panel on the Assessment of Key Legislation and the Acceleration of Fundamental Change. Available online: https://www.parliament.gov.za/storage/app/media/Pages/2017/october/High_Level_Panel/HLP_Report/HLP_report.pdf (accessed on 27 January 2021).
10. National Planning Commission. *National Development Plan 2030 Our Future-Make It Work*; National Planning Commission, The Presidency: Pretoria, South Africa, 2012; ISBN 9780621411805.
11. African Union Commission. *Agenda 2063: The Africa We Want*; African Union Commission: Addis Ababa, Ethiopia, 2015; ISBN 9789295104235.
12. United Nations. *Transforming Our World: The 2030 Agenda for Sustainable Development*; United Nations: New York, NY, USA, 2015; ISBN 9780874216561.
13. Enemark, S.; McLaren, R.; Lemmen, C. *Fit-For-Purpose Land Administration Guiding Principles*; United Nations Habitat Global Land Tools Network: Nairobi, Kenya, 2015.
14. Whittal, J. Fiscal Cadastral Systems Reform A Case Study of the General Valuation Project 2000 in the City of Cape Town. Ph.D. Thesis, University of Calgary, Calgary, AB, Canada, 2008.
15. Carlsson, S. Informatics, School of Economics and Management, L.U. A Critical Realist Perspective on IS Evaluation Research. In Proceedings of the 13th European Conference on Information Systems, Information Systems in a Rapidly Changing Economy, ECIS 2005, Regensburg, Germany, 26–28 May 2005.
16. Çağdaş, V.; Stubkjær, E. Design research for cadastral systems. *Comput. Environ. Urban Syst.* **2011**, *35*, 77–87. [CrossRef]
17. McGann, M.; Blomkamp, E.; Lewis, J.M. The rise of public sector innovation labs: Experiments in design thinking for policy. *Policy Sci.* **2018**, *51*, 249–267. [CrossRef]
18. Buchanan, R. Worlds in the Making: Design, Management, and the Reform of Organizational Culture. *She Ji J. Des. Econ. Innov.* **2015**, *1*, 5–21. [CrossRef]
19. Hull, S.A.; Whittal, J. Achieving Success and Sustainability Through Significance: A Cross-Case Analysis of Cadastral Systems Development. In Proceedings of the FIG Working Week 2020: Smart Surveyors for Land and Water Management, Amsterdam, The Netherlands; 2020; pp. 10–14.
20. Kingwill, R. Square Pegs in Round Holes: The competing faces of land title. In *Untitled: Securing Land Tenure in Urban and Rural South Africa*; Hornby, D., Kingwill, R., Royston, L., Cousins, B., Eds.; University of KwaZulu-Natal Press: Pietermaritzburg, South Africa, 2017; pp. 235–282.
21. Enemark, S.; Bell, K.C.; Lemmen, C.; McLaren, R. *Fit-For-Purpose Land Administration*; FIG & World Bank: Copenhagen, Denmark, 2014; ISBN 9788792853103.
22. Enemark, S.; McLaren, R. Fit-for-Purpose Land Administration: Developing Country Specific Strategies for Implementation. In Proceedings of the 2017 World Bank Conference on Land and Poverty, Washington, DC, USA, 20–24 March 2017; pp. 1–18.
23. Borras, S.; Franco, J.C. Contemporary discourses and contestations around pro-poor land policies and land governance. *J. Agrar. Chang.* **2010**, *10*, 1–32. [CrossRef]
24. Zevenbergen, J.; Augustinus, C.; Antonio, D.; Bennett, R. Pro-poor land administration: Principles for recording the land rights of the underrepresented. *Land Use Policy* **2013**, *31*, 595–604. [CrossRef]
25. Barry, M. Fit-for-purpose land administration—Administration that suits local circumstances or management bumper sticker? *Surv. Rev.* **2018**, *50*, 383–385. [CrossRef]
26. Hull, S. A Framework for Guiding Cadastral Systems Development in Customary Land Rights Contexts. Ph.D. Thesis, University of Cape Town, Cape Town, South Africa, 2019.
27. Carlsson, S.A. Towards an information systems design research framework: A critical realist perspective. In Proceedings of the First International Conference on Design Science Research in Information Systems and Technology, Claremont, CA, USA, 24–25 February 2006; pp. 192–212.
28. Searle, J.R. *The Construction of Social Reality*; The Free Press: New York, NY, USA, 1997; ISBN 0029280451.
29. Checkland, P. *Systems Thinking, Systems Practice: Includes a 30-Year Retrospective*; John Wiley & Sons, Ltd.: Chichester, UK, 1999; ISBN 0-471-986062.
30. Mabesa, M.J. The Cadastral System of Lesotho: Challenges and Opportunities. M.Phil Thesis, University of Cape Town, Cape Town, South Africa, 2011.
31. Augustinus, C. Catalysing global and local social change in the land sector through technical innovation by the United Nations and the Global Land Tool Network. *Land Use Policy* **2020**, *99*, 105073. [CrossRef]
32. Scott, J.; Carrington, P.J. *The SAGE Handbook of Social Network Analysis*, 1st ed.; SAGE Publications Ltd.: New York, NY, USA, 2011; ISBN 9781847873958.

33. Brousseau, É.; Glachant, J.-M. *New Institutional Economics: A Guidebook*; Cambridge University Press: New York, NY, USA, 2008; ISBN 9780521700167.
34. Van der Merwe, A.; Gerber, A.; Smuts, H. Guidelines for Conducting Design Science Research in Information Systems. In *ICT Education. SACLA 2019. Communications in Computer and Information Science*; Springer: Cham, Switzerland, 2020; Volume 1136, pp. 163–178. ISBN 978-3-030-35629-3_11.
35. Buchanan, R. Human Dignity and Human Rights: Thoughts on the Principles of Human-Centered Design. *Des. Issues* **2001**, *17*, 35–39. [CrossRef]
36. Hull, S.; Whittal, J. Human rights in tension: Guiding cadastral systems development in customary land rights contexts. *Surv. Rev.* **2019**, *51*, 97–113. [CrossRef]
37. Guba, E.G. ERIC/ECTJ Annual Review Paper: Criteria for assessing the trustworthiness of naturalistic inquiries. *Educ. Commun. Technol.* **1981**, *29*, 75–91.
38. Woods, M.; Paulus, T.; Atkins, D.P.; Macklin, R. Advancing qualitative research using Qualitative Data Analysis Software (QDAS)? Reviewing potential versus practice in published studies using ATLAS.ti and NVivo, 1994–2013. *Soc. Sci. Comput. Rev.* **2016**, *34*, 597–617. [CrossRef]
39. Friese, S. *Qualitative Data Analysis with ATLAS.ti*, 2nd ed.; SAGE Publications Ltd.: London, UK, 2014; ISBN 978-1-44628-204-5.
40. Bringer, J.D.; Johnston, L.H.; Brackenridge, C.H. Using Computer-Assisted Qualitative Data Analysis Software to Develop a Grounded Theory Project. *Field Methods* **2006**, *18*, 245–266. [CrossRef]
41. Koeva, M.; Bennett, R.; Gerke, M.; Crommelinck, S.; Stöcker, C.; Crompvoets, J.; Ho, S.; Schwering, A.; Chipofya, M.; Schultz, C.; et al. Towards innovative geospatial tools for fit-for-purpose land rights mapping. *Int. Arch. Photogramm. Remote Sens. Spat. Inf. Sci. ISPRS Geospatial Week.* **2017**, *42*, 37–43. [CrossRef]
42. Hull, S.; Whittal, J. Human rights and land in Africa: Highlighting the need for democratic land governance. In *Human Rights Matters*; Corrigan, T., Ed.; IntechOpen: London, UK, 2021; ISBN 978-1-83968-874-4. Available online: https://www.intechopen.com/online-first/human-rights-and-land-in-africa-highlighting-the-need-for-democratic-land-governance (accessed on 3 May 2021).
43. Department of Rural Development and Land Reform (DRDLR). *The Communal Land Tenure Policy Framework*; DRDLR; Pretoria, South Africa, 2013. Available online: https://www.drdlr.gov.za/sites/Internet/ResourceCenter/DRDLR%20Document%20Centre/Communal_Land_Tenure_Policy_Framework.pdf (accessed on 3 May 2021).
44. South Africa. *White Paper on South African Land Policy*; Government Printer: Pretoria, South Africa, 1997; ISBN 0514308869.
45. Abubakari, Z.; Richter, C.; Zevenbergen, J. Evaluating some major assumptions in land registration: Insights from Ghana's context of land tenure and registration. *Land* **2020**, *9*, 281. [CrossRef]
46. Van Asperen, P.; Hendriks, B.; Zevenbergen, J. Scaling up pro-poor land recordation: Findings and consequences of three peri-urban cases from Sub-Saharan Africa. *Afr. J. Land Policy Geospat. Sci.* **2019**, *2*, 13–39.
47. Akrofi, E.O. Assessing Customary Land Administration Systems for Peri-Urban Land in Ghana. Ph.D. Thesis, University of Cape Town, Cape Town, South Africa, 2013.
48. South Africa. *Green Paper on Land Reform*; Government Printer: Pretoria, South Africa, 2011; pp. 3–13.
49. Bennett, R.M.; Alemie, B.K. Fit-for-purpose land administration: Lessons from urban and rural Ethiopia. *Surv. Rev.* **2016**, *48*, 11–20. [CrossRef]
50. Deininger, K.; Selod, H.; Burns, A. *The Land Governance Assessment Framework*; The World Bank: Washington, DC, USA, 2012; ISBN 9780821387580.
51. Zein, T. Fit-For-Purpose land administration: An implementation model for cadastre and land administration systems. In Proceedings of the Land and Poverty Conference 2016: Scaling up Responsible Land Governance, Washington, DC, USA, 14–18 March 2016.
52. Flores, C.C.; Tan, E.; Buntinx, I.; Crompvoets, J.; Stöcker, C.; Zevenbergen, J. Governance assessment of the UAVs implementation in Rwanda under the Fit-For-Purpose land administration approach. *Land Use Policy* **2020**, *99*, 104725. [CrossRef]
53. Association for Rural Advancement (AFRA). Traditional Courts Bill Uses Apartheid Laws to Subjugate Communities. *Cape Times*, 21 September 2012; 11.
54. Archary, L. Strengthening relative rights of people working the land. In Proceedings of the National Land Tenure Summit, Presentation in Johannesburg, Johannesburg, South Africa, 4–6 September 2014.
55. COGTA. *Traditional and Khoi-San Leadership Bill*; Minister of Cooperative Governance and Traditional Affairs: Cape Town, South Africa, 2015.
56. Hull, S.; Whittal, J. Towards a framework for assessing the impact of cadastral development on land rights-holders. In Proceedings of the FIG Working Week 2016 Recovery from Disaster, Christchurch, New Zealand, 2–6 May 2016; pp. 1–21.
57. Hall, R.; Mbilinyi, M.; Rusimbi, M.; Omeje, K.; Plaut, M.; Gonzales, D. Briefings. *Rev. Afr. Polit. Econ.* **2005**, *32*, 621–651. [CrossRef]
58. Hall, R. Land Reform Policy Discussion Document; South African History Online. 2012. Available online: https://www.sahistory.org.za/archive/land-reform-policy-discussion-document-june-2012 (accessed on 3 May 2021).
59. Balas, M.; Murta, J.; Matlava, L.; Marques, M.R.; Joaquim, S.P.; Carrilho, J.; Lemmen, C. A Fit-For-Purpose Land Cadastre in Mozambique. In Proceedings of the 2017 World Bank Conference on Land and Poverty, Washington, DC, USA, 20–24 March 2017; p. 26.

60. Moreri, K.; Fairbairn, D.; James, P. Issues in developing a fit for purpose system for incorporating VGI in land administration in Botswana. *Land Use Policy* **2018**, *77*, 402–411. [CrossRef]
61. Cousins, B.; Cousins, T.; Hornby, D.; Kingwill, R.; Royston, L.; Smit, W. *Will Formalising Property Rights Reduce Poverty in South Africa's 'Second Economy'? Questioning the Mythologies of Hernando de Soto*; PLAAS Policy Br No. 18; Programme for Land and Agrarian Studies (PLAAS), University of the Western Cape: Cape Town, South Africa; pp. 1–6.
62. Hull, S.A.; Kingwill, R.; Fokane, T. *An Introduction to Land Administration*; LandNNES: Cape Town, South Africa, 2020; Available online: https://www.researchgate.net/publication/344162012_An_Introduction_to_Land_Administration_Notes_to_accompany_the_video_series_produced_for_LandNNES (accessed on 3 May 2021). [CrossRef]
63. Lewis, L. *10 Vital Considerations for Software Implementation: How to Ensure Success*; Thomson Reuters: Toronto, ON, Canada, 2009; Available online: https://tax.thomsonreuters.com/site/wp-content/pdf/aumentum/10-Considerations-3S-Whitepaper.pdf (accessed on 3 May 2021).
64. Kotter, J.P. *Leading Change*; Harvard Business Review Press: Boston, MA, USA, 1996.
65. Thaler, R.H.; Sunstein, C.R. *Nudge: Improving Decisions about Health, Wealth, and Happiness*, 2nd ed.; Penguin Books: London, UK, 2009; ISBN 9780143115267.
66. Barnes, G.; Griffith-Charles, C. Assessing the formal land market and deformalization of property in St. Lucia. *Land Use Policy* **2007**, *24*, 494–501. [CrossRef]
67. Whittal, J. A New Conceptual Model for the Continuum of Land Rights. *S. Afr. J. Geomat.* **2014**, *3*, 13–32.
68. Lemmen, C. *The Social Tenure Domain Model: A Pro-Poor Land Tool*; International Federation of Surveyors (FIG): Copenhagen, Denmark, 2010; pp. 1–21.
69. McLaren, R. Crowdsourcing support of land administration—A partnership approach. International Federation of Surveyors. December 2011. Available online: https://www.fig.net/resources/monthly_articles/2011/december_2011/RobinMcLarenPaperCadastre20InnsbruckConference02092011v3.pdf (accessed on 3 May 2021).
70. Oldfield, J.; Van Oosterom, P.; Beetz, J.; Krijnen, T.F. Working with open BIM standards to source legal spaces for a 3D cadastre. *ISPRS Int. J. Geo Inf.* **2017**, *6*, 351. [CrossRef]
71. Williams-Wynn, C.; Africa, S. Applying the Fit-For-Purpose Land Administration Concept to South Africa: Will it Work? In Proceedings of the FIG Working Week 2020: Smart Surveyors for Land and Water Management, Amsterdam, The Netherlands, 10–14 May 2020; pp. 1–15.
72. Barry, M.; Roux, L. A change based framework for theory building in land tenure information systems. *Surv. Rev.* **2012**, *44*, 301–314. [CrossRef]
73. Burns, A.; Grant, C.; Nettle, K.; Brits, A.; Dalrymple, K. *Land Administration Reform: Indicators of Success, Future Challenges*; Land Equity International: Wollongong, Australia, 2006.
74. Williams-Wynn, C. Land rights: What people want. In Proceedings of the FIG Working Week 2015, Sofia, Bulgaria, 17–21 May 2015; pp. 1–16.
75. Clark, M.; Luwaya, N. *Communal Land Tenure 1994–2017: Commissioned Report for High Level Panel on the Assessment of Key Legislation and the Acceleration of Fundamental Change*; An initiative of the Parliament of South Africa; Land Accountability Research Centre (LARC): Cape Town, South Africa, 2017. Available online: https://www.parliament.gov.za/high-level-panel (accessed on 3 May 2021).
76. Ntsebeza, L. Chiefs and the ANC in South Africa: The reconstruction of tradition. In *Land, Power & Custom: Controversies Generated by South Africa's Communal Land Rights Act*; Claasens, A., Cousins, B., Eds.; UCT Press: Cape Town, South Africa, 2008; pp. 238–261.
77. De Satgé, R.; Cartwright, K.; Kingwill, R.; Royston, L. (Eds.) The Role of Land Tenure and Governance in Reproducing and Transforming Spatial Inequality: Commissioned Report for High Level Panel on the Assessment of Key Legislation and the Acceleration of Fundamental Change. 2017. Available online: https://www.parliament.gov.za/storage/app/media/Pages/2017/october/High_Level_Panel/Commissioned_Report_land/Commissioned_Report_on_Spatial_Inequality.pdf (accessed on 3 May 2021).
78. Adams, M.; Sibanda, S.; Turner, S. Land tenure reform and rural livelihood in Southern Africa. What is land tenure reform? *Nat. Resour. Perspect.* **1999**, *39*, 1–15.
79. CDE. *Land Reform in South Africa: Getting Back on Track*; The Centre for Development and Enterprise: Johannesburg, South Africa, 2008.
80. Cousins, B.; Hall, R. *Rights without Illusions: The potential and Limits of Rights-Based Approaches to Securing Land Tenure in Rural South Africa*; Working Paper 18 PLAAS; Programme for Land and Agrarian Studies (PLAAS), University of the Western Cape: Cape Town, South Africa, 2011.
81. Asiama, K.; Bennett, R.; Zevenbergen, J. Participatory land administration on customary lands: A practical VGI experiment in Nanton, Ghana. *ISPRS Int. J. Geo Inf.* **2017**, *6*, 186. [CrossRef]

land

MDPI

Article

Geospatial Tool and Geocloud Platform Innovations: A Fit-for-Purpose Land Administration Assessment

Mila Koeva [1,*], Mohammed Imaduddin Humayun [2], Christian Timm [2], Claudia Stöcker [1,3], Sophie Crommelinck [1], Malumbo Chipofya [1], Rohan Bennett [4,5] and Jaap Zevenbergen [1]

1 Faculty of Geo-Information Science and Earth Observation (ITC), University of Twente, 7514 AE Enschede, The Netherlands; claudia.stoecker@uni-muenster.de (C.S.); sophie.crommelinck@gmail.com (S.C.); m.c.chipofya@utwente.nl (M.C.); j.a.zevenbergen@utwente.nl (J.Z.)
2 Hansa Luftbild, Nevinghoff 20, 48147 Münster, Germany; humayumo@hansaluftbild.de (M.I.H.); timm_chr@web.de (C.T.)
3 Institute for Geoinformatics, Westfälische Wilhelms-Universität Münster, Heisenbergstrasse 2, 48149 Münster, Germany
4 Swinburne Business School, Swinburne University of Technology, Hawthorn, VIC 3122, Australia; rohanbennett@swin.edu.au
5 Kadaster Netherlands, The Dutch Cadastre, Land Registry, and National Mapping Agency, 7300 GH Apeldoorn, The Netherlands
* Correspondence: m.n.koeva@utwente.nl; Tel.: +31-53-487-4410

Abstract: The well-recognized and extensive task of mapping unrecorded land rights across sub-Saharan Africa demands innovative solutions. In response, the consortia of "its4land", a European Commission Horizon 2020 project, developed, adapted, and tested innovative geospatial tools including (1) software underpinned by the smart Sketch maps concept, called SmartSkeMa; (2) a workflow for applying unmanned aerial vehicles (UAV); and (3) a boundary delineator tool based on the UAV images. Additionally, the consortium developed (4) a platform called Publish and Share (PaS), enabling integration of all the outputs of tool sharing and publishing of land information through geocloud web services. The individual tools were developed, tested, and demonstrated based on requirements from Rwanda, Kenya, Ethiopia, and Zanzibar. The platform was further tested by key informants and experts in a workshop in Rwanda after the AfricaGIS conference in 2019. With the project concluding in 2020, this paper seeks to undertake an assessment of the tools and the PaS platform against the elements of fit-for-purpose land administration. The results show that while the tools can function and deliver outputs independently and reliably, PaS enables interoperability by allowing them to be combined and integrated into land administration workflows. This feature is useful for tailoring approaches for specific country contexts. In this regard, developers of technical approaches tackling land administration issues are further encouraged to include interoperability and the use of recognized standards in designs.

Keywords: fit-for-purpose; land tenure; land administration; UAV; feature extraction

Citation: Koeva, M.; Humayun, M.I.; Timm, C.; Stöcker, C.; Crommelinck, S.; Chipofya, M.; Bennett, R.; Zevenbergen, J. Geospatial Tool and Geocloud Platform Innovations: A Fit-for-Purpose Land Administration Assessment. *Land* **2021**, *10*, 557. https://doi.org/10.3390/land10060557

Academic Editors: Stig Enemark, Robin McLaren, Christiaan Lemmen and Francisco Manzano Agugliaro

Received: 14 April 2021
Accepted: 19 May 2021
Published: 26 May 2021

1. Introduction

Land tenure security supports orderly land dealings, poverty reduction, dispute minimization, and overall sustainable development. Delivering "tenure security for all" is one of the implied objectives of the sustainable development goals (SDGs)—target 1.4—set by the United Nations (UN) [1]. It can influence household income, food security, and equality [2,3]. However, millions of people-to-land relationships are still not recorded and remain unknown to governments [4].

This situation, unfortunately, is a fact in Sub-Saharan African countries [5]. With conventional cadastral surveying and mapping approaches, full completion would take decades, if not centuries [6]. Furthermore, there are many legal, social, and historical differences between the countries: it is hard to develop a unique scalable approach.

Across the 2000s and 2010s, significant effort was undertaken to improve the situation. In support of the development of an efficient land administration system, the Land Administration Domain Model (LADM), aimed at establishing a common ontology of land administration concepts, was proposed [7,8]. In addition, the Fit-For-Purpose (FFP) approach was developed, encouraging participatory, innovative, and scalable methods that match the needs [9] and challenges of a specific country context [4]. One of the key principles of FFP was to use "general" rather than "fixed" boundaries, using participatory mapping based on image interpretation [10]. Good examples of implementation of the FFP for land administration are found in Ethiopia, Rwanda, Namibia, and Indonesia, among others [11,12]. FFP includes principles that cover spatial, legal, and institutional country aspects [13]. The seven main elements based on the FFP concept [14] that the land administration system should take into account are:

- **Flexible** in the spatial data capture process to provide information about the different uses and occupations of the land;
- **Inclusive** in the extension to cover all types of tenure and all types of land;
- **Participatory** in the manner to capture and use data, ensuring community support;
- **Affordable** operation for the government and society to use it;
- **Reliable** regarding the information, it should be authoritative and updated;
- **Attainable** to create a system within a short time frame and with the available resources;
- **Upgradable** regarding improvement over time to respond to social and legal needs as well as economic opportunities.

In response to the need for innovative solutions based on FFP principles, the "its4land" H2020 project was initiated [15,16]. The project aimed to develop technologies that consider the needs and readiness of the users that are implementable and scalable. They include (1) smart Sketch maps—also known as SmartSkeMa; (2) a UAV workflow and ortho generator; (3) a boundary delineator tool based on the UAV images; and (4) the Publish and Share (PaS) platform providing geocloud services.

In the following section, the development of the tools is briefly described, referring to the publications over the years [5,16–19]. More emphasis is put on the description of the PaS platform, which was the last to be developed in the design sequence and has therefore undergone less formal assessment. Subsequently, the methods for assessing the tools and the platform against the seven fit-for-purpose elements are described. In Sections 4–6, from an academic perspective, a critical results assessment, feedback, and ways forward, in terms of future research and scaling activities, are provided. From a practical standpoint, it is suggested that the work at hand can help in improving the geospatial tool and geocloud platform solutions, and that focus should be placed on ensuring interoperable tools to support the successful upscaling to the its4land toolbox (suite of tools) and Fit-For-Purpose Land Administration (FFPLA) more generally.

2. Background: "its4land" Tools and Platform

2.1. SmartSkeMa

SmartSkeMa was developed to document land rights information, taking into account the local knowledge of the communities. SmartSkeMa tool includes: (1) a developed domain model that uses the land tenure characteristics as described by communities that live with them; (2) a spatial model based on hand-drawn sketchmaps; (3) a method for recognizing and georeferencing the sketchmaps and embedding them into the land information system (spatial datasets) [20]. The tool can be used in two ways (1) by overlaying the vector data on top of the Sketch maps; or (2) by aligning the sketches with an orthophoto image [21]. In addition, the nonspatial information can be processed via local domain model (LDM) which is connected to the Land Administration Domain Model (LADM). Spatial information is extracted via the object detection techniques. An example of the workflow and its implementation can be found in [15,16].

2.2. UAVs

The UAVs solution involved developing a workflow to incorporate the tasks of choice of the vehicle, investigation of the regulations and policies, pilot training and certification, flight permissions, data capture, processing, and quality assessment. During the project, numerous flights and missions were completed and assessed depending on the country aim. In Kenya, where pastoralism is a dominant land use and spatial information is not mapped numerous flights have been done and scientific comparison of the results was performed. More details on the data acquisition in September 2018 Kajiado and March 2018 Mailua with RGB sensors using DT18, a fixed-wing UAV and DJI Phantom 4 can be found in [16]. In Rwanda, the aim was cadastral map updating using UAV data, since the country supported by international donors, mapped over 12 M parcels in 2007/2008. Therefore, detailed explanations on the application of UAV data acquisition in Rwanda 2017 and 2018 can be found in [22]. In addition, in collaboration with the World Bank a joint survey and data quality assessment was performed in Zanzibar in July 2019. Moreover, the its4land team developed a tool called "OthoGenerator" which is based on the open-source image processing software called "OpenDroneMap" [23]. The tool integrates the processes of generating point clouds, digital surface models, and orthomosaics from images. This tool is integrated into the PaS platform, where the user must specify the required resolution of the output, overlap, context of the scene, and the mode for georeferencing.

2.3. Boundary Delineator

The "boundary delineator," or also called in some of the publication "automatic feature extraction" tool, was developed as an independent QGIS plugin. It incorporates a boundary extraction machine learning method based on RGB satellite, aerial, or UAV images. The methodology includes image segmentation, boundary classification, and interactive delineation [24–27]. The source code is available for download [25]. Numerous examples on the application of the tool have been shared for Kenya and Rwanda in [28–30]. A detailed explanation of the sequence of the work, with the individual tools integrated into the platform, for a case study in Ethiopia, is explained in [15].

2.4. PaS Geocloud Platform

The most recently developed tool in the its4land suite was the "PaS" platform. It aims to support workflows in projects related to land administration [31]. It was premised on the idea that while land administration systems worldwide may differ in detail, most of them have in common some generalizable requirements, functions, and outputs [12,32]. In this regard, a typical definition of land administration refers to "the processes of determining, recording and disseminating information about the ownership, value, and use of land when implementing land management policies" [31]. Every land administration system requires a generalized spatially related reference system as a basis. In contemporary systems, this spatial reference is provided and maintained by a GIS database. PaS focuses on providing the spatial reference for a land administration system.

The PaS platform offers a core set of features that can be customized and extended. The functionalities of PaS are primarily targeted at independent software vendors (ISV) or integrators to create services or applications for end-users. PaS provides a set of high-level geocloud-based services for developers of land administration systems to use or integrate spatial references for tenure registration. Using these services allows a vendor to concentrate on functionalities required by their customer instead of re-implementing existing solutions for common problems. These services are based on the concepts introduced by LADM. To improve or assist land administration workflows and tasks, four different usage models for the platform were developed:

- **Application:** A self-contained application for an end-user which uses the Application **Programming Interface (API)** provided by the platform.
- **Integration:** PaS functionality is used to host a Land Administration workflow, or parts of it.

- **Tools:** Applications with self-contained functionality, which use the API for integration with other tools or applications.
- **Platform extension:** Adding new core functionalities to the platform.

The platform is capable of hosting and integrating tools and data to facilitate land tenure recording services and applications. This allows the complementary tools and methods, developed in the its4land project—which use images captured by UAVs [33–36], qualitative data processing using Sketch maps [21], and Boundary Delineation [18,37–40]—to be integrated coherently into existing or new land administration workflows. For seamless integration with existing systems, PaS is implemented on standards such as Representational State Transfer (REST) [41], LADM, and Open Geospatial Consortium (OGC) Web Services (OWS). It is composed of technical components shown in Figure 1.

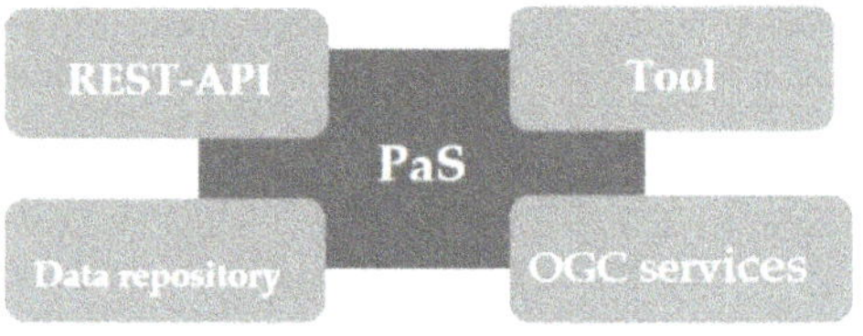

Figure 1. Components of the PaS platform.

In more detail, these PaS elements are:

- A public REST over HTTP API that allows tools and applications to interact with the PaS platform. The choice of an HTTP API allows applications to be developed in a wide variety of programming languages;
- A tool runtime environment for Smart Sketchmaps, UAV image processing, boundary delineation, and other tools. Since computing resource usage of some of the tools can be quite demanding, they are started on-demand via the API;
- A data repository for alphanumeric, geo, binary, and image data. Data are added or manipulated via the API;
- OGC services for data dissemination. These allow usage of GIS such as QGIS to access maps and query data via services such as WMS/WFS.

The implementation of the PaS platform follows a toolbox approach (integration of tools) and provides a framework consisting of standard APIs and services used by all other its4land tools. From this toolbox, users can select those its4land tools fitting their tasks best. All components developed for PaS are open-source (https://github.com/its4land/publish-and-share, accessed on 24 May 2021). The components can be hosted in a cloud environment such as Amazon Web Services, Microsoft Azure, or a private cloud environment.

The development of the PaS platform was one of the key exploitable results of the its4land project. The utility of the platform in implementing abstract concepts from LADM in a programmatic manner enables building usable land administration systems on top of it. Additionally, the platform also demonstrated how innovative tools serving a niche purpose could be tied together, enabling the use of innovative methods where traditional land tenure recording methods fall short or require immense effort.

The geocloud services provided by PaS are based on the conceptual model introduced by LADM. These concepts are abstract, and their concrete meaning depends on country-specific interpretation. By providing a means to define and implement concepts from LADM, PaS can provide the necessary information to a land administration system that is needed for legal registration. A developer familiar with LADM will recognize concepts such as SpatialUnits or AdminSources and can map them to the structures and localized meanings needed by a specific land administration system.

The current version of PaS focuses on the capture of data for subsequent registration in a land administration system. Therefore, only a subset of LADM is implemented. In practice, this means that PaS can handle all those LADM concepts which deal with the evidence of land rights and thus enables a legally secure registration of rights by a land administration system. The following concepts of LADM are supported in PaS as shown in Table 1 below:

Table 1. Concepts from LADM supported by PaS.

LADM Concept	Usage in PaS
LA_SpatialUnit	LA_SpatialUnit is the spatial reference in an LADM-based Land Administration System (LAS). It is used in PaS because it is the spatial reference for any kind of tenure registration. The interpretation of LA_SpatialUnit provided by PaS is subject of the context, including the legal framework in which PaS is used. Example 1: A SpatialUnit is a parcel that was created by the *Boundary Delineator* tool. Example 2: The *SmartSkeMa* tool produces spatially demarcated interest in land, which is treated as LA_SpatialUnit.
LA_Level	LA_Level can be used to group LA_SpatialUnits with a geometric or thematic coherence. Example: Distinguish SpatialUnits, which represent different types of community land, such as seasonal pastures.
LA_BoundaryFaceString	LA_BoundaryFaceString forms the outside of LA_SpatialUnit in a 2D geometric representation. It represents a general or fixed boundary. According to the principles of Fit-For-Purpose it is treated in PaS as a general boundary. A land administration system can use boundary face strings to create SpatialUnits as needed in the specific implementation. Example: The *Boundary Delineator* tool produces general boundaries, which are managed as BoundaryFaceStrings in PaS.
LA_SpatialSource	LA_SpatialSource documents the evidence of a spatial unit or a boundary face string. It can either be an input or output of a step in the workflow. A LA_SpatialSource can be any kind of document, such as orthomosaics, images, surveying sketches PaS does not restrict it to a specific type. How a SpatialSource is interpreted forms part of the use of PaS, in a specific project or implementation. Example 1: The SketchMaps used in *SmartSkeMa* are treated as SpatialSources Example 2: Orthoimages produced by the *UAV Ortho Generator* are treated as LA_SpatialSource as well. They are used by the *Boundary Delineator* a starting point to delineate boundaries.
LA_AdministrativeSource	LA_AdminSource documents the evidence of an interest in land. This includes rights, restrictions, responsibilities, and the involved parties. The documents can be any kind of document and files—from a scanned contract to a recorded narrative description of alternative concepts of land rights. According to Fit-for-Purpose principles, this addresses inclusive and participatory dimensions since it allows a formalized documentation of evidence of interests in land in a wide range of ways. The interpretation of a LA_AdminSource Document is part of the use of PaS, in a specific project. This depends highly on the legal framework of the country where the project is conducted. The legally valid registration itself is done in a LAS. Example: *SmartSkeMa* captures information about land rights and land usage based on community-related ontologies. The LA_AdminSources information is stored in a structured form in PaS. Furthermore, it is linked LA_SpatialUnits which are created by *SmartSkeMa* as well.

The PaS platform offers developers and system integrators a Public API on top of which custom applications can be built or platform features integrated into existing land administration systems. The API provides an implementation for LADM concepts such as spatial sources and spatial units, among others, such that they can be used in a programmatic fashion in an information system. Linking different concepts, as required by the implemented workflow, can be performed by the API. A parcel can be linked to an orthoimage or a Sketch as their SpatialSource. It can also be linked to a file that documents the ownership of that parcel. During the legal registration in a LAS, the user of the LAS can access this information and relationships to prove the evidence easily.

The API itself is a REST API [41] and is served over HTTP(S). Implementing a REST API makes integration on any platform or programming language seamless. Figure 2 shows the main user interface of the online documentation of the Public API (Public API is used here to differentiate from an internal private API used to extend the platform itself). The document acts as a reference for developers to interact with the platform to perform tasks such as creating instances of concepts and adding and querying information to the instances. In addition to this, the platform website (https://platform.its4land.com, accessed on 24 May 2021) offers additional documentation and guides for developers, integrators and other users to familiarize themselves with its concepts and usage.

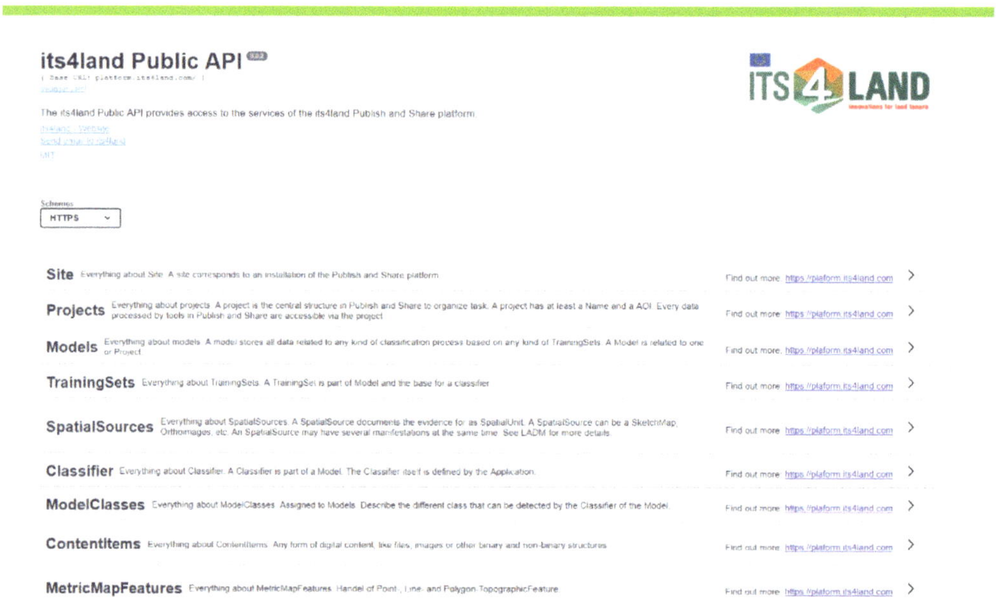

Figure 2. The main user interface of the online documentation of the Public API.

The Public API lends itself to multiple usage models discussed previously. These include an end-user who interacts using only a GUI client, a developer wishing to extend their legacy land administration system with features of PaS, or wanting to extend and customize the platform itself. The API is designed to achieve these workflows by following the OpenAPI 2.0 (https://spec.openapis.org/oas/v2.0.html, accessed on 24 May 2021) standard for specification, which allows the generation of interactive documentation that includes examples.

The platform's Data Dissemination Interface (DDI) provides an alternative means of accessing spatial data stored in PaS. Compared to the Public API, the DDI allows limited read-only access to data via OGC services such as WMS/WFS. The DDI is useful to access LADM spatial concepts such as spatial units or boundary face strings. DDI also provides high-performance access to orthoimages and other raster data stored in PaS via WMS services. The main advantage of the DDI is the ease of use. Nearly every popular GIS Tool can use OGC services out of the box. This allows lightweight data access by workflows developed by the application or integration usage model. The Public API requires more effort on the client-side to handle requests and responses. However, on the other hand, the Public API allows better control, such as advanced filtering capabilities compared to the WFS interface. If data should be manipulated, the use of the Public API is required. Table 2 provides an overview of the available spatial data classes in PaS and their access services in DDI.

Table 2. Spatial data classes in PaS (X indicates support).

Data	WMS	WFS	Remarks
Spatial Unit	X	X	Only 2D polygon profile
Boundary Face String	X	X	
Metric Map Feature	X	X	
Orthomosaic	X		
Other Raster Data	X		
Other Vector Data	X	X	

2.4.1. Tool-Based Workflow

Tools and services built using features provided by PaS enable innovative workflows, which can be integrated into a land administration system. As depicted in Figure 3, the components provided by the platform support a workflow from capture to dissemination of data, including the intermediate steps, the output of which is ready for consumption by a land administration system.

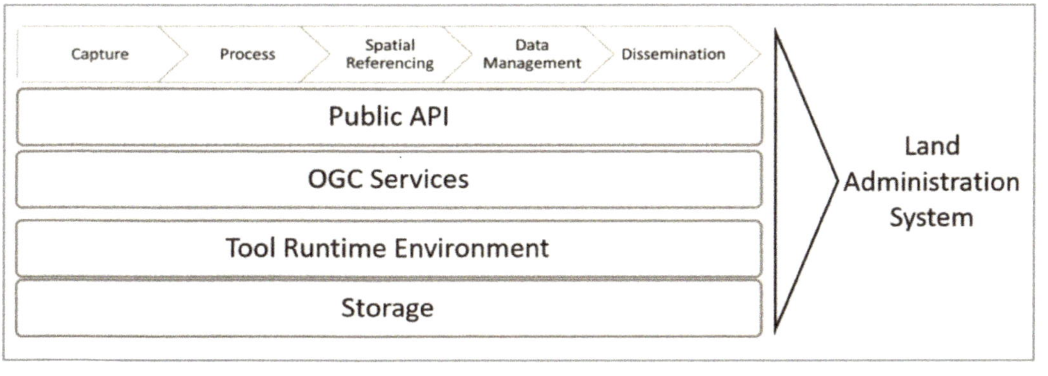

Figure 3. Abstract workflow of PaS.

To add a tool, a developer creates an application in a programming language of their choice and uses the REST API to communicate with PaS. Tools may be made available and run on the platform itself. In this case, the tool is packaged using Docker (https://www.docker.com/, accessed on 24 May 2021), added to the PaS server and registered on the platform by the administrator. The its4land tools—UAV Orthogenerator and SmartSkeMa were made available this way. The alternative is for the tool to run outside the PaS environment, which was the model used by the Boundary Delineator tool, since it was a QGIS plugin. In both cases, the PaS API provides various endpoints to read and write data. Tools available and registered in the PaS environment may be started/stopped and their execution status queried via the tools endpoint. A PaS-based workflow consists of executing tools at various stages, often using the output of one tool as the input to another.

The its4land project demonstrated how innovative tools could be orchestrated to provide a coherent workflow to capture and process spatial information from non-traditional sources. This workflow used all the developed tools—UAV Ortho Generator, SmartSkeMa, and the Boundary Delineator. Likewise, a client web application used the public platform API to provide a user interface to carry out different steps in the workflow.

A typical workflow consists of the following steps, all of which are performed via the API:

1. Create a project context by specifying the project metadata—name, description bounding box, tag, external links.
2. Add resources to the project (Figure 4). Resources can be of the type Spatial Source or DDI Layers (served via OGC). Additionally, tool-specific resources such as validation sets required by the Boundary Delineator tool can also be added. The tool developers specify what kind of resources are supported and what metadata are required for execution. Figure 4 shows this being accomplished via the client web application.
3. Once resources have been added, tools can be started within the context of the project by providing the required parameters. The tools can access the requisite resources within the project context, process them and add the output back to the project. It is not a requirement for all tools to run on the cloud platform. For instance, the Boundary Delineator tool was developed as a QGIS plugin and runs locally on the user's workstation. The input and output data for such tools, however, are optionally stored on the platform when used in conjunction with complementary tools.
4. Different its4land tools produce different outputs. For e.g., SmartSkeMa produces spatial units, while the Boundary Delineator produces BoundaryFaceString—both concepts from LADM. Other tools such as the UAV Ortho Generator produces an orthomosaic (DDI Layers), which are used as inputs for the next tool.

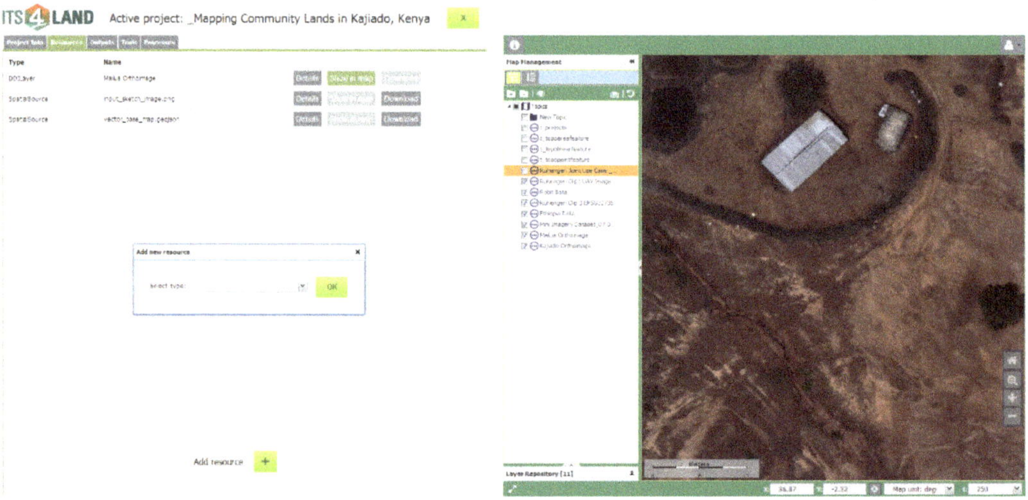

Figure 4. Client Web Application—Adding Resources.

The UAV Ortho Generator tool is based on and allows the creation of orthomosaics from imagery captured by unmanned aerial vehicles (UAVs). ODM is open source and freely available. However, in most projects, the creation of orthomosaics involves extensive image processing. Depending on the workload, the limited processing and storage capacities of PCs and mobile devices can prove insufficient. Offloading such workload to the cloud, where processing can be scaled up on-demand, is ideal. The UAV Ortho Generator of the PaS platform provides this solution and allows the processing of large datasets even though local processing capacities are limited. Upon starting the tool, the status of the execution can be queried via the API. Once the orthomosaic is generated, it is added as a resource within the project context. An internally carried out step here involves optimizing the resulting orthomosaic in the Cloud Optimized GeoTIFF (COG) format for cloud-based access (https://www.cogeo.org/, accessed on 24 May 2021). The next step in the workflow is the use of the Boundary Delineator tool via QGIS.

The Boundary Delineator fetches data from the project context and assists the user in interactively delineating visible boundaries from imagery. Once this process is complete, the georeferenced boundaries are uploaded as BoundaryFaceString in the project context.

The SmartSkeMa tool complements the Boundary Delineator and can collect non-visible boundary information from sources such as Sketch maps. In addition to this, it can capture additional information such as LADM rights, restrictions, and responsibilities. Like all tools, SmartSkeMa can exchange data with PaS within the context of a project. Instances of SmartSkeMa may also be launched from PaS, allowing existing data to be accessed by multiple users. After capturing and annotating data, SmartSkeMa produces SpatialUnits and AdminSource documents. Such information for community-owned lands in Kenya being displayed by the web client is shown in Figure 5.

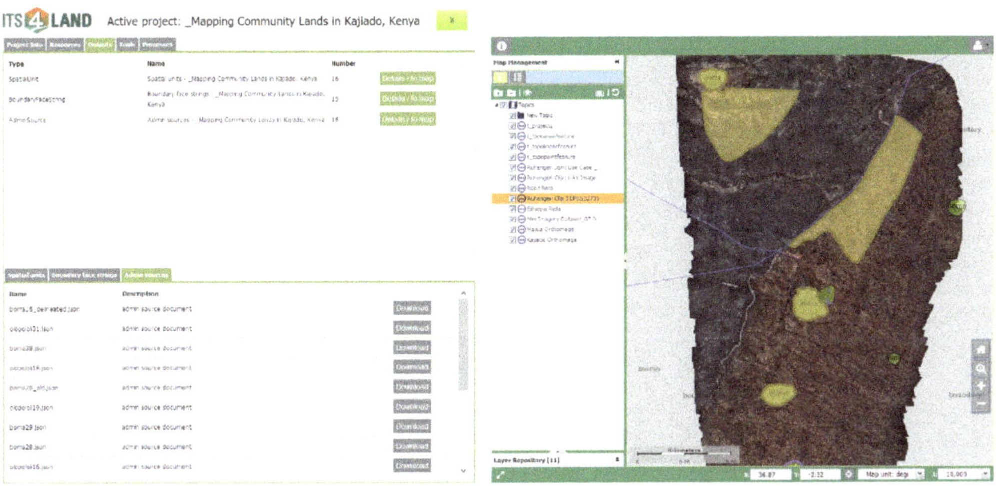

Figure 5. Result of Processing by the SmartSkeMa tool.

Output results after processing by the tools are accessible via the platform API or DDI. These use popular standards such as OGC services for spatial information or well-supported formats for information interchange on the web such as JSON/GeoJSON, resulting in ease of access by other applications and services.

PaS focuses on workflows that are related to spatial processes in an LAS. PaS is not so beneficial in supporting transfer processes (such as buying or inheritance of properties): such processes must follow a well-defined legal framework and an LAS is optimized in handling those processes for a specific legal context. Currently, this is a shortcoming of PaS, in its goal to provide a flexible platform based on generic LADM concepts.

Referring to the core land administration processes [42], the typical scenarios where PaS is beneficial are:

1. Formally titling land—Especially in developing countries, where no LAS exists, the formal titling of land is one of the first steps to implement an LAS. Formally titling of land requires tasks such as:
    - Capturing base data, such as aerial photographs, as a spatial source and a basis for later parceling
    - Capturing documents and evidence for rights, restrictions or responsibilities as admin sources
    - Adjudication process with spatial unit and admin source as result, for formal registration

2. PaS can host the necessary workflows for these tasks and provide the results to the LAS for formal registration.
3. Forming new interests in the cadaster (such as subdivision or merging)—To apply changes to the cadaster, information must be gathered in advance, which documents the evidence of the changes. These are documents that describe the target situation in the cadaster, but could be also the target situation itself, which is registered afterwards by the LAS.
4. Determining boundaries—Determining new boundaries in a cadaster requires a complex workflow and several documents to confirm the correctness of the new boundaries (like survey sketches or orthophotos). This workflow can be implemented on PaS and the results can be provided as a boundary face string to the LAS, which creates parcels according to regulatory standards in the LAS.

2.4.2. Example Workflow

To better illustrate how PaS facilitates Land Administration workflows, Figure 6 offers a glimpse of how it was applied to a proof-of-concept scenario in Ruhengeri, Rwanda. In this scenario, an orthophoto of the study area was created using data collected from UAVs. The orthophoto was used to create precise geometries from visible boundaries using the Boundary Delineator. SmartSkeMa made use of hand-drawn maps collected via participatory mapping. This resulted in less precise geometries, but it allowed recording of information on the right restrictions and responsibilities (RRR)S formed the glue, allowing both tools and users access to the data, which was its design purpose.

(a) Aerial image obtained from the UAV Orthogenerator

(b) Viewing output of Boundary Delineator in QGIS

(c) Using SmartSkeMa to add RRR information

(d) Viewing cumulative output in the PaS GUI client

Figure 6. Example workflow using PaS and tool in Ruhengeri, Rwanda.

Figure 6a, depicts a georeferenced aerial orthophoto image of a study area in an urban environment. To generate this orthophoto image, the raw aerial photograph dataset was uploaded to PaS and the UAV Orthogenerator tool was started by setting a few required parameters. Once the tool completed the image processing procedure, the resulting orthophoto image was made available via WMS or directly in a GeoTIFF raster format. The latter was used by the Boundary Delineator in QGIS (Figure 6b) to demarcate visible boundaries. The resulting BoundaryFaceStrings are saved in PaS. SmartSkeMa (Figure 6c) provided a means to record non-visible RRR information and save the resulting spatial units back to PaS. Storage and exchange of data between the tools was made possible via the PaS API. A demo GUI client (Figure 6d) enabled access to the data uploaded to the platform via a modern web browser, for visualization and editing. Similar scenarios for chosen areas in Kenya and Ethiopia also demonstrated this integrated approach.

In summary, the its4land project developed a suite of independent, innovative tools for undertaking specific land administration activities, but, via PaS, also provided an open platform to enable the integration of outputs from those independent tools, with a view to enabling mixing and matching of the tools, for a given project context. Until now, while the individual geospatial tool innovations have been formally evaluated, they have not been assessed together against FFP elements or criteria. Moreover, evaluation of PaS as a standalone platform, in terms of usability and functionality, let alone its ability to enable integration of outputs from other tools, and FFP for that matter, has yet to be undertaken. Therefore, there exists good motivation to undertake such an evaluation. The following section explains the methods for enabling this evaluation.

3. Methods

In this section, the methods for assessment of the individual tools and the integrative platform (PaS), with respect to FFP elements, are explained. In this regard, the overarching research paradigm can be considered to be pragmatist, i.e., commencing from the starting point that there many means to respond to a particular set of real-world requirements and basing the solution assessment on the integration of concepts and methods from a range of applied sciences. In this case, those applied sciences are the geospatial sciences, information systems, and computer sciences. More specifically, the methods in this paper gain some inspiration from action-research and reflexivity-based methodologies [43], although they cannot be said to be complete or formal applications of those approaches.

First, the individual data collection and analysis tools—SmartSkeMa, UAVs, Boundary Delineator—were tested during their development and demonstrated across Eastern Africa. Fieldwork, workshops, semi-structured interviews, and focus group discussions were completed for all tools in Kenya in 2017 and 2018, where 58 land administration stakeholders from local government institutions, non-governmental organizations, private companies, and national institutions participated [22]. Two workshops were organized; one at the local government in Kajiado and one at the Regional Centre for Mapping of Resources for Development (RCMRD) in Nairobi. Both followed the same structure: presentation by the facilitators, splitting the participants into groups for detailed activities, demonstrations, and discussions on the individual tools. Independently, SmartSkeMa testing was done in Ethiopia in 2017 with 20 local stakeholders. To collect data for analyzing UAV workflows, three different UAVs (Inspire 2 DJI, e FireFLY6, and DT18) were tested and demonstrated in Rwanda in 2017 [19]. At the same location, the previously collected UAV-based orthomap was used to test a participatory mapping approach along with presentations about the results [44]. In addition, UAV data collection and workshops were carried out in Zanzibar in 2019. Overall, 33 people from local and national government participated during those workshops (16 in Zanzibar and 17 in Rwanda) [45].

Second, the overarching PaS platform, the latest development of its4land was assessed. It was demonstrated during Rwanda AfricaGIS in December 2019. Furthermore, a workshop with 4 experts from ESRI Rwanda was organized to test the functionalities of the platform and to obtain their feedback.

Third, and for the specific aims of this paper, i.e., the assessment of the individual tools and the platform against the FFP elements reported, in addition to the information obtained during the above-mentioned events, feedback from 15 key informants and experts was gathered via an online survey. The background of the selected experts is shown in Appendix A. From them, 3 experts were involved in the development and testing of the tools, and 4 were directly involved in the testing and demonstration of PaS during Africa GIS 2019 in Rwanda. The remaining 8 experts were well familiar with the functionalities of the tools and the platform. The selection of the participants was intentionally made having in mind their familiarity with these particular tools and platform since in this way, more critical views and feedback is obtained. Given the close (and sometimes direct) links between the participants in this evaluation work and the tool/platform development, the potential for bias needs to be and is hereby acknowledged.

The experts were asked to fill a survey rating the platform and the tools with respect to the seven FFP elements (Figure 7). Provision for qualitative feedback was also provided.

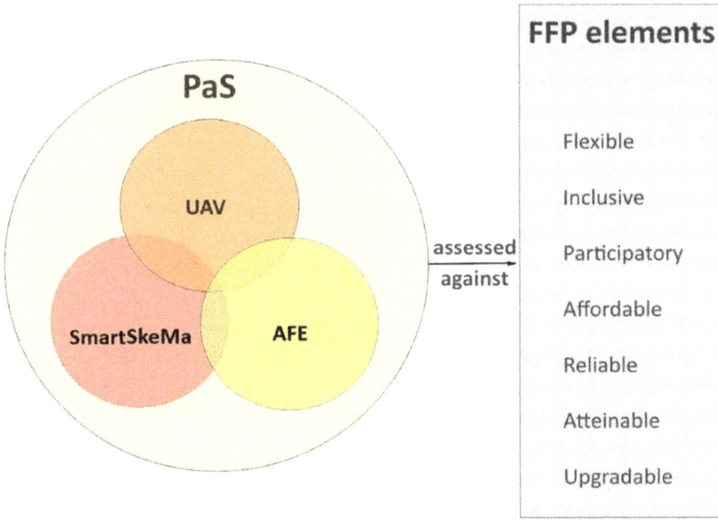

Figure 7. Assessment method.

A clarification on how to understand these elements for the specific context of the tools was initially provided to the experts (Annex2). Based on the results from the survey, answers to the following questions could therefore be ascertained, each in the context of FFP:

- What value does PaS bring (in terms of FFP)?
- How does the PaS improve the performance of the other its4land tools (with respect to FFP)?
- When do the individual tools perform better on their own?
- What can be improved in the future (in terms of FFP support)?

After obtaining the results, the authors that were involved in the development of the tools and the platform did the initial result screening and analysis. The answers which differ significantly between the participants have been analyzed, and an explanation of the reasoning is provided. For this task, the level of familiarity of the participants with the concrete tool was taken into account. Therefore, more trust was ascribed the values given by the expert than voted being more familiar. Some participants wrote extensive comments to explain their answers which helped us with the assessment. From one side, the low number of participants can be seen as a limitation of this research. However, if there were more participants without a deep understanding of the tools, the obtained feedback would have been global and not so critical. Therefore, even with the limited number of participants, the answers from the survey provide enough information for the analysis.

4. Results

In this section, the results from the assessment of the platform and the individual tools with respect to FFP elements are presented. The boxplot method for graphical representation grouping the numerical data through their quartiles is selected. The box shows the interquartile range (IQR) containing 50% of the scores between the first (Q1) and the third quartile (Q3). The middle line of the box represents the median and the mean value is denoted with a cross (x). The whiskers display the values 1.5 times below Q1 and 1.5 times above Q3, respectively. Scores falling outside the whiskers are considered outliers. The longer the box the more scattered the data. The shorter the box the less scattered is the data.

4.1. Evaluation Results of the Individual Geospatial Tools (SmartSkeMa, UAVs, Boundary Delineator)

The results of the tools are combined in one chart to compare them easier and to split them from the one for PaS. Out of the 15 participants in the survey, seven persons reported being somewhat familiar with SmartSkeMa, four mentioned that they were very familiar, three respondents were ambivalent, while one person was not at all familiar with it (Figure 8). The one unfamiliar respondent also chose the not applicable option for all FFP elements and was not included in the analysis.

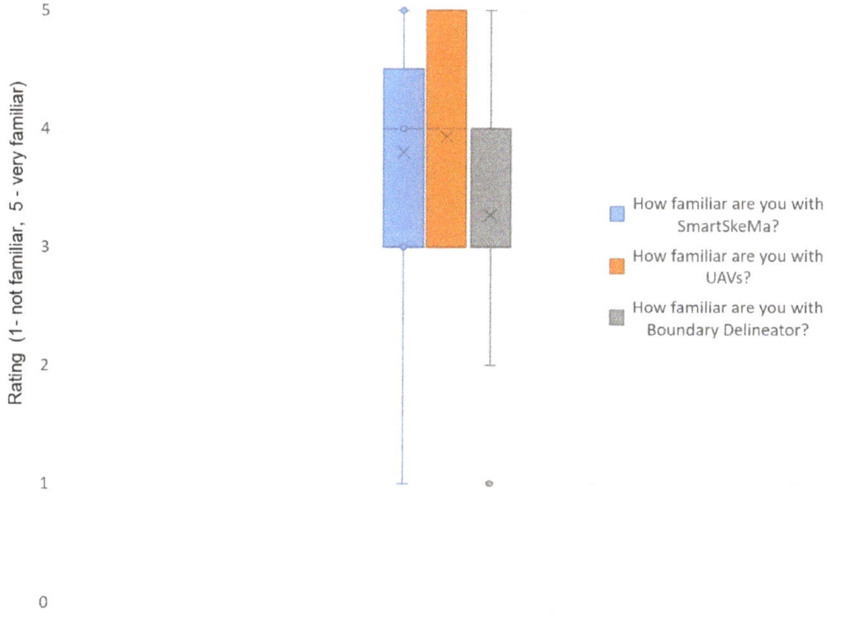

Figure 8. Familiarity of the participants with the tools and the platform.

SmartSkeMa was rated high on the flexibility, inclusivity, participatory nature, and upgradability elements of the FFP (see Figure 9). These ratings agree with the design principles behind SmartSkeMa's development. The fact that interactions with SmartSkeMa are based on physical interaction via sketching makes it a very participatory tool. This explains the high rating on the participatory element (8 out of 14 participants with score of 5 and 2 with score of 4 and 5 with score of 3). Not surprisingly, most participants could not decide whether the tool met the FFP reliability criteria. One participant noted that SmartSkeMa "SmartSkeMa is not suited for authoritative data, but for [up-to-date] data [it is] very usable." Another participant observed that the tool might not meet "legal precision and accuracy requirements". One can interpret these data as indicating that the tool has some reliability with respect to the production of up-to-date data but is not reliable for the production of authoritative data.

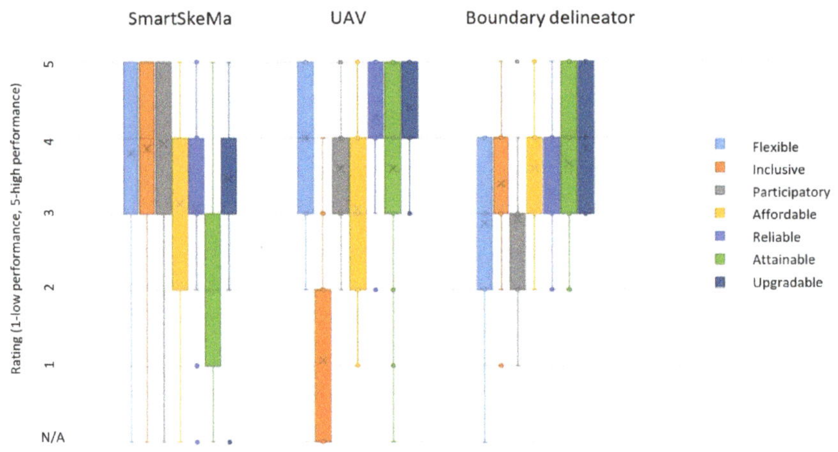

Figure 9. Assessment of the tools against the FFP elements.

The elements of affordability and attainability received the worst ratings (7 of the participants rated reliability with a score of 3, meaning they did not know whether SmartSkeMa meets the reliability criteria for FFP). Affordability was the worst-rated element, with 6 participants responding that it did not meet the FFP criteria for affordability. The reason for this rating is not clear at present, but one possible explanation is that participants considered it a human resource-intensive approach.

In contrast to the other tools, all respondents showed a medium to high familiarity with UAVs (see Figure 9) as a data collection technology for land administration. This is not surprising given that UAVs are increasingly applied to various surveying and mapping tasks during recent years. As indicated in Figure 9, UAVs score notably high in terms of flexibility, reliability, attainability, and upgradability. The availability of different UAV platforms and sensor combinations, and highly customizable flight plans, and various means of ground-truthing, cater to a high level of flexibility. Furthermore, it appears that the high resolution of generated orthophotos is perceived as reliable and authoritative data, which is easy to upgrade with an increased temporal or spatial resolution. The participatory and affordable elements were rated with medium performance. Particularly affordability stands out as the primary concern shared by half of the respondents. It reflects the high costs typically involved in UAV data collection processes, including the purchase of equipment, ground-truthing measurements, import fees, charges for flight permissions, etc. Out of the eight elements, "inclusive" was the only one rated not applicable for UAV workflows: UAV data can only be used to create base-maps and does not provide further information on existing people-to-land relationships.

For the boundary delineation tool, 15 participants provided input to our study. As shown in Figure 9, most participants (6) have medium experience with the tool. Likewise, Figure 9 shows how the participants rated the degree to which the tool fulfills each of the FPP criteria. The tool scores high (5) in terms of being attainable and upgradable. This can be explained by the fact that the tool is open source and that it can be further developed. The tool scores low (2) to medium (3) in terms of being flexible, inclusive, participatory, affordable, and reliable. This can be explained by the fact that the tool can be used to delineate visible boundaries only. In addition to its dependency on visible boundaries, the tool requires trained GIS staff to be implemented as well as image data as an input to locate the visible boundaries.

4.2. Evaluation Results of PaS

Most experts who participated in the assessment rated themselves as having a high degree of familiarity (7 nos.), with PaS as indicated in Figure 10. When it came to judging the platform with respect to the FFP elements (Figure 10), the element of upgradability was rated the highest, followed by flexibility and attainability. This can be attributed to PaS's ability to support different tools and features via the API, use of OGC services and LADM. Experts judged the proof-of-concept connectivity between the tools demonstrated here at the R&D level as having the potential for upgrading/extending to respond to future needs (see Upgradable in Appendix B-1). This is where it could be inferred that PaS's ability enables it to connect different tools to make possible more complex land administration workflows. Complexity here refers to combining usage of tools as opposed to using them singularly, for example, the combination of UAVs and use of SmartSkeMa, or UAVs combined with automatic boundary detection, or perhaps some other combination. The inclusivity and affordability elements garnered low-medium ratings overall. This indicates not all participants are convinced PaS can store different types of land tenure in an affordable manner.

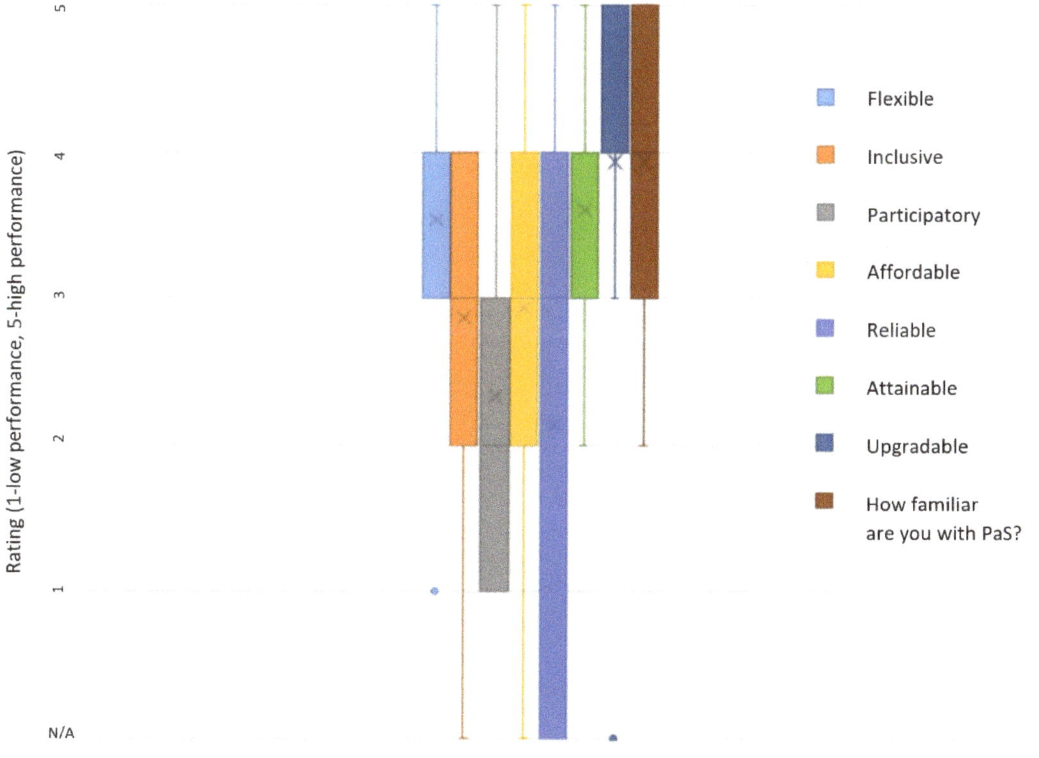

Figure 10. Assessment of PaS against the FFP elements and familiarity with the platform.

PaS is not meant for data capture and cannot offer guarantees regarding the authoritativeness and reliability of data, which are sourced elsewhere. The survey made this point clear, and this could explain the reason for the low ranking of reliability. It is also possible that the infrastructure and expertise needed to operate the platform has affected the reliability perception.

The participatory element fared poorly. Given the nature of PaS this is difficult to assess since the participatory element is more of a procedural dimension in FFP and refers to how different groups participate and contribute to the registration of land and rights. The development and use of the platform is not participatory, in a manner that someone interested in land can use it directly. Rather, it can be used to create tools and applications which allow for participation.

As per the comments of one of the participants, the nature of PaS as a development platform for LADM-based applications means every FFP goal can be achieved by an application that uses it. Some criticism was also leveled at the platform requiring high technical expertise to operate, its dependency on a cloud service provider, and the fact that testing its workability by the wider public is restricted due to its experimental nature. Future work in this direction will benefit by addressing these concerns.

5. Discussion

This section is organized as follows. First, an overall assessment of the suite of its4land tools, including (and with a specific emphasis on) PaS, against FFPLA elements is provided, based on the results presented. Second, the interactions between PaS and the other tools, in terms of any further strengths, weaknesses, opportunities, and threats, are provided. Third, areas for further developments based on the results are presented.

5.1. Overall FFPLA Assessment

Overall, the its4land suite of tools and the associated PaS platform appear to align well with the elements of FFPLA, at least according to the experts surveyed, and albeit with some variation between tools with regards to specific FFPLA elements. This is perhaps to be expected, given the broad definitions ascribed to the FFPLA elements and the potential for different interpretations among survey participants. Interpretation of the elements is also challenged by the fact that the elements also describe the social and legal context for the implementation of a LAS, but only the technical aspects are taken into consideration here. In this regard, it is perhaps most interesting to look at the extremes in responses, and here, UAVs scoring low on "inclusiveness", SmartSkeMa scoring low on "attainability", and the Boundary Delineator scoring lower on "participatory", speak to the tools still requiring a high level of geospatial technical insights to understand and use, and by association, still needing a layer of simplification and increased levels of usability if the aim is for lay-people to use those tools themselves.

In contrast to the other tools developed in the its4land project, the PaS platform is not meant as a tool for an in-field surveyor or community mapper. While it was also seen to adhere or align to many of the FFPLA elements, at least in terms of the experts surveyed, it is interesting that overall it scored lower on most of the FFPLA elements—tending to score near "2" at the lower quartile, whereas the other tools tended to bottom out at "3". Looking at the more extreme results, PaS scores lower on reliability, inclusiveness, affordability, and participation. The elements on inclusiveness and participation are perhaps best explained via potential differences in interpretation by respondents, if not practicalities of implementation: while the other tools provide a direct and self-contained benefit for an end-user in the area of land administration for specific tasks, in practice, PaS would have a degree of separation from users in the field, or even local land officers, in that PaS is more for developers and system integrators, and actors who create a land administration workflows. For these kinds of users, PaS can be beneficial. That said, it is understandable that PaS could be misinterpreted or understood as "out of the hands" of land administration practitioners, let alone communities, and therefore less "participatory" or "inclusive" (noting anyway that "inclusiveness" relates to types of tenure included in a system, not the participants doing the mapping).

Perhaps of more interest or concern is that PaS scored lower on reliability and affordability. The reliability perception could be to do with the expert experiences with PaS, but is more likely to do with the fact that it requires a level of IT and internet infrastructure maturity to function. In the locations where PaS is intended for use, these infrastructures are often not reliable, particularly outside major urban centers, although it should be noted that this situation, particularly via mobile communications networks, is steadily improving. The affordability element links to this: PaS application does and would require mature levels of IT investment and the associated skill sets needed to maintain them. This not only applies to the land administration function within a jurisdiction but all government services generally.

These potential negatives aside, via those surveyed, it can be seen that PaS is agreed to provide a targeted and widely accepted development model and a platform concept for the specific needs of the land administration sector. There is currently no platform available on the market that addresses these specific needs and combines the land administration concepts with a state-of-the-art API.

Following the platform approach, the utility of PaS highly depends on the availability of tools, which provide high-level building blocks for land administration workflows. Although the tools could operate alone, the combination of the tools allowed more integrated workflows.

5.2. PaS-Tool Interactions FFPLA Assessments

Attention now turns to consider how PaS can integrate and interoperate with the other tools, and specifically, what added value these combinations might bring—in terms of enabling what we characterize as complex land administration workflows.

First, SmartSkeMa was designed to be used independently as a data collection tool. However, as with any other data collection, to make use of its outputs, it must be used in conjunction with other tools. Deployment of SmartSkeMa within PaS showed that several of the poorly rated FFP criteria could be realized when SmartSkeMa is combined with the other its4land tools. On the upgradability question, one participant in the survey mentioned that SmartSkeMa needs to be upgraded "in such a way that it should work as a standalone system" by supporting independent database access capabilities and the ability for cloud deployment. This is precisely where PaS can play a key role in a project based on its4land tools. It can also be argued that PaS can allow SmartSkeMa data to be used together with authoritative data, especially for planning purposes. In this case, the reliability aspect is addressed by reference to the authoritative data and by the fact SmartSkeMa tenure data are likely to be more up to date than official data collected at longer intervals. This would be achieved without losing most of the other FFP attributes of SmartSkeMa itself.

Evidently, there are scenarios where SmartSkeMa performs better or at least not worse when used without the PaS platform. For example, PaS does not address the affordability concerns rated poorly by the survey participants. PaS can further negatively affect the attainability of SmartSkeMa by requiring an internet connection and the upload of base map data to the platform. This a double-aged sword—on the one hand, PaS allows access to base data produced by other teams and tools while, on the other hand, constraining the number of situations in which SmartSkeMa can be used.

Second, On UAVs, the battery endurance and the productivity of UAVs have increased, leading to larger areas that can be mapped. Consequently, more significant amounts of data are collected, and more substantial computations are needed to process these images, which consumer-grade laptops and PCs can hardly accomplish. In this aspect, the functionalities of the PaS platform are highly beneficial for UAV image processing overcoming the bottleneck of low computational power. In particular, cloud computing allows the speeding up of image processing, if needed. Furthermore, the last mile in terms of dissemination, visualization, and data sharing of large UAV-based orthophotos can be reached by the PaS platform using embedded WMS services and web-based data visualization. The use of the PaS platform and its UAV Ortho Generator tool would significantly improve the affordability and attainability of UAV-based data acquisition for two reasons. First, less money would need to be spent on expensive image processing software. Second, less technical expertise would be required by local staff implementing the UAVs as a data acquisition tool.

Third, the integration of the boundary delineation tool into the PaS platform is not mandatory, but beneficial. Since all the source code is publicly available (See: https://github.com/its4land/delineation-tool, accessed on 24 May 2021), it can be used on its own. However, the tool requires image data as input that can be processed and stored on the PaS platform. Such image data can come from UAVs and thus be combined with another tool's output via the platform. Furthermore, the output of the tool can be stored and viewed as boundary facestrings on the platform. These boundaries can then be used in the SmartSkeMa process to visually compare hand-drawn and visible boundaries from both tools. The tool could be improved via the platform when especially its first part (image segmentation) would be done online on UAV images. This result could then be used in the QGIS plugin Boundary Delineation (https://plugins.qgis.org/plugins/BoundaryDelineation/, accessed on 24 May 2021), which represents the second mandatory part of the tool. At the time of writing, the image segmentation requires extensive IT knowledge and is less user-friendly than the QGIS plugin. If the image segmentation could be done on the platform, less processing power would be required from the user, and the entire tool experience would be more user-friendly.

5.3. Opportunities for Further PaS Development

Looking ahead and taking into account the inputs from the survey respondents, it appears that the three complementary tools developed in the its4land project show the potential for further development and improvement in several ways. These include: adding more tools to the platform; adding more concepts from the LADM model, such as the LA_RRR concept; adding more spatial profiles (like textual or survey Sketch-based) to the LA_SpatialUnit API; developing new business models, such as an online app store system for tools; using tools such as Terraform (https://www.terraform.io/, accessed on 24 May 2021) and Chef (https://www.chef.io/, accessed on 24 May 2021) to make PaS independent of a specific cloud platform; this would include the possibility to operate PaS also on a private cloud; bundle the hyper-converged infrastructure of PaS with standard hardware to a self-contained system that can be operated in the field without the need for permanent internet connection. This will address the affordability concerns.

6. Conclusions

This paper aimed to contribute to the broad challenge of mapping unrecorded land rights across sub-Saharan Africa. It provided an update on the results of the "its4land" project, a European Commission Horizon 2020 project aimed at developing, adapting, and testing innovative geospatial tools for the purposes of enabling alternate approach must land rights mapping in that region. Specifically, the project developed (1) software that underpinned by the smart Sketch maps concept, called SmartSkeMa; (2) a workflow for applying unmanned aerial vehicles (UAV); and (3) a semi-automatic feature extraction (AFE) tool. Additionally, the consortium developed (4) a platform called Publish and Share (PaS), enabling integration of all the outputs of tools and that could share and publish land information through a geocloud web service approach. This particular paper aimed at assessing these tools against FFPLA elements for the tools and platform individually, but also taking more holistic viewpoints.

Although the tools and platform development processes had previously been undertaken in Rwanda, Kenya, Ethiopia, and Zanzibar, this paper sought to use key informants and experts, including those who had taken part in the tool development, to obtain a more reflexive and post-project perspective. In this regard, it provides an alternate perspective on the its4land results and constitutes a further and novel contribution.

Overall, the results show that individually, each of the tools can be said to adhere or deliver upon FFPLA elements, albeit with areas of improvement available in all cases. Each of the tools can function and deliver outputs independently and reliably. That said, each is seen to have its own challenges or weaknesses in the regard, usually linked to the level of technical maturity of the technology generally, the level of technical acumen needed to use or maintain the use of the tool, and/or the level of cost involved. Interestingly, each tool and the platform is also seen to have its own strengths and weaknesses: it is not that there are FFPLA elements that scored high or low uniformly across all tools.

On PaS, it was shown to be complementary, if not underpinning, providing enhancements but also potentially enabling the development of more complex land administration workflows by combining usage of tools and data—that could be tailored to the specific needs of given country contexts. The viability, in terms of cost-benefit needs further validation though through practice and scaled implementation. That said, it is important to emphasize that there is currently no platform available on the market that addresses the specific needs and combines the land administration concepts with a state-of-the-art API that PaS enables.

In terms of further work, aside from examining further the limitations against specific FFPLA elements for each of the tools and the platform, now that PaS is established as an integrative and interoperable platform, there lies the potential to add further tools and extensions and to consider ways of scaling its use through business model considerations and larger-scale piloted usage.

Author Contributions: Conceptualization, M.K., S.C., M.C., C.S., M.I.H., C.T., R.B., methodology, M.K., S.C., M.C., C.S., M.I.H., C.T., R.B., formal analysis, M.K., S.C., M.C., C.S., M.I.H., C.T., R.B., writing—original draft preparation M.K., S.C., M.C., C.S., M.I.H., C.T., R.B. final version review M.K., S.C., M.C., C.S., M.I.H., C.T., R.B., J.Z. All authors have read and agreed to the published version of the manuscript.

Funding: The research described in this paper was funded by the research project "its4land," which is part of the Horizon 2020 program of the European Union, project number 687828. Funding of the publication costs for this article has kindly been provided by the School of Land Administration Studies, University of Twente, in combination with Kadaster International, the Netherlands.

Institutional Review Board Statement: Not applicable.

Informed Consent Statement: Not applicable.

Data Availability Statement: Not applicable.

Acknowledgments: Its4land team would like to also express their acknowledgments to their colleagues for their support during the fieldworks and workshops and the experts that were part of the survey.

Conflicts of Interest: The authors declare no conflict of interest.

Appendix A

N	Background of the Experts
1	Photogrammetry and Remote Sensing, Cadastre
2	Spatial Knowledge Representation
2	Geography, UAV
4	Geo-information, automatic feature extractionfrom images
5	Land administration and Management
6	Land administration and Management
7	Land administration and Management
8	Geoinformatics
9	Spatial Knowledge Representation
10	Software engineer
11	Software engineer
12	GIS and geo-data
13	Geo-consulting
14	GIS, Geo-information, land administration
15	GIS, software developer

Appendix B

1. Rate the PaS according to the FFP elements. In the context of PaS, they mean the following:

Flexible: To what extent can PaS be used to store and view information about different uses and land occupations?

1(− −) Cannot store or view such information

5(++) Supports storing/viewing information about all uses and occupations of land.

Inclusive: To what extent can PaS be used to store and view information about different land tenure types?

1(− −) Cannot store or view land tenure information

5(++) Supports storing and viewing all types of land tenure information

Participatory: To what extent does PaS support community engagement and participation in storing and distributing land administration data?

1(− −) Not designed for community engagement and participation

5(++) Fully supports community engagement and participation

Affordable: Is PaS affordable as a platform to build land administration systems for government and society?

1(− −) Not affordable

5(++) Very affordable

Reliable: Is the data provided by PaS authoritative and reliable?

*(note that PaS is not meant for data capture)

1(− −) Not authoritative or reliable

5(++) Highly authoritative or reliable

N/A: PaS does not provide reliable data on its own

Attainable: Can PaS support creating land administration systems with new and innovative methods to capture land use and tenure?

1(− −) Impossible in a short time with available resources

5(++) Highly achievable in a short time with available resources

Upgradable: Can PaS be upgraded and improved to respond to social and legal needs as well as economic opportunities?

1(− −) It is not upgradable

5(++) It is highly upgradable

2. Rate the SmartSkeMa according to the FFP elements. In the context of SmartSkeMa they mean the following:

Flexible: To what extent can SmartSkeMa be used to capture information about different uses and occupations of land?

1(− −) SmartSkeMa can capture information about ONLY ONE use or occupation of land

5(++) SmartSkeMa can capture information about ALL uses and occupations of land

Inclusive: To what extent can SmartSkeMa be used to capture different types of land tenure?

1(− −) SmartSkeMa can capture ONLY ONE type of land tenure

5(++) SmartSkeMa can capture ALL types of land tenure

Participatory: To what extent does/can SmartSkeMa support community engagement and participation in the data capture process?

1(− −) SmartSkeMa does/can NOT support community engagement/participation

5(++) SmartSkeMa does/can fully support community engagement/participation

Affordable: How do you rate the affordability of SmartSkeMa as a land tenure data capture tool from the perspective of the government or civil society organizations?

1(− −) SmartSkeMa is NOT affordable for the government and civil society

5(++) SmartSkeMa is affordable for the government and civil society

Reliable: To what extent can SmartSkeMa be used to generate data that is authoritative and up to date?

1(− −) SmartSkeMa CANNOT be used to produce authoritative and up-to-date data

5(++) SmartSkeMa CAN be used to produce authoritative and up-to-date data

Attainable: To what extent can SmartSkeMa be set up and put into operation within a short time-frame and with limited resources?

1(− −) SmartSkeMa CANNOT be quickly set up and put into operation
5(++) SmartSkeMa CAN be quickly set up and put into operation
Upgradable: To what extent can SmartSkeMa be upgraded and improved to respond to social and legal needs as well as economic opportunities?
1(− −) SmartSkeMa CANNOT be upgraded
5(++) SmartSkeMa CAN be upgraded

3. Rate the UAVs according to the FFP elements. In the context of UAV, they mean the following:

Flexible: To what extent can UAV be used to capture images of different uses and occupations of land?
1(− −) UAV data can capture only one use or occupation of land
5(++) UAV data can capture all uses and occupations of land
Inclusive: N/A for UAVs—UAV cannot collect land tenure data
Participatory: To what extent support UAVs community engagement and participation in the data capture process?
1(− −)UAV do not support community engagement
5(++) UAV fully support community engagement
Affordable: To what extent can UAVs be afforded by the government and society as a collection tool for aerial images?
1(− −) UAV are not affordable for the government and society
5(++) UAVs are affordable for the government and society
Reliable: To what extent provide UAV imagery authoritative and reliable base data?
1(− −) UAV imagery are not authoritative and reliable
5(++) UAV imagery are authoritative and reliable
Attainable: To what extent can UAVs be used to collect aerial base data in a short time with limited resources?
1(− −) UAV do not allow for quick data collection of aerial imagery
5(++) UAV allow for quick data collection of aerial imagery
Upgradable: To what extent can UAVs be used to provide upgraded and improved base data to respond to social and legal needs as well as economic opportunities?
1(− −) UAVs cannot be used to upgrade existing data
5(++) UAVs can be used to upgrade existing data

4. Rate the AFE according to the FFP elements. In the context of AFE they mean the following:

Flexible: To what extent can AFE be used to capture different uses and occupations of land?
1(− −) AFE can capture only one use or occupation of land
5(++) AFE can capture all uses and occupations of land
Inclusive: To what extent can AFE be used to capture different types of land tenure
1(− −) AFE can capture only one type of land tenure
5(++) AFE can capture all types of land tenure
Participatory: To what extent can AFE be used participatory by engaging the community?
1(− −) AFE cannot be used in a participatory manner
5(++): AFE can fully be used in a participatory manner
Affordable: To what extent is AFE affordable for the government and society
1(− −) AFE is not affordable for the government and society
5(++) AFE is freely affordable for the government and society
Reliable: To what extent are the boundaries delineated with AFE reliable?
1(− −) AFE cannot capture reliable boundary information
5(++) AFE can capture reliable boundary information
Attainable: To what extent is a system for boundary capture based on AFE attainable in a short time with limited resources?
1(− −) AFE does not allow quick boundary delineation in any system

5(++) AFE allows quick boundary delineation in any system

Upgradable: To what extent can AFE be upgraded and improved to respond to social and legal needs as well as economic opportunities?

1(− −) AFE cannot be upgraded

5(++) AFE can be upgraded

References

1. Nations, U. Home | Sustainable Development. Available online: https://sdgs.un.org/ (accessed on 25 May 2021).
2. Why Secure Land Rights Matter. Available online: https://www.worldbank.org/en/news/feature/2017/03/24/why-secure-land-rights-matter (accessed on 15 March 2021).
3. Lakner, C.; Mahler, D.G.; Negre, M.; Prydz, E.B. *How Much Does Reducing Inequality Matter for Global Poverty?* The World Bank: Washington, DC, USA, 2019.
4. Bennett, R.M.; Alemie, B.K. Fit-For-Purpose land administration: Lessons from urban and rural ethiopia. *Surv. Rev.* **2016**, *48*, 11–20. [CrossRef]
5. Koeva, M.N.; Crommelinck, S.; Stöcker, C.; Crompvoets, J. Its4land—Challenges and opportunities in developing innovative geospatial tools for fit-for-purpose land rights mapping. In Proceedings of the FIG Congress 2018, Istanbul, Turkey, 6–11 May 2018; International Federation of Surveyors (FIG): Copenhagen, Denmark, 2018; pp. 1–17.
6. Zevenbergen, J.; Augustinus, C.; Antonio, D.; Bennett, R. Pro-Poor land administration: Principles for recording the land rights of the underrepresented. *Land Use Policy* **2013**, *31*, 595–604. [CrossRef]
7. Lemmen, C.; van Oosterom, P.; Bennett, R. The land administration domain model. *Land Use Policy* **2015**, *49*, 535–545. [CrossRef]
8. International Organization for Standardization (ISO). Geographic information—Land Administration Domain Model (LADM). Available online: https://www.iso.org/standard/51206.html (accessed on 12 July 2018).
9. Ho, S.; Pattyn, V.; Broucker, B.; Crompvoets, J. Needs assessment in land administration: The potential of the nominal group technique. *Land* **2018**, *7*, 87. [CrossRef]
10. Rahmatizadeh, S.; Rajabifard, A.; Kalantari, M.; Ho, S. A framework for selecting a fit-for-purpose data collection method in land administration. *Land Use Policy* **2018**, *70*, 162–171. [CrossRef]
11. De Vries, W.; Bennett, R.; Zevenbergen, J. Toward responsible land administration. In *Advances in Responsible Land Administration*; CRC Press: Boca Raton, FL, USA, 2015; pp. 3–14.
12. Zevenbergen, J.; De Vries, W.; Bennett, R. *Advances in Responsible Land Administration*; CRC Press: Padstow, UK, 2015; ISBN 9781498719599.
13. Enemark, S.; McLaren, R.; Lemmen, C.; Antonio, D.; Gitau, J. *Guiding Principles for Building Fit-For-Purpose Land Administration Systems in Developing Countries*; UN-Habitat: Nairobi, Kenya, 2016.
14. Enemark, S.; Bell, C.K.; Lemmen, C.; Mclaren, R. *Fit-For-Purpose Land Administration*; International Federation of Surveyors (FIG): Copenhagen, Denmark, 2014; ISBN 9788792853103.
15. Koeva, M.; Stöcker, C.; Crommelinck, S.; Chipofya, M.; Kundert, K.; Schwering, A.; Sahib, J.; Zein, T.; Timm, C.; Humayun, M.; et al. Innovative geospatial solutions for land tenure mapping. *Rwanda J. Eng. Sci. Technol. Environ.* **2020**, *3*, 34–49. [CrossRef]
16. Koeva, M.; Stöcker, C.; Crommelinck, S.; Ho, S.; Chipofya, M.; Sahib, J.; Bennett, R.; Zevenbergen, J.; Vosselman, G.; Lemmen, C.; et al. Innovative remote sensing methodologies for Kenyan land tenure mapping. *Remote Sens.* **2020**, *12*, 273. [CrossRef]
17. Stöcker, C.; Nex, F.; Koeva, M.; Gerke, M. Uav-Based cadastral mapping: An assessment of the impact of flight parameters and ground truth measurements on the absolute accuracy of derived orthoimages. *ISPRS Int. Arch. Photogramm. Remote Sens. Spat. Inf. Sci.* **2019**, *42*, 613–617. [CrossRef]
18. Koeva, M.; Bennett, R.; Gerke, M.; Crommelinck, S.; Stöcker, C.; Crompvoets, J.; Ho, S.; Schwering, A.; Chipofya, M.; Schultz, C.; et al. Towards innovative geospatial tools for fit-for-purpose land rights mapping. *ISPRS Int. Arch. Photogramm. Remote Sens. Spat. Inf. Sci.* **2017**, *42*, 37–43. [CrossRef]
19. Stöcker, C.; Ho, S.; Koeva, M.N.; Nkerabigwi, P.; Schmidt, C.; Zevenbergen, J.A.; Bennett, R. Towards UAV-based land tenure data acquisition in Rwanda. In Proceedings of the FIG Congress 2018, Istanbul, Turkey, 6–11 May 2018; International Federation of Surveyors (FIG): Copenhagen, Denmark, 2018.
20. Schwering, A.; Wang, J.; Chipofya, M.; Jan, S.; Li, R.; Broelemann, K. SketchMapia: Qualitative Representations for the alignment of sketch and metric maps. *Spat. Cogn. Comput.* **2014**, *14*, 220–254. [CrossRef]
21. Chipofya, M.; Jan, S.; Schultz, C.; Schwering, A. Towards smart sketch maps for community-driven land tenure recording activities. In Proceedings of the 20th Conference on Geo Information Science AGILE, Wageningen, The Netherlands, 9–12 May 2017.
22. Casiano Flores, C.; Tan, E.; Buntinx, I.; Crompvoets, J.; Stöcker, C.; Zevenbergen, J. Governance assessment of the UAVs implementation in Rwanda under the fit-for-purpose land administration approach. *Land Use Policy* **2020**, *99*. [CrossRef]
23. Drone Mapping Software—OpenDroneMap. Available online: https://www.opendronemap.org/ (accessed on 14 April 2021).
24. Crommelinck, S.; Bennett, R.; Gerke, M.; Koeva, M.N.; Yang, M.Y.; Vosselman, G. SLIC superpixels for object delineation from UAV data. *ISPRS Ann. Photogramm. Remote Sens. Spat. Inf. Sci.* **2017**, 9–16. [CrossRef]
25. Crommelinck, S. Delineation-Tool GitHub. Available online: https://github.com/its4land/delineation-tool (accessed on 13 May 2019).

26. Crommelinck, S.; Yang, M.Y.; Koeva, M.N.; Vosselman, G. Towards automated cadastral boundary delineation from UAV data. *arXiv* **2018**, arXiv:1709.01813.
27. Crommelinck, S.; Koeva, M.; Yang, M.Y.; Vosselman, G. Robust object extraction from remote sensing data. *arXiv* **2019**, arXiv:1904.12586.
28. Crommelinck, S.; Bennett, R.; Gerke, M.; Nex, F.; Yang, M.Y.; Vosselman, G. Review of automatic feature extraction from high-resolution optical sensor data for UAV-based cadastral mapping. *Remote Sens.* **2016**, *8*, 689. [CrossRef]
29. Crommelinck, S.; Höfle, B.; Koeva, M.N.; Yang, M.Y.; Vosselman, G. Interactive cadastral boundary delineation from UAV images. *ISPRS Ann. Photogramm. Remote Sens. Spat. Inf. Sci.* **2018**, *4*, 81–88. [CrossRef]
30. Xia, X.; Koeva, M.; Persello, C. Extracting cadastral boundaries from UAV images using fully convolutional networks. In Proceedings of the IGARSS 2019 IEEE International Geoscience and Remote Sensing Symposium, Yokohama, Japan, 28 July–2 August 2019; IEEE: Piscataway, NJ, USA, 2019; pp. 2455–2458.
31. Xia, X.; Persello, C.; Koeva, M. Deep fully convolutional networks for cadastral boundary detection from UAV images. *Remote Sens.* **2019**, *11*, 1725. [CrossRef]
32. United Nations Economic Commission for Europe (UNECE). *Land Administration Guidelines. With Special Reference to Countries in Transition*; United Nations Economic Commission for Europe: New York, NY, USA, 1996.
33. Enemark, S. The Land Management Paradigm for Institutional Development. In Proceedings of the Expert Group Meeting on Incorporating Sustainable Development Objectives into ICT Enabled Land Administration Systems, Centre for Spatial Data Infrastructures and Land Administration the Land Management Paradigm for Institutional Development, Melbourne, Australia, 9–11 November 2005.
34. Stöcker, C.; Bennett, R.; Nex, F.; Gerke, M.; Zevenbergen, J. Review of the current state of UAV regulations. *Remote Sens.* **2017**, *9*, 459. [CrossRef]
35. Stocker, C.; Koeva, M.; Bennett, R.; Zevenbergen, J. Evaluation of UAV-based technology to capture land rights in Kenya: Displaying stakeholder perspectives through interactive gaming. In Proceedings of the 2019 Annual World Bank Conference on Land and Poverty, "Catalyzing Innovation", Washington, DC, USA, 25–29 March 2019.
36. Koeva, M.; Muneza, M.; Gevaert, C.; Gerke, M.; Nex, F. Using UAVs for map creation and updating. *Surv. Rev.* **2018**, *50*, 312–325. [CrossRef]
37. Stöcker, C.; Nex, F.C.; Koeva, M.N.; Gerke, M. Quality assessment of combined IMU/GNSS data for direct georeferencing in the context of UAV-based mapping. *ISPRS Ann. Photogramm. Remote Sens. Spat. Inf. Sci.* **2017**, 355–361. [CrossRef]
38. Crommelinck, S.; Koeva, M.; Yang, M.; Vosselman, G. Application of deep learning for delineation of visible cadastral boundaries from remote sensing imagery. *Remote Sens.* **2019**, *11*, 2505. [CrossRef]
39. Crommelinck, S.; Bennett, R.; Gerke, M.; Yang, M.; Vosselman, G. Contour detection for UAV-based cadastral mapping. *Remote Sens.* **2017**, *9*, 171. [CrossRef]
40. Crommelinck, S.; Höfle, B.; Koeva, M.N.; Yang, M.Y.; Vosselman, G. Interactive cadastral boundary delineation from UAV images. In Proceedings of the ISPRS TC II Mid-Term Symposium "Towards Photogrammetry 2020", Riva del Garda, Italy, 4–7 June 2018; Volume 4, pp. 81–88.
41. Fielding, R.T. *Architectural Styles and the Design of Network-Based Software Architectures*; University of California: Irvine, CA, USA, 2000; Volume 54.
42. Williamson, I.; Enemark, S.; Wallace, J.; Rajabifard, A. *Land Administration for Sustainable Development*; ESRI Press Academic: Redlands, CA, USA, 2010; p. 487.
43. Unger, E.M.; Bennett, R.; Lemmen, C.; de Zeeuw, K.; Zevenbergen, J.; Teo, C.H.; Crompvoets, J. Global policy transfer for land administration and disaster risk management. *Land Use Policy* **2020**, *99*. [CrossRef]
44. Stöcker, C.; Ho, S.; Nkerabigwi, P.; Schmidt, C.; Koeva, M.; Bennett, R.; Zevenbergen, J. Unmanned aerial system imagery, land data and user needs: A socio-technical assessment in Rwanda. *Remote Sens.* **2019**, *11*, 1035. [CrossRef]
45. Stöcker, C.; Nex, F.; Koeva, M.; Gerke, M. High-Quality UAV-based orthophotos for cadastral mapping: Guidance for optimal flight configurations. *Remote Sens.* **2020**, *12*, 3625. [CrossRef]

land

MDPI

Article

SmartSkeMa: Scalable Documentation for Community and Customary Land Tenure

Malumbo C. Chipofya [1,*], Sahib Jan [2] and Angela Schwering [2]

1 Faculty of Geo-Information Science and Earth Observation (ITC), University of Twente, Hengelosestraat 99, 7514 AE Enschede, The Netherlands

2 Institute for Geoinformatics, Westfälische Wilhelms-Universität Münster, Heisenbergstrasse 2, 48149 Münster, Germany; s_jan001@uni-muenster.de (S.J.); schwering@uni-muenster.de (A.S.)

* Correspondence: m.c.chipofya@utwente.nl; Tel.: +31-53-489-4760

Abstract: According to the online database landmarkmap, up to an estimated 50% or more of the world's habitable land is held by indigenous peoples and communities. While legal and procedural provisions are being made for bureaucratically managing the many different types of tenure relations in this domain, there continues to be a lack of tools and expertise needed to quickly and accurately document customary and indigenous land rights. Software and hardware tools that have been designed for documenting land tenure through communities continue to assume a parcel-based model of land as well as categories of land relations (RRR) largely dimensionally similar to statutory land rights categories. The SmartSkeMa approach to land tenure documentation combines sketching by hand with aerial imagery and an ontology-based model of local rules regulating land tenure relations to produce a system specifically designed to allow accurate documentation of land tenure from a local perspective. In addition, the SmartSkeMa adaptor which is an OWL-DL based set of rules for translating local land related concepts to the LADM concepts provides a more high-level view of the data collected (i.e., what does this concept relate to within the national LADM profile?) In this paper we present the core functionalities of SmartSkeMa using examples from Kenya and Ethiopia. Based on an expert survey and focus groups held in Kenya, we also analyze how the approach fairs on the Fit-for-Purpose Land Administration tools scale. The results indicate that the approach could be beneficial in scaling up mapping of community and customary lands as well as help reduce conflict through its participatory nature.

Keywords: customary land tenure; participatory mapping; fit-for-purpose land administration; land recordation tools; semantic technologies; land information system

Citation: Chipofya, M.C.; Jan, S.; Schwering, A. SmartSkeMa: Scalable Documentation for Community and Customary Land Tenure. *Land* **2021**, *10*, 662. https://doi.org/10.3390/land10070662

Academic Editors: Stig Enemark, Robin McLaren and Christiaan Lemmen

Received: 30 April 2021
Accepted: 16 June 2021
Published: 22 June 2021

1. Introduction

Much of the world's habitable land is held by indigenous peoples and communities. Surprisingly, there is a dearth of tools and expertise needed to quickly and accurately document customary and indigenous land rights. By contrast a lot of investment is put into implementation and improvement of statutory land administration systems as witnessed by many recent projects in the domain (LIFT in Ethiopia and the Rwanda LTR for example). This is ironic considering landmarkmap's estimate that of the nearly 70% undocumented tenures on land in the world a significant proportion is likely to cover up to 50% of the world's habitable areas (http://landmarkmap.org).

One possible explanation for the skewed investment towards securing state sanctioned forms of land rights is what James Scott referred to as the process of legibility making [1] (pp. 33–36). Legal land rights categories are often created to simplify (and make uniform) the relations of tenure on land to render them amenable to administration. This simplification inevitably entails a tearing down of some existing local social norms and cultural practices. As both Scott [1] and Abubakari et al. [2] note, this legibility making often meets some forms of resistance. State systems can therefore co-exist or compete with existing norms, and may come into direct conflict with them.

Considering the impact of good land governance and tenure security on many of today's pressing problems (including environmental sustainability, food security, and economic and social development—the global Sustainable Development Goals (SDGs) acknowledge the impact of tenure security on all these aspects and more). it seems reasonable to consider supporting a symbiotic coexistence of state and local systems of tenure administration. The need for legibility making at the human scale is quickly being replaced by intelligent computer systems. Such systems are able to deal with complexities largely out of the reach of the average human expert. They interpret huge amounts of complex information and provide summarized information in a form that human experts can use and act on. We propose a deeper focus of using such systems in the domain of land administration. SmartSkeMa is a system that exemplifies how such high-tech solutions can make a difference. The role of these technologies should not be to help replace customary ways of managing access to and use of land but to make the systems of rules, norms, and ways of knowing that govern customary land tenure legible from a state perspective as is. In the guiding principles for country implementation of Fit-for-purpose land administration, it is proposed that lands with customary tenure "can be included in the formal system by demarcating the outer boundaries while retaining the community institutions that allocate and manage individual and household plots, with the option to register these land rights as the need arises" [3], (pp. 26–27). Systems such as SmartSkeMa can be used for registration of land rights within the inner boundaries of the customary land.

This paper presents the SmartSkeMa land data documentation approach built around the SmartSkeMa software system. SmartSkeMa is a portable cross-platform tool that allows people to document their land tenure rights using the familiar process of sketching or tracing and using concepts from their everyday experiences of exercising those rights. It achieves this fit by applying methods from Computer Vision and Artificial Intelligence. A sketch map is automatically digitized and (approximately) georeferenced using either numerical or logical approximation. Once the sketch is digitized and georeferenced social information can be attached to the detected objects or parcels. The social information must conform to a set of concepts actually used by the community where the tenure documentation is being conducted. The relevance of using predefined concepts is that the data can then be automatically interpreted using relations between those concepts. This is where the Artificial Intelligence methods of knowledge representation and reasoning are used to give life to cultural knowledge and information. It is possible, for example, to infer that an individual's right to use a particular piece of land under either or both the customary and legal regimes using information about the social relations in which she participates (e.g., marriage, living in a particular community, etc.)

SmartSkeMa was developed as part of the "its4land" H2020 project [4,5]. The its4land project developed four technological solutions for land tenure data acquisition. The tools were required address actual needs and readiness of their target user groups. They were also expected to be implementable and scalable within the contexts in which they were expected to be used. The guiding principles for the tools' development therefore included consideration of the seven Fit-For-Purpose (FFP) elements [6] which recommend that tools be: 1. **Flexible** enabling the capture of information about the different uses and occupations of the land; 2. **Inclusive** by covering all types of tenure and all types of land 3. **Participatory** in the manner that data are captured and used 4. **Affordable** to use for government and society at large; 5. **Reliable** for acquiring authoritative and updated information; 6. **Attainable** within a short timeframe and with the available resources; 7. **Upgradable** in response to social, legal, and economic needs and opportunities. We present an evaluation of SmartSkeMa based on an expert survey and focus groups held in Kenya. We then discuss the results in the context of the seven FFP elements as a frame of reference.

The next section gives a brief overview of the state of the art in alternative land tenure documentation tools that can be used in customary tenure regimes. We also note the structure of conventional land administration systems through reference to the ISO 19152:2012 standard Land Administration Domain Model (LADM) [7,8]. SmartSkeMa has the ability to interpret local, i.e., customary, land concepts in the language of the LADM. This is described in more detail in Section 3 where the authors present the SmartSkeMa workflow and provide a quasi-tutorial of how SmartSkeMa works. In Section 4 we present the evaluation of SmartSkeMa. We conclude the paper with final remarks in Section 5.

2. Background

SmartSkeMa is a fit-for-purpose land tenure documentation approach. This section provides a context for land tenure documentation with SmartSkeMa. We give an overview of a few fit-for-purpose land registration and land tenure documentation approaches and outline how the SmartSkeMa approach differs from them.

2.1. Land Tenure Relations

In most official land administration institutions, the mapping of land units and the registration of the interests of them are carried out by separate functions (e.g., Departments or even Ministries). In project based, systematic land registration, on the other hand, these processes are often combined into a single stream-lined process. The distinction is between sporadic and systematic registration of land [9] (pp. 349–354).

Regardless of the 'fashion' in which land registration is executed, there are still common steps that can be generally considered to be part of a land registration process. With respect to SmartSkeMa the most important of these include (i) formalization of land tenure concepts and relations, (ii) mapping land units including parcels but also other identifiable land resources, (iii) recording of the parties involved, and (iv) documentation of the relations between parties and land units—in LA terms, rights, restrictions, and responsibilities [9] (pp. 34, 38).

In the formal land administration domain the formalization of land tenure concepts is achieved by the presence of relevant laws and statutes to govern and secure the exercise of rights on land as well as stipulate the restrictions and responsibilities on holders of land rights [9] (p. 213). A title to land is the evidence of a person's rights to the land [10] (p. 109). There are different classes of land titles by which a party may hold land rights including Freehold, Leasehold, Perpetual use rights, etc., and they vary in their interpretation from jurisdiction to jurisdiction. There are also differences in the extent to which land may be exploited and/or alienated depending on the title and jurisdiction. For example, during a field visit to Ethiopia, we learned from local communities around Bahir Dar that there are restrictions on land development in rural Ethiopia which stipulate that no permanent structures may be constructed in certain types of agricultural plots, a restriction not present in analogous scenarios in other countries like Kenya.

Underlying the different classes of land title is the form of tenure they represent. Williamson et al. [9] list a possible categorization of tenures based on observations of "processes used by a society to stabilize its access to land and resources" [9] (p. 333, Table 12.2). They make a distinction between formal and informal tenure but also list tenures including leasehold, freehold (i.e., private ownership), and customary, traditional, indigenous, and native tenures. As alluded by many authors [1,11,12] the subject of land tenure is complex. Malinowski [10] (pp. 317–320) seems to point out, based on his experiences in the Trobriands, that tenure on land is at the same time negotiated through social, cultural, religious, economic constructs as it plays a role in their construction. SmartSkeMa's domain models are designed to help bring out the "true" nature of land tenure in the customary, traditional, and informal settings. Where official land administration systems are given a generic structure using the LADM [7], customary or traditional or indigenous or informal land tenures can be approximated by SmartSkeMa's domain models.

2.1.1. LADM

The LADM provides a standardized global vocabulary for land administration. It specifies the general elements expected within a Land Administration System arranged in three main packages covering parties, rights, and tenure information (basic administrative units, RRRs) and spatial units. The spatial units package has a surveying and representation subpackage [7].

In the LADM land is represented as spatial units. Rights are modelled by a combination of an RRR and a basic administrative unit which is a conceptual object through which specific rights are related to spatial units. While traditionally spatial units were considered to be well defined polygonal geographic features, the advent of the fit-for-purpose paradigm required the inclusion of more flexible spatial representations of spatial units. Lemmen et al. [13], introduced the notion of levels to allow the representation of spatial units using different representational forms in the LADM. In addition to points, lines and polygons [13] also describe text-based and sketch-based spatial units. However, in its raw form, this information cannot be automatically interpreted by a land administration system: a sketch in raster format or a text in ASCII format has no spatial reference and its contents (unless properly annotated) cannot be understood and [spatially] operated on by a computer. SmartSkeMa supports the spatial representation of non-precise geographic features such as those included in sketch maps, textual descriptions, and similar data sources using qualitative spatial representations.

2.1.2. Customary, Traditional, Indigenous, and Native Tenures

The LADM provides a conceptual model of land tenure information. Modelling of land administration processes is outside the scope of the model. Those processes support all required transactions to keep the data up to date. The LADM concept of a Basic Administrative Unit (BAU) facilitates the abstraction the rights-to-land relation such that a customary, informal, or any other social tenure relationship also can be recorded [7] (p. 2). Through the BAU a unique or homogeneous set of rights, restrictions, or responsibilities can be associated with a group of spatial units treated as a single administrative entity. Examples of the application of this structure to a customary tenure system are provided in Annex C of [7].

Tenure, however, is much more complex and dynamic than the models represented within modern Land Administration Systems. Malinowski in his seminal chapters (Ch. XI and XII) on Land Tenure in the Trobriand Islands noted that (notwithstanding Colonization as a wholesome phenomenon in its own right) the focus on legal categories may have been the source of many land related problems in the British colonies [11]. The observation is not unique. C.K. Meek's Land Law and Custom in the Colonies [12] is full of examples of simplifications in land tenure representation that had catastrophic outcomes.

The distinction between the de facto and the de jure is that the former is practical and the latter legible [1]. The illegibility of de facto customary tenures makes them inappropriate or ill-structured for the land management functions of land administration officials [1]. However, this is not a truism. The incapacity of officialdom to grasp and control de facto organic systems of land tenure is the main reason for which official systems abstract from the multitude of rules and rule systems to a few fixed categories of tenure (hence, Seeing like a State). Although modern technology has made approximating such complex and various tenure systems more practical, one is hard pressed to find information systems and tools that support the documentation and execution of local practices related to land tenure.

2.1.3. STDM

Initiatives to capture customary tenure in modern LAS include the Social Tenure Domain Model or STDM [14]. STDM refers both to a tool and the land tenure domain model it implements. The STDM model is based on the continuum of rights concept which holds that land tenure is not a unitary concept but exists on a formality-informality spectrum [14,15]. STDM is related to the LADM and most of its classes have counterparts in the LADM. This relationship is also described in Annex I of the LADM specification [7]. The idea of the spectrum, which is a simplifying model, allows different kinds of tenure to be arranged according to a fixed dimension—i.e., the extent to which tenure is enforced by state apparatuses. Rights in STDM are represented by a variety of tenure categories and it supports extension of the data model through the addition of new custom categories. For example, if special rights hold for mountain side areas then a user could add "mountainside-parcel" as a spatial unit type. However, it is not possible to infer the special rights from the types of parcels to which they apply or from other rights. That is, STDM records the static aspects of land tenure. In SmartSkeMa on the other hand, rights are transient (i.e., dynamic) [16]. A right exists contingent on other conditions which may be physical or abstract in nature, as is illustrated in Section 3.2.5 below.

2.2. *Land Tenure Mapping*

Both the collection of tenure information and the enforcement of laws to protect land rights depend on an adequately clear identification of the land involved. Many approaches have been employed for identifying land in the records and on the ground. Here we look at only a few approaches which we consider as a continuum to which the SmartSkeMa approach belongs.

2.2.1. Ground Surveys in Sporadic Land Registration

In traditional land administration the most commonly used mapping approach is the ground survey in which the positions of boundary points for a parcel or other unit of land are recorded using bearings and distances or as coordinates. Ground surveys are mostly used for sporadic land registration. The challenge here is the cost factor. Professional fees, equipment costs, and time are billed at a premium. This is impractical for systematic large scale cadastral mapping—e.g., in Cambodia a communal land titling project started in 2013 cited a 40 USD per hectare cost [17].

2.2.2. GPS and Orthophoto Maps in Systematic Land Registration

Systematic land registration requires more flexible and cost-effective approaches. One approach is to use low-cost location measuring instruments such as GPS/GNSS devices. While high-end GPS or GNSS receptor instruments are sometimes used with electronic survey instruments such as the TotalStation, for large-scale projects the low-end cheaper versions are a more practical solution. As demonstrated in the Rwanda LTR project aerial image printouts can be deployed en masse because they can be operated with little to no training reducing the human resource and hardware cost [18].

Another commonly used cadastral mapping approach involves the use of aerial imagery. The LTR projects in Rwanda and Ethiopia both used aerial (orthorectified) imagery for identification of visible boundaries. The procedure for this process is often standard. Landowners/users identify their boundaries in an image resolving differences through site visits where necessary. Boundary features marked on the aerial image are digitized in a GIS. Two approaches are prevalent. A GIS operator may look at the boundary markings on the image and mark the corresponding points in the GIS using the original correctly projected version of the image as a background guide. This usually results in many errors. The digitizer may perceive markings wrongly, he/she may have jittery hands that cause boundary points to be off from where he/she intended them, or he/she may introduce his/her own interpretations to resolve apparent ambiguities (do the hedges forming a boundary belong to the left or right side parcel?). A much more stable approach involves marking the printed orthomap with known ground control points. In this case the digitization is performed on top of a rectified version of the marked and then scanned aerial image. This means that the GIS operator simply follows the ink as marked out in the field. This somewhat reduces the human error inherent to manual digitization. The Village Land Use Planning project in Tanzania used this digitization technique for large-scale recurrent village planning exercises [19]. The advantage of drawing boundaries over orthorectified aerial images as compared to taking boundary Coordinates on the ground using GPS is that the mappers do not need to visit every boundary corner in case of difficult terrain.

2.2.3. Open Tenure and SOLA

Open Tenure is an open source software application that supports mapping of land parcels or territories (demarcation), association of land users and claims to mapped parcels, and generation of summaries and reports including maps [20]. It runs on a mobile device but can be connected to the SOLA registry server which provides centralized storage and management of land tenure information [20]. The authors claim that SOLA provides functions and services provided by a typical land office. This suggests that SOLA is generic cadaster and registry system that does not take specific account of the peculiarities of different land tenure regimes. For customary tenure, flexibility, participation, and ownership are key values. These values should be part of the key design principles guiding the development of any tool that targets customary or other non-statutory land tenure regimes.

2.2.4. Sketching Land

It may be argued that sketching is the preferred technique of mapping for land sector non-governmental organizations (NGOs). For NGOs hand drawn sketch maps are a particularly useful tool in community mapping as they facilitate collaborative map creation among participants involved in the exercise [21,22]. Tenure mapping in these settings focusses on the community's own knowledge and consensus is achieved in a participatory manner. This is exemplified by the works of such organizations as Namati through their numerous mapping guides including the most recent Community Facilitators Guide to the Community Land Act [23] which aims to support communities in Kenya to register their community lands pursuant to the provisions of the Community Land Act (2016) of Kenya. The sketching process in the manuals focuses on the organizational aspect of the exercises: who will participate in the mapping and to what extent, how will negotiations be structured, how should the data be stored and used, etc. SmartSkeMa is a candidate tool to provide operational support to make processes such as Namati's easier to implement in a digital environment. Rugema et al. [22] also present a tool for sketch mapping but using a digital pen. By contrast SmartSkeMa is based on ink and paper sketching.

2.3. Summary

The report on accuracy assessment of unconventional mapping tools published by Spatial Collective [24] indicates that simple handheld GPS devices as well as consumer smartphones with embedded GPS receivers can meet the requirements in certain conditions. Spatial Collective's study particularly showed that with little training local citizens of even remote communities were able to conduct mapping by themselves. This increases participation but still has technological and skills barriers. Besides, only one person can command a handheld device at a time. An alternative is to use sketch mapping. But when sketch maps are stored as is, as paper or raster images, they will invariably lack the contextual details required to make them interoperable with other data sources. Other approaches go further and involve the manual digitization of the sketch maps into a GIS system, enriching the data with a geographic context [20]. However, even in this case the data may fail to accurately capture significant social values and norms that govern human relationships on land. Land registration tools for customary, indigenous or other informal tenure regimes must therefore support both the capture of both the spatial and social/legal aspects of land information, and tool developers must take seriously accuracy on all dimensions of land information, not only the spatial one.

3. SmartSkeMa

The primary objective of the SmartSkeMa system is to provide a community focused land tenure documentation tool. It is composed of several components which come together to provide three main functions:

1. Digitize sketch maps of land parcels, territories, and/or resources;
2. Approximately georeference geometries in a sketch map via simple linear or quadratic regression model or using qualitative spatial relations such as left-of, inside, overlap;
3. Capture local land tenure concepts and facts and translate them into statutory tenure terms using the LADM as the intermediating domain model.

As a software SmartSkeMa has two distinct data processing workflows with overlapping functions based on the type input map used. In the first workflow sketching is done over a georeferenced map, usually an aerial orthophoto. In the second, no background map is used during sketch; the sketch is produced freehand. The two workflows then proceed with analogous steps 1. Vectorize, 2. Align, 3. Add Land Tenure Attributes (*Data Processing in SmartSkeMa* in the third lane of Figure 1). Users can also review the data and go back to correct some data before exporting it for use with other information systems or software.

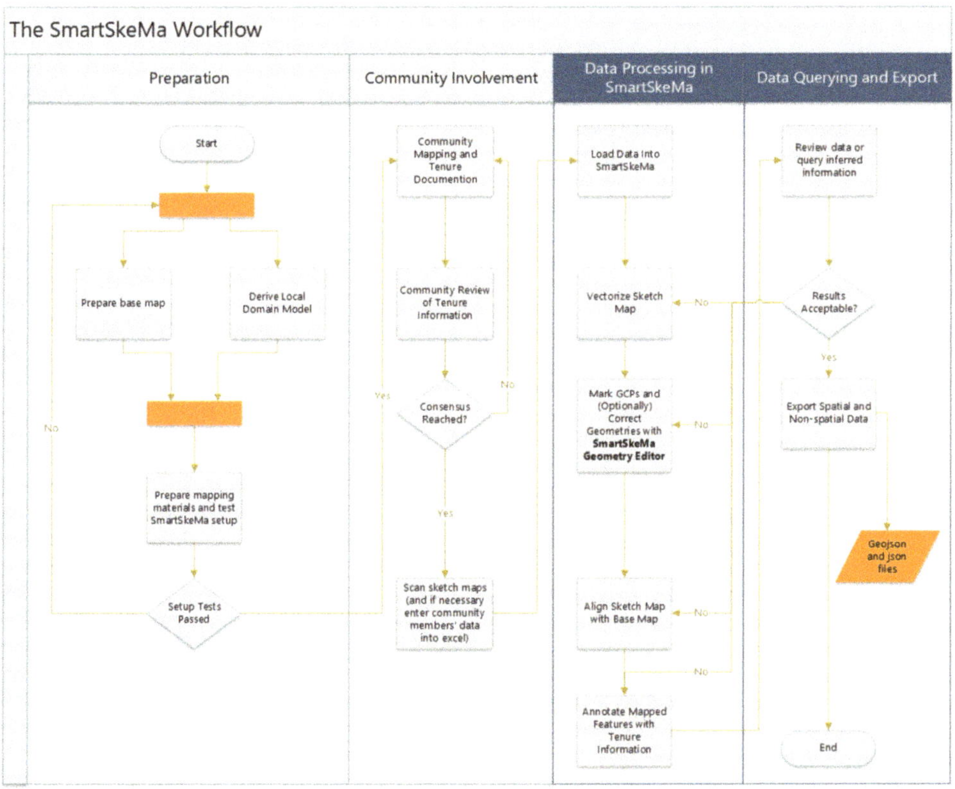

Figure 1. The SmartSkeMa Tenure Documentation Workflow.

Outside the software, a user must perform steps that fit into what we call the SmartSkeMa Tenure Documentation Workflow (see Figure 1). We return to the inner workings of SmartSkeMa in Section 3.2. In the next subsection we provide an overview of the SmartSkeMa Tenure Documentation Workflow.

3.1. The SmartSkeMa Tenure Documentation Workflow

The SmartSkeMa Tenure Documentation workflow can be integrated with other processes such as for example those promoted by Namati or other organizations. Thus, SmartSkeMa is designed to support processes that are already being applied at scale. It allows non-professionals to participate in the data digitalization process, potentially reducing time and labor cost. The key role that the steps in the current workflow play is to ensure the successful execution of each subprocess of the SmartSkeMa data processing workflow.

The first step is to prepare the system for a land data acquisition project or for a community to use on an ongoing basis (the *Preparation* lane in Figure 1). A geospatial expert is required for this step. Base reference data must be acquired and set up in the system. These can be orthophotos or vector maps that will be used for georeferencing the input sketch map data. For a given location a set of at least four ground control points (GCP) must be identified and saved to a geojson file. In addition, a domain model of the local land tenure concepts must be developed to be used during data capture and export—this can occur in parallel as seen in Figure 1.

Once everything is setup data can be collected in the community. For the sketching it is recommended to print the base map or image on an A1 or A0 size sheet (see Figure 2a). In case this is not feasible, smaller sections of the whole map can be printed on A3 sheets. Using a base map allows more precision for spatial alignment. Each printout must contain at least four visible ground control points. In addition, data sheets for recording personal and tenure information of community members must be prepared. The structure of these data sheets can be varied (there is no prescribed format) but they must be able to record personal details, relations between people and land objects, and any special conditions that apply to these relations. As we will see such conditions might be qualifying conditions or required dependencies for a relation to validly exists according local norms.

Field data collection (2nd lane Figure 1) can follow any procedure provided maps are drawn with a clear marker. In the current version sketching must be done on a transparent sheet lain on top of the base map printout (Figure 2c,e). The GCPs must be marked, preferably with a differently colored pen and pen thicknesses should be relatively similar. Field data can be scanned with a scanner for better quality results although a camera can also be used to take a photo the sketch. With digital versions of all their field data at hand (e.g., as in Figure 2d,f), the user can then proceed to produce vectorized, georeferenced and interoperable records.

3.2. The SmartSkeMa Data Processing Workflow

SmartSkeMa is browser-based software application that can be installed as a standalone system on a user's local computer or be accessed remotely over the internet. For purposes of portability, SmartSkeMa has been packaged using a tool called Docker (https://www.docker.com/) which allows the tool to be run in different operating systems including Windows, Linux, or MacOS. The user interface is accessed as a web page in a browser. For the technical reader, the backend components are written mostly in python with parts written in Java and the C programming language. It integrates semantic reasoning using the owlapi and the HermiT owl reasoner. The semantic reasoner allows complex social rules to be represented and reasoned over. The reasoner also uses predefined rules to interpret concepts in the local domain models as LADM concepts (see below for more details).

The user experiences the data processing workflow as six sequential steps. The following subsections describe these steps.

Figure 2. Data collection for SmartSkeMa processing. (**a**) Base map must be printed, preferably on A0 sheet. Mapping exercises can be performed in different formats e.g., in community workshop (**b**) or between small number of neighbors (**c**). Names of community members must be put into an excel sheet either directly in the field or after field work (**d**). After fieldwork the map must be scanned into a digital raster format such as *.png or *.jpeg (**e**,**f**).

3.2.1. Load Data

Once SmartSkeMa is started, the user must select the type of data processing workflow they want to execute. This is necessary for SmartSkeMa to determine the internal functions to use when processing the input data as well as the user interface elements it will need to display the results. The user is then prompted to upload all the data required depending on the selected workflow (Figure 3a). The final loaded data as they appear in SmartSkeMa can be seen in Figure 3b on the bottom.

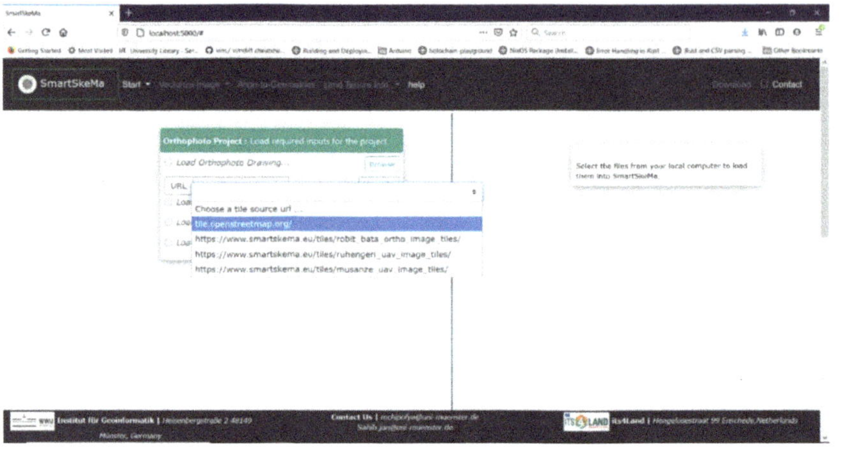

(a)

(b)

Figure 3. SmartSkeMa starts on a blank page. (**a**) Under the Start button users can select the type of data processing workflow they want to execute before uploading their data. In this figure, the user is specifying a source URL for the base reference data. (**b**) When spatial data are loaded in SmartSkeMa the sketch map raster image is loaded on the left panel and base map image with reference points overlayed as black dots is loaded in the panel on the right-hand-side of the window.

3.2.2. Vectorize the Sketch

Vectorizing the sketch map requires only the click of a button. In the background, one of two processes is invoked depending on the selected workflow. In the case of a sketch map with GPCs and no other symbols included, the system first separates image into layers of different colors. Then each layer is processed separately computing first the corner vertices of each geometry (Figure 4). From the corner vertices and the ink extracting the connections between the vertices to form topological structures called a doubly connected edge list (DCEL) [25] (pp. 29–33). The final polygons are obtained from this DCEL structure.

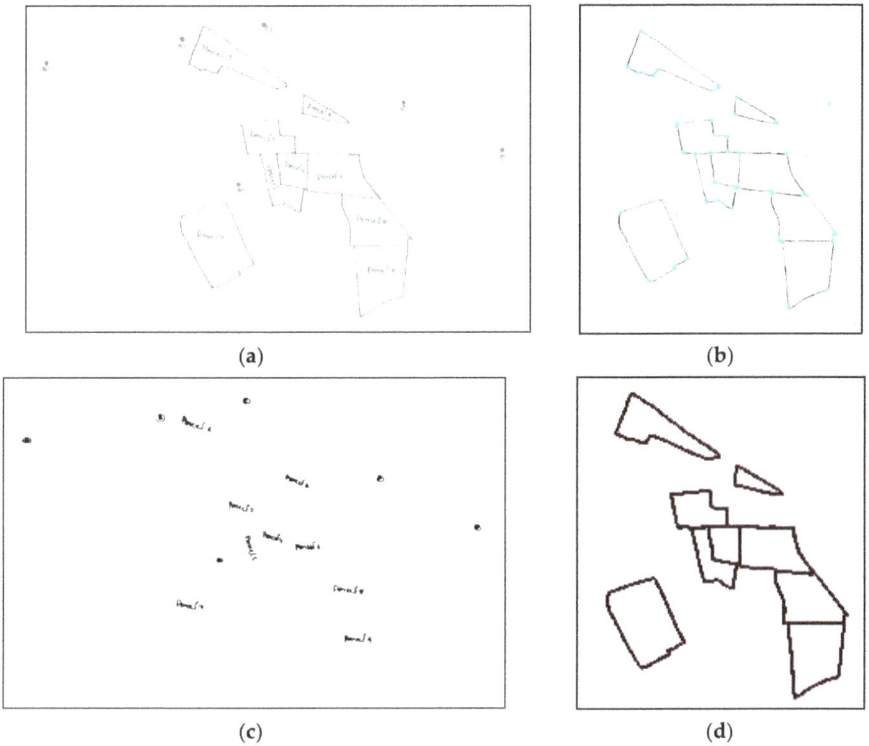

(a) (b) (c) (d)

Figure 4. Automatic vectorization of an input vector map. (**a**) The input map, (**b**) corner vertices overlaid on polygon layer of input map, (**c**) text and markers extracted from the input map, (**d**) edges/lines of the polygons extracted shown in red.

3.2.3. Label Ground Control Points in Sketch Map and Edit Geometries If Necessary

Figure 5 shows the sketch map after vectorization and alignment. SmartSkeMa provides an editor function for the user to be able to adjust the geometries manually if necessary (editor buttons appear above the sketch map after vectorization is complete as seen in the left-hand image in Figure 5). The editor is also used to annotate the GCPs in the sketch map with the correct labels or identifiers so that they can be aligned correctly.

Figure 5. Vectorized sketch overlayed on original sketch map image and georeferenced version overlayed on base map.

3.2.4. Align the Sketch Map with the Base Map: Approximate Georeferencing

Georeferencing is also achieved by the user with one click of a button. The user clicks on Align-to-Geometries and in the background SmartSkeMa transforms the coordinates of the polygons in the vectorization of the sketch map into the coordinate system of the base map using the GCPs as anchor points.

Here we use a simple polynomial regression model. That means GCPs are paired: one from the sketch map to one from the base map. Each GCP has a x- and a y-coordinate. Assuming linearity in the relationship between the sketch map and base map, one forms a systems of simple linear equations relating the left-hand side coordinates to the right-hand side ones with two unknowns that become the coefficients of the transform from the coordinate space of the sketch map to the coordinate space of the base map. This is similar to defining an affine coordinate transformation between two planar geographic coordinate projection systems [26]. The map in the left-hand side panel in Figure 5 shows the georeferenced vector layer derived from the loaded sketch map.

In case a base map wasn't used during sketching and therefore no nearly affine relationship exists between the sketch map coordinates and the base map coordinates an alternative approach is taken [27]. First qualitative spatial relations are computed using the methods reported in Schwering et al. [28]. These relations include observable facts such as the fact that a feature is contained *inside or outside, near or far, adjacent or disjoint,* etc. relative to another feature. The types of relations used depend on the topological form of the feature representation (0-dimensional, 1-dimensional, or 2-dimensional). Using the set of relations between features in the sketch map and the set of relations between features in the base map together with any feature type or feature identifying information (e.g., a specific *school* identified by its name), the most likely feature correspondences between the two maps can be computed. The algorithms used in this process have been reported in Chipofya [29] and Chipofya et al. [30,31]. The output is the set of spatial relations between the elements of the two maps where the algorithm has determined some elements as spatially coinciding with each other.

3.2.5. Enter Land Tenure Relations (Rights, Responsibilities, Restrictions)

The last task involving data manipulation that the user must perform is to add social and tenure information associated with the mapped features if any. SmartSkeMa provides a simple form for this which allows only drag and drop interactions (Figure 6). Basic data about persons and institutions concerned are already parsed into the domain model upon upload in the first step of the workflow (Section 3.2.1). These pre-parsed data are shown at the top row in the form in Figure 6. The user selects individuals, groups of individuals, institutions (incl. clans and families) drag them to the Selected Related Elements row. S/he must then choose the tenure relations to declared (under Activities and Statuses) and drag them to the Selected Related Elements row. S/he can also drag conditional facts into the last position of the Selected Related Elements row as explained at the beginning of this section. The ⊞ button confirms the selected relation. This is repeated for as many relations as there are to be declared.

On the user side everything works as if they are simply manipulating text objects. However, the system will reject any relation that is invalid according to the local domain model defined for that particular session of SmartSkeMa. This is intended to be a domain model representing the local or cultural rules relating to land within the community being mapped.

So how does this domain model look and where does come from? The domain model is a Web Ontology Language (OWL) model. It is simply a conceptual tree structure where the nodes represent the concepts of interest together specification about binary relations between any concepts in the tree. The relationship between child and parent nodes in the tree is that the child node is a subconcept of the parent.

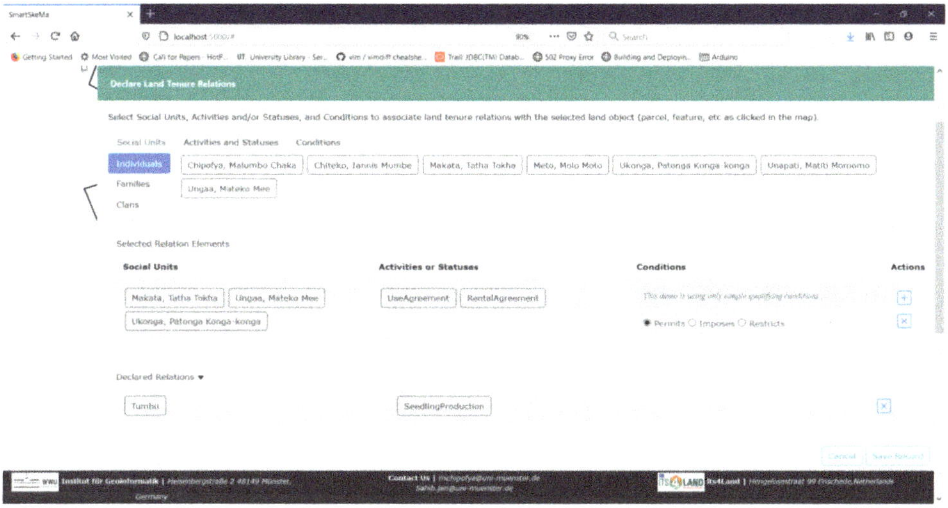

Figure 6. Form for declaring social relations and associations recorded during data collection in the field.

Figure 7 shows the top two levels of the concept tree of a local domain model used during one of our studies called the MSKDM [32]. The domain model has high level classes *Activity*, *SocialUnit*, *Livestock*, and *EnvironmentalCharacteristic*. As an example, Maasai culture revolves around their semi-nomadic pastoralism. Thus, a Maasai clan and its members can be represented as *SocialUnit* instances. Grazing their *Livestock* is an *Activity* which can be carried out in a geographic region that is modeled as an *EnvironmentalCharacteristic*. The significance of having geographic regions modelled as *EnvironmentalCharacteristic* is that in this particular domain the most significant aspect of tenure is not the specific geographic location but rather its ecological properties. More examples based on the MSKDM are presented in Karamesouti et al. [32] and Chipofya et al. [16].

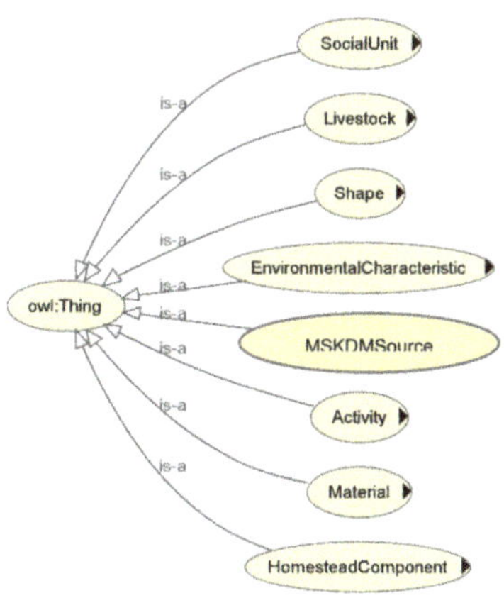

Figure 7. Extract of owl concept-tree from the MSKDM [32]. The top concept is called *Thing* and is a super-concept of all other concepts in the model. Each lower concept in the tree is a specialized version of its parent concept.

The reasoning process uses a series of rules and the predefined information of infer new information. Figure 8 shows a schematic of how the non-membership status of an individual to a family can be led to the conclusion that the individual has no right to conduct a particular activity on a particular piece of land. In this case it may assumed that the parcel is the family's pasture. The rule applied behind the scenes is:

hasValidatedExclusiveRightTo(SocialUnit: x, Activity: y)

and

isNotMemberOf(SocialUnit: z, SocialUnit: x)

implies hasNoRight(z, y).

Data entered in SmartSkeMa can be translated into LADM complaint objects—i.e., objects that are instances of LADM classes. This part is also based on the reasoning capabilities described above. The idea is to interpret certain instances as Parties, other instances as RRR, others as Basic Administrative Units, Spatial Units, etc. For example, the Activity *WaterAnimalsRanch1* can be interpreted as a right with *Kashu Family* being the party. A detailed description of the design and workings the local domain model can be found in [16,32]. The rules for mapping from local domain model concepts to LADM concepts are detailed in [16].

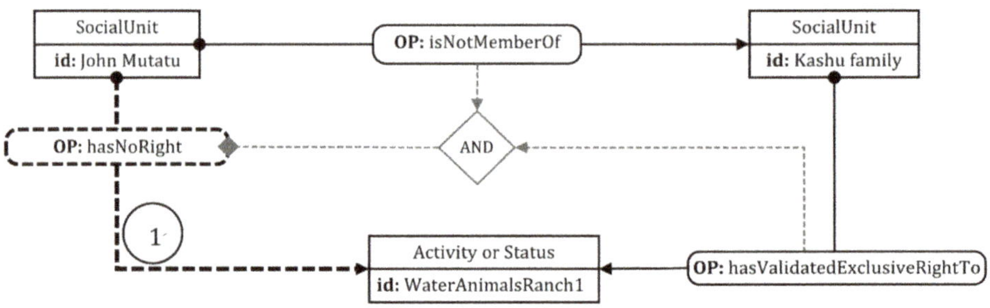

Figure 8. Schematic of reasoning process by which the reasoner infers new facts from input data in the domain model [32]. Rectangles represent concept instances, rounded rectangles represent relations between concept instances, and diamond shapes are logical connectors that take one or more relations as inputs and it outputs other concept relations. Here the output is the relation *hasNoRight* between the SocialUnit instance *John Mutatu* and the Activity instance *WaterAnimalsRanch1*.

3.2.6. Download the Data to a Chosen Location

After all processing is complete, the data can be downloaded to a folder on the local computer and be used in further processes.

3.3. *Summary*

SmartSkeMa implements a variety of functions designed to make the digital processing of data obtained through participatory mapping in customary and informal setting more streamlined, simple, accurate. However, it is more than that. By attempting to represent local norms, rules, and recognized social relations, SmartSkeMa brings the prospect of making customary land rights mainstream.

4. Expert Impressions of SmartSkeMa

During the development phase of SmartSkeMa Land Administration Experts from several countries were engaged to gain an understanding of how experts judged the value proposition of SmartSkeMa. At workshops in Kajiado and Nairobi, Kenya, 20 experts answered questions regarding how and where they believed SmartSkeMa could be applied in Kenya.

In this section we present results from the workshop. In the discussion we compare the results from the study to another study where participants were asked to rank SmartSkeMa, among three other tools, on each of the seven elements of the FFP-LA paradigm [33].

4.1. Materials and Methods

The workshops were organized in three sessions with a roughly equal number of participants in each participating in each session. The participants were invited to the workshop to obtain a mix of land sector actors with different roles and expertise. The group consisted of:

1. Kenyan government officials from the departments of surveys and physical planning, the land registrar's office, and the National Land Commission;
2. Land Administration researchers from four Universities in Kenya;
3. Private companies involved in the land sector in Kenya;
4. Non-governmental Organizations involved in land rights mapping and advocacy.

In each session a demonstration of SmartSkeMa was presented to the participants. This included a presentation of the preparation steps, the data collection process, and a display of the tools used in the field. The demonstration was based on data collected in a field study in Kajiado, Kenya. Participants were allowed to ask questions about the SmartSkeMa Tenure Documentation workflow and the tool's functionality.

The second part of the session comprised of discussions about the applicability of SmartSkeMa for community land documentation in Kenya. The discussions were guided by a set of questions which participants were first requested to answer independently on a questionnaire before sharing their responses with the group. The discussion was then used to seek clarification on the responses shared by individual participants. The written responses consisted a Yes/No/Maybe option and free text part for verbose descriptive answers to justify the Yes/No/Maybe response.

4.2. Results

The results presented here focus on four questions. The verbose responses were coded by two researchers as recommended for qualitative analysis [34] and then combined to obtain the summary justifications given by each respondent. Participants were asked to indicate whether the tool could:

1. be used in conjunction with standard land administration systems at county or national level;
2. be used independently by community-based organizations, national civil society organizations, faith-based organizations working in the land sector;
3. enable communities to register and govern their lands and natural resources locally according to local cultures and customs.

Pie charts summarizing the responses are shown in Figure 9. In response to question 1, 2 out of 20 respondents thought SmartSkeMa could not be used in that context, 4 indicated that there may be some use for it but also raised doubts (response: May be), while 14 responded with a 'Yes'. The justification given for the 'No' responses were related to low expected accuracy of the mapped points. The most common justifications for the 'Yes' responses were (i) interpretation of the actual relations on the ground, (ii) mainstreaming or capturing the community perspective, (iii) the participatory nature of the tool, and (iv) the ability to map shared or communal resources. These correspond well to the response to the next question presented below.

For question 2 the one respondent who responded 'No' gave the reason that "Ground truthing is necessary and the basemap [must be] well done to show what one has collected in the field—especially on legal [aspects]". This is of course anti-thetic to the assumptions behind SmartSkeMa's design. 17 participants responded with 'Yes' indicating conflict resolution, planning, capturing attribute and legal data as potential applications. 'Maybe' respondents expressed that its application would be require professional guidance.

The reasons given for the 'Yes' responses to question 3 were quite varied but the dominant themes included

1. potential to reduce conflicts;
2. use in resource use/management and planning (also at the community level);

3. And related to the above, that it is useful for planning purposes.

Unsurprisingly, those who answered 'No' mentioned that there were already preexisting legal restrictions on permissible land uses and that the outputs would not meet legal thresholds for surveys.

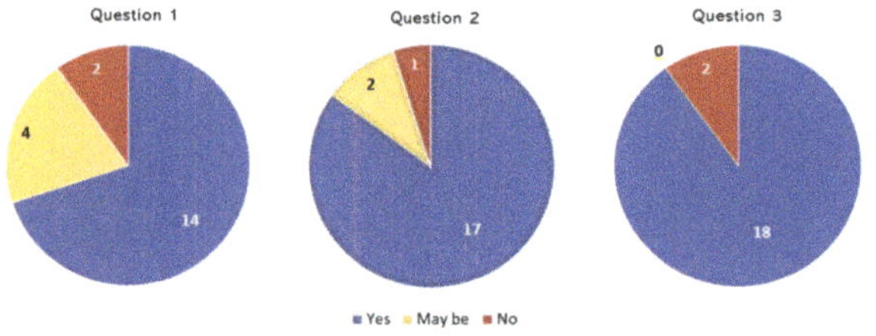

Figure 9. Expert opinions on whether or not SmartSkeMa could be used (i) in conjunction with standard Land Administration systems, (ii) by non-governmental land sector organizations, and (iii) to enable land tenure self-governance by communities.

Respondents were also asked to mention areas where they saw SmartSkeMa being applied within their domain of work. The responses were summarized into categories and the most common categories are shown in Figure 10. The most prevalent application areas fell under the category of "Participatory capture of community view on land (mapping/appraisal/consultation)". However, respondents also mentioned application areas which we grouped under the following categories with high frequency (7): conflict resolution, land use planning, and documentation of varied tenure information.

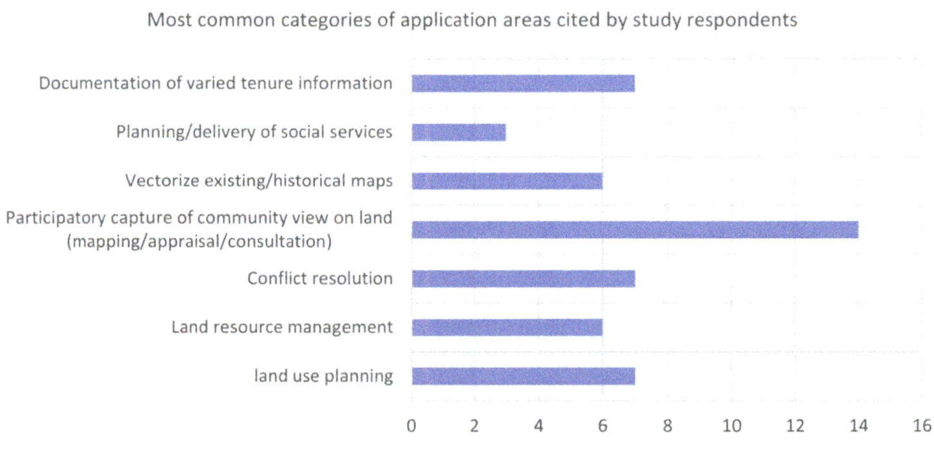

Figure 10. Expert opinions on the possibility of using SmartSkeMa in conjunction with standard land administration systems.

4.3. Discussion

The observations made by respondent in the study reported above were matched with a separate survey [33] conducted in January 2021 where respondents were asked to rank SmartSkeMa on each of the Fit-For-Purpose dimensions of Flexibility, Inclusivity, Participation, Affordability, Reliability, Attainability, and Upgradability.

Of the 14 participants who took part in the latter survey 10 rated SmartSkeMa with a score of at least 4 on a 5-point scale for the dimensions Flexibility, Inclusivity, and Participation while 7 participants rated the Reliability dimension with a score of 3 (don't know) indicating, among other things, that data produced with the tool could not be used as "authoritative data" and would not meet "legal precision and accuracy requirements"—see [33] for more details. This result should not be surprising. Rahmatizadeh et al. [35] in a study on parameters that affect how fit-for-purpose spatial data collection tools are selected also found that over 50% of respondents considered accuracy to be one of the two most important criteria.

The results above are significant as they highlight what the expert community considers as the weak points for the SmartSkeMa system and its workflows: it has not yet been proven to meet legally stipulated accuracy thresholds for geolocation (i.e., spatial) measurement in any jurisdiction. This is significant to attain government buy-in and acceptance of data collected with the tool. Experiments by Asiama et al. [36] in the Dagbon region of Ghana found that although the accuracy of mapping with a GPS equipped mobile phone ranged between 1 m and 5 m there were not many disputes between neighbors regarding the position of mapped parcel borders. This could be explained by the fact that during the mobile mapping all neighbors were always present [36].

Buy-in can also be achieved through acceptance by the community as a whole including local authorities such as chiefs and community/village heads men and women. The Global Land Tool Network and its partners, for example, achieved a community level of acceptance in their land rights documentation projects in Zambia [37] and Uganda [38] through engagement of the chiefs and the community. The projects used STDM as the software tool to support mapping and attribute capture.

The strength of SmartSkeMa as shown in both studies is in its participatory nature and its ability to capture culturally relevant local land tenure categories. It is our impression that all other positive ratings obtained from this study directly follow from these two attributes of the tool. Juxtaposing the criticism of poor accuracy with the positive evaluation for participation one notices a discrepancy between the legal requirements for recognition of land records and the needs of communities whose tenure on land is not legally recognized.

On one hand mapping the vast areas covered by customary land tenure regimes using conventional methods within any reasonable time frame is a near impossible task. Simplifications are possible by engaging teams of non-professional mappers as done in Rwanda. But this solves only one of the problems. The solution must also account for local custom. As such legal requirements for the recognition of customary land right must accept lower thresholds for the measurement accuracy of the maps used. Conflicts, which are what most land professionals fear would result from low accuracy maps, can be dealt with using existing conflict resolution mechanisms within community structures.

5. Conclusions

There are several tools for documenting customary land rights. Most of these tools are designed to be easy and cheap to use. Most are also designed to work with a variety of land rights. The challenge, however, is to be able to document land rights and their dynamics as they truly function. In customary settings this is an essential gap as recorded land rights may differ significantly with the actual experiences of land users. SmartSkeMa attempts to close the gap by modeling land tenure concepts as closely as possible. This allows SmartSkeMa to support both the legibility of customary land tenure to government authorities and the preservation of the customs within which the tenure relations operate. Preserving customary rights to land requires also preserving customary ways of allotting, negotiating, and exercising those rights. Otherwise, the entire notion of customary land tenure itself becomes a shell or a cover for replacing customary tenure with statutory tenure.

In this paper we introduced the functionalities of SmartSkeMa, presented a workflow for using it, and described some the underlying technical details. SmartSkeMa has functions to make participatory mapping and recordation of land rights accessible to non-professionals. Setting it up, however, may require a good understanding of several diverse skills. An OWL ontology needs to be set up, base map data need to be obtained, and field work materials need to be prepared. Once the setup is done, however, only a little training may be required to allow a community to independently document their lands.

Overall SmartSkeMa received positive feedback from two separate groups of experts in two different studies. SmartSkeMa's strength is in its ability to include detailed local knowledge into a land information database. Many respondents to our first study noted that SmartSkeMa had the potential reduce land related conflicts. This is a point to be investigated further in actual use of the tool in practice.

Going forward there are two directions for advancing our work. First, we will eliminate the need for using a transparent overlay in order to trace over a map. One approach to do so is to apply image registration on the scanned field sketch with its original. If good approximate alignment can be achieved, then taking the image differences would eliminate original patterns leaving only the sketched ink. The other extension has to do with facilitating communities' self-governance of land tenure. We believe this could facilitated by user friendly tools built on top peer-to-peer overlay networks and decentralized systems, including blockchain networks.

Author Contributions: Conceptualization, M.C.C., S.J. and A.S.; methodology, M.C.C. and S.J.; software, M.C.C. and S.J.; formal analysis, M.C.C.; data curation, M.C.C.; writing—original draft preparation, M.C.C.; writing—review and editing, M.C.C. and S.J.; visualization, M.C.C. and S.J.; supervision, A.S.; project administration, A.S. and M.C.C.; funding acquisition, A.S. and M.C.C. All authors have read and agreed to the published version of the manuscript.

Funding: Funding of the publication costs for this article has kindly been provided by the School of Land Administration Studies, University of Twente, in combination with Kadaster International, the Netherlands. The research was partially funded by the research project "its4land," which is part of the Horizon 2020 program of the European Union, project number 687828.

Institutional Review Board Statement: Not applicable.

Informed Consent Statement: Not applicable.

Data Availability Statement: Not applicable.

Acknowledgments: The authors acknowledge the contributions by other members of the its4land project to the successful completion of the fieldworks and workshops.

Conflicts of Interest: The authors declare no conflict of interest. The funders had no role in the design of the study; in the collection, analyses, or interpretation of data; in the writing of the manuscript, or in the decision to publish the results.

References

1. Scott, J.C. *Seeing Like a State: How Certain Schemes to Improve the Human Condition Have Failed*; Yale University Press: New Haven, USA, 1998.
2. Abubakari, Z.; Richter, C.; Zevenbergen, J. Making Space Legible across Three Normative Frames: The (Non-) Registration of Inherited Land in Ghana. *Geoforum* **2020**, *108*, 217–226. [CrossRef]
3. United Nations Human Settlements Programme, UN-Habitat. Fit-For-Purpose Land Administration: Guiding Principles for Country Implementation. 2016. Available online: www.unhabitat.org (accessed on 25 November 2020).
4. Koeva, M.; Stöcker, C.; Crommelinck, S.; Chipofya, M.; Kundert, K.; Schwering, A.; Sahib, J.; Zein, T.; Timm, C.; Humayun, M.; et al. Innovative Geospatial Solutions for Land Tenure Mapping. *Rwanda J. Eng. Sci. Technol. Environ.* **2020**, *3*, 3. [CrossRef]
5. Koeva, M.; Bennett, R.; Gerke, M.; Crommelinck, S.; Stöcker, C.; Crompvoets, J.; Ho, S.; Schwering, A.; Chipofya, M.; Schultz, C.; et al. Towards innovative geospatial tools for fit-for-purpose land rights mapping. In Proceedings of the International Archives of the Photogrammetry, Remote Sensing and Spatial Information Sciences-ISPRS Archives, Moscow, Russia, 15–17 May 2017; Volume 42.
6. Enemark, S.; Bell, K.; Lemmen, C.; McLaren, R. *Fit-for-Purpose Land Administration*; International Federation of Surveyors (FIG): Copenhagen, Denmark, 2014.
7. ISO/TC 211 Geographic Information/Geomatics. *ISO 19152:2012 Geographic Information—Land Administration Domain Model (LADM)*; International Organization for Standardization: Geneva, Switzerland, 2012.
8. Lemmen, C.; Van Oosterom, P.; Bennett, R. The Land Administration Domain Model. *Land Use Policy* **2015**, *49*, 535–545. [CrossRef]
9. Williamson, I.; Enemark, S.; Wallace, J.; Rajabifard, A. *Land Administration for Sustainable Development*; ESRI Press: Redlands, CA, USA, 2010.
10. United Nations Economic Commission for Europe. *Land Administration Guidelines: With Special Reference to Countries in Transition*; Report: ECE/HBP/96; United Nations: New York, NY, USA; Geneva, Switzerland, 1996.

11. Malinowski, B. *Coral Gardens and Their Magic. Vol I: Soil-Tilling and Agricultural Rites in the Trobriand Islands*; George Allen & Unwin: London, UK, 1935.
12. Meek, C.K. *Land Law and Custom in the Colonies*, 2nd ed.; Oxford University Press: London, UK, 1949.
13. Lemmen, C.; Oosterom, P.V.; Thompson, R.J.; Hespanha, J.; Uitermark, H. The modelling of spatial units (parcels) in the land administration domain model (LADM). In Proceedings of the XXIV FIG International Congress 2010: Facing the Challenges—Building the Capacity, Sydney, Australia, 11–16 April 2010.
14. Lemmen, C.; Augustinus, C.; van Oosterom, P.; van der Molen, P. The social tenure domain model: Design of a first draft model. In Proceedings of the FIG Working week; 2007.
15. Lemmen, C.H.J.; Augustinus, C.; du Plessis, J.; Laarakker, P.; de Zeeuw, K.; Saers, P.; Molendijk, M. The operationalisation of the Continuum of Land Rights at Country Level. In *Linking Land Tenure and Use for Shared Prosperity, Proceedings of the Annual World Bank Conference on Land and Poverty, Washington, DC, USA, 23–27 March 2015*; p. 27.
16. Chipofya, M.; Karamesouti, M.; Schultz, C.; Schwering, A. Local Domain Models for Land Tenure Documentation and their Interpretation into the LADM. *Land Use Policy* **2020**, *99*, 105005. [CrossRef]
17. FAO and MRLG. Challenges and Opportunities of Recognizing and Protecting Customary Tenure Systems in Cambodia, Bangkok, Thailand. Licence: CC BY-NC-SA 3.0 IGO. 2019. Available online: http://www.fao.org/3/CA1034EN/ca1034en.pdf (accessed on 1 December 2019).
18. African Natural Resources Center. *Land Tenure Regularization in Rwanda: Good Practices in Land Reform—Case Study*; African Development Bank: Abijan, Ivory Coast, 2016. Available online: https://www.afdb.org/fileadmin/uploads/afdb/Documents/Publications/anrc/AfDB_RwandaLandReform_FA.pdf (accessed on 30 March 2021).
19. National Land Use Planning Commission. *Tools and Spatial Technologies for Village Land Use Planning: A Practitioner's Manual for Active Community Engagement*; Report of the National Land Use Planning Commission; National Land Use Planning Commission: Dar Es Salaam, Tanzania, 2018. Available online: https://www.nlupc.go.tz/uploads/publications/en1565469142-Tools%20and%20Spatial%20Technologies%20for%20Village%20Land%20Use%20Planning%20-%20Practitioner%E2%80%99s%20Manual%20for%20Active%20Community%20Engagement%20-%20FINAL.pdf (accessed on 13 December 2020).
20. Open Source Software for Recording Tenure Rights. Available online: http://www.fao.org/tenure/activities/administration/recording-of-rights/software/en/ (accessed on 21 March 2021).
21. Knight, R.; Brinkhurst, M.; Vogelsang, J. The Community Land Protection Facilitator's Guide; Namati: 2016. Available online: https://namati.org/resources/community-land-protection-facilitators-guide/ (accessed on 25 November 2020).
22. Rugema, D.M.; Verplanke, J.J.; Lemmen, C.H.J. Digital pen method. In *Advances in Responsible Land Administration*; Zevenbergen, J.A., de Vries, W.T., Bennett, R.M., Eds.; CRC Press: Boca Raton, FL, USA, 2015; pp. 131–144, ISBN 978-1-4987-1961-2.
23. Vogelsang, J.; Mbarak, H.; Wario, H.; Arach, D. Facilitator Guide: Kenyan Community Land Act; Namati: 2020. Available online: https://namati.org/resources/facilitator-guide-kenya-community-land-act/ (accessed on 25 November 2020).
24. Spatial Collective. Putting Community and Rights: On the Map in Southern Kenya. Report to the Omidyar Network. 2018. Available online: https://www.landportal.org/library/resources/putting-community-and-rights-map-southern-kenya (accessed on 21 March 2021).
25. De Berg, M.; Cheong, O.; van Kreveld, M.; Overmars, M. *Computational Geometry: Algorithms and Applications*, 3rd ed.; Springer: Berlin/Heidelberg, Germany, 1997; ISBN 978-3-540-77973-5.
26. Iliffe, J.; Lott, R. *Datums and Map Projections for Remote Sensing, GIS and Surveying*, 2nd ed.; Whittles Publishing: Dunbeath, Scotland, 2000; pp. 109–122.
27. Chipofya, M.; Jan, S.; Schultz, C.; Schwering, A. Towards Smart Sketch Maps for Community-driven Land Tenure Recording Activities. In Proceedings of the 20th AGILE International Conference on Geographic Information Science, Wageningen, The Netherlands, 9–12 May 2017.
28. Schwering, A.; Wang, J.; Chipofya, M.; Jan, S.; Li, R.; Broelemann, K. SketchMapia: Qualitative Representations for the Alignment of Sketch and Metric Maps. *Spat. Cogn. Comput.* **2014**, *14*, 220–254. [CrossRef]
29. Chipofya, M. Matching qualitative constraint networks with online reinforcement learning. In *GCAI 2016, 2nd Global Conference on Artificial Intelligence, Berlin, Germany, 29 September–2 October 2016*; Benzmueller, C., Sutcliffe, G., Rojas, R., Eds.; EPiC Series in Computing EasyChair; EasyChair: Manchester, UK, 2016; Volume 41, pp. 266–279.
30. Chipofya, M.; Schultz, C.; Schwering, A. A metaheuristic approach for efficient and effective sketch-to-metric map alignment. International. *J. Geogr. Inf. Sci.* **2015**, *30*, 405–425. [CrossRef]
31. Chipofya, M.; Schwering, A.; Binor, T. Matching Qualitative Spatial Scene Descriptions á la Tabu. In *Advances in Soft Computing and Its Applications*; Lecture Notes in Computer Science; Castro, F., Gelbukh, A., González, M., Eds.; Springer: Berlin/Heidelberg, Germany, 2013; Volume 8266.
32. Karamesouti, M.; Schultz, C.; Chipofya, M.; Jan, S.; Galeano, C.E.M.; Schwering, A.; Timm, C. The Maasai of southern Kenya Domain Model of Land Use. *ISPRS Ann. Photogramm. Remote Sens. Spat. Inf. Sci.* **2018**, *4*. [CrossRef]
33. Koeva, M.; Humayun, M.I.; Timm, C.; Stöcker, C.; Crommelinck, S.; Chipofya, M.; Bennett, R.; Zevenbergen, J. Geospatial Tool and Geocloud Platform Innovations: A Fit-for-Purpose Land Administration Assessment. *Land* **2021**, *10*, 557. [CrossRef]
34. Bengtsson, M. How to plan and perform a qualitative study using content analysis. *NursingPlus Open* **2016**, *2*, 8–14. [CrossRef]
35. Rahmatizadeh, S.; Rajabifard, A.; Kalantari, M.; Ho, S. A framework for selecting a fit-for-purpose data collection method in land administration. *Land Use Policy* **2018**, *70*, 162–171. [CrossRef]

36. Asiama, K.; Bennett, R.; Zevenbergen, J. Participation, innovative approaches and customary cadastres: A practical experiment in Nanton, Ghana. In Proceedings of the Annual World Bank Conference on Land and Poverty, Washington, DC, USA, 19–23 March 2018.
37. Antonio, D.; Gitau, J.; Sylla, O. Innovative Customary Land Governance in Zambia: Experiences, Lessons Learned and Emerging Impacts. In Proceedings of the FIG Working Week 2019, Geospatial Information for a Smarter Life and Environmental Resilience, Hanoi, Vietnam, 22–26 April 2019.
38. Antonio, D.; Makau, J.; Mabala, S. Piloting the Social Tenure Domain Model in Uganda. Innovative Pro-Poor Land Tools under Implementation. GIM International Magazine. 2013. Available online: https://www.gim-international.com/content/article/piloting-the-social-tenure-domain-model-in-uganda (accessed on 1 June 2021).

land

MDPI

Article

Good Practices in Updating Land Information Systems that Used Unconventional Approaches in Systematic Land Registration

Mireille Biraro [1,*], Jaap Zevenbergen [2] and Berhanu Kefale Alemie [1]

[1] Institute of Land Administration, Bahir Dar University, Bahir Dar P.O. Box 79, Ethiopia; berhanu.kefale@bdu.edu.et

[2] Faculty of Geoinformation Science and Earth Observation (ITC), University of Twente, 7522 NB Enschede, The Netherlands; j.a.zevenbergen@utwente.nl

* Correspondence: biraro.mireille@ines.ac.rw

Abstract: To properly govern people-to-land relationships, there is a need to formally recognize land rights, and for this to bring recognizable societal change, the established Land Information System (LIS) has to be updated continuously. Though existing literature suggests different parameters to consider when updating an LIS, little is said on how countries are doing this, especially when unconventional approaches through systematic land registration were initially used. This paper comes up with recommendable good practices where the suggested needs for updating land records were made workable. Nine countries with similar data collection procedures for the initial registration were selected based on literature study; questionnaires designed and distributed to LIS experts from each country using internet; and the collected data were analyzed qualitatively. Fortunately, all the case countries possess infrastructure supporting land records updating (procedures, mobilization means, institutional and legal frameworks, and so on). For the majority, the systems are simplified; registration fees are reasonable; services are decentralized; the database is accessible by citizens and local officers; staff are trained; the system effectiveness is assessed; and the political support is offered. However, the procedures are long, data sharing is inexistent, financial and technical sustainability is uncertain, and many different institutions are involved in the registration. Whilst updating used to appear as a forgotten activity, good practices now exist.

Keywords: good practices; updating land records; systematic land registration; unconventional approach

Citation: Biraro, M.; Zevenbergen, J.; Alemie, B.K. Good Practices in Updating Land Information Systems that Used Unconventional Approaches in Systematic Land Registration. *Land* **2021**, *10*, 437. https://doi.org/10.3390/land10040437

Academic Editor: Robin McLaren

Received: 18 March 2021
Accepted: 15 April 2021
Published: 20 April 2021

1. Introduction

Land registration emerged throughout the history of humanity for publicity of land records. This helps landholders and land administrators to know what belongs to whom and for the State to know all land units liable for taxation or other services [1]. Countries are using different methods when establishing their Land Information System (LIS) as there is no 'one-fit-for-all type' approach. LIS refers to a system made of land records; human and technical resources; and appropriate procedures and techniques to collect, analyze, maintain, disseminate, and use this information [1,2]. These records on land can be collected through sporadic or systematic land registration depending on the purpose, on available technologies, and/or on who is funding the activity (donors, government, or landholders) [3]. The systematic land registration is often compulsory, done plot by plot to cover the entire area under registration, regardless of the wishes of landholders [2,4]. Sporadic land registration, on the other hand, can either be voluntary (at the initiative of the landholder) or compulsory (during land transfer) [2]. Systematic registration is strongly recommended in the case of first registration [1] as it allows to have a complete database for the area under registration, low cost per unit of registration (as the registration covers many units simultaneously [1]), less recorded disputes (since there is a possibility to appeal and the registration is done in the presence of local community [3]), and a faster process (as photogrammetric techniques are often used to demarcate boundaries). However, the initial cost may be high due to the fact that resources are needed for photogrammetric mapping, to equip registration offices, train staff, hire external experts to design the procedures and proceedings, and so on [1,3]. That is why this approach is usually adopted through government- or donor-funded projects [3].

Systematic land registration is the one that is mostly recommended to developing countries [1,4,5] due to insufficient financial and human resources and yet the property rights have to be registered. As this registration type covers a vast area and involves many people, unconventional approaches that are cheap and faster are suggested instead of using standardized methods that are longer and expensive. Contemporary philosophies in the area suggest the fit for purpose [6] and pro-poor land recordation [7] approaches when recording land as they are flexible, inclusive, participatory, affordable, and upgradable. These unconventional approaches have become popular in the developing world. Kenya, Nepal, and Thailand are the oldest cases [1,3,4,8]; while Ethiopia, Namibia, and Rwanda are the recent ones [6,9–12]. Though there are some particularities in each program, in all these cases, a considerable number of parcels was systematically registered through the use of unconventional approaches.

Nowadays, keeping the collected land records up to date seems to be more challenging than establishing an LIS. Though existing literature suggests different parameters to consider when updating an LIS, little is said on how countries are doing this, especially when unconventional approaches through systematic land registration were initially used. This paper provides good practices used to update an LIS that is originally developed through unconventional approaches and systematic land registration.

The paper contains six sections. The first one gives a background followed by the methodology used for data collection. Theoretical perspectives on updating come as the third section whereas a description of the land registration programs in the case study countries follows. The fifth section presents the analysis and the discussion done on the findings and the conclusion comes as the last and sixth section.

2. Materials and Methods

The selection of countries that have systematically registered their land by using unconventional approach was done through an extensive systematic literature review. The snowballing approach, often used in systematic literature review [13,14], helped to find publications on various land registration programs. The first entry point was to find a set of publications done on land registration programs and to perform a backward iteration (looking in the list of the initial reference) and a forward iteration (identifying references citing the initial references) [13]. 'Fit for purpose land administration: guiding principles for countries implementation' [5] and 'Securing Africa's Land for Shared Prosperity: a program to scale up reforms and investments' [11] are two publications that helped in this exercise. Seventeen (17) countries were found in the first round, some referred to as those that successfully implemented a national land registration using an unconventional approach [5] while others conducted land governance reforms by using faster and cost effective approaches [11]. To find more about the registration programs in each of these countries, backward and forward iterations were performed by using the references provided in the two publications.

The second entry point was the cadastral template website (www.cadastretemplate.org (accessed on 6 May 2020)) as it provides extensive information on the historical development of cadastral systems in different countries. By the time of accessing this website, fifty-seven (57) countries around the world had uploaded information concerning the development of their cadastral system. As this paper focuses on countries that systematically registered land, a selection was done and only fifteen (15) were kept (as they indicated that they used this approach).

From the two rounds, thirty-two (32) countries were identified. As Ethiopia appeared twice, the total number became thirty-one (31) countries. For all these countries, more publications were searched in order to find out when the initial land registration was done, for what purpose, and how land records were collected. The main sources of documentation were World Bank and FIG publications as experiences from these registration programs are often shared in their annual conferences. Additionally, a general search on World Wide Web through google scholar was done in order to access international journals that published work on the registration programs from the selected countries. Throughout the reading, two more countries that systematically registered their land were added making a total of thirty-three (33) countries (see the Appendix A for the list).

Guided by the question of how the land records were initially collected, a further refinement was done. Among the thirty-three (33) countries, eleven (11) were kept. These are countries that did a systematic land registration (a considerable number of parcels were registered), the registration program was not a pilot project, and unconventional approaches were used. Regarding the use of unconventional approaches, not all the criteria provided by Enemark [6] and Zevenbergen [7] could be assessed through literature. That is why a few criteria were applied in the assessment being *participatory and flexible mapping* (which does not require high professional training); *affordability* (no expensive process); and the *co-management of the process* (involvement of local citizens) [6,7]. Though Armenia and Vietnam fulfilled the selection criteria, they were dropped at the end as the contacted land experts were unable to participate in the research. The final nine (9) countries for which the land registration programs were explored in depth are listed in Table 1.

Table 1. Explored land registration programs.

#	Country	Land Registration Program
1	Ethiopia	1st and 2nd level rural land registration (in 1998 and 2013 until now)
2	Kenya	Registration of Native Reserves (1965–1966 until now)
3	Rwanda	Land Tenure Regularization (2009 to 2013)
4	Thailand	Land Titling Program (1984 to 2004)
5	Kyrgyzstan	Registration of Land and Real Estates (2000 to 2007)
6	Namibia	Communal Land Registration (2003 up to now)
7	Nepal	Systematic registration (1965 to 1995)
8	Mexico	Certification of *Ejido* (communal land) (1992 to 2006)
9	Cambodia	Land Management and Administration Project (2002 to 2013)

For these countries, contact was made with LIS professionals working in the land sector and researchers who published on the registration programs in order to know mechanisms in place to update the collected land records. Most literature suggests sixteen (16) parameters to consider when updating an LIS sustainably. Accordingly, a questionnaire with both open-ended and closed questions was designed. Questions were asked about the identified registration programs, the existing process for updating the collected land records, and the technical structure of the database storing these records. The questionnaire was distributed to at least two (2) LIS experts for each country using internet (email, survey monkey or a virtual video discussion). The selection was purposively done, privileging those who could potentially provide the needed information. Snowballing also helped to reach more experts as those contacted linked to others. Out of the seventy-eight (78) people contacted via email, forty-five (45) replied among which twenty-eight (28) responded to the questionnaire. The collected data are analyzed qualitatively.

3. Theoretical Perspectives on Updating Land Records

Land records are very dynamic, as people-to-land relationships do change regularly, and need to be kept up to date. The term updating refers to the recording of any change from the real world into the system as soon as it occurs. It is a daily exercise that requires adequate level of trained professionals and communities willing to contribute to the updating process [15].

Capturing the real-time change of land rights into the database is critical as without it, the land system loses relevance and is replaced by an informal system [3]. Without regular updating, there is a high risk of going back to the previous situation, and even worse. Outdated land records may be used to cheat those who are less knowledgeable, thus creating a critical situation for the Land Administration System (LAS). The later refers to an infrastructure of implementing land policies and land management strategies in support of sustainable development [3]. LIS is a fundamental infrastructure for LAS [3] as it allows to secure land rights, undertake land/property valuation and taxation, plan and control the use of land, and implement utilities/infrastructures and construction planning [6].

When the database is up to date, records on land tenure allow a secured land market: Potential buyers may check whether they are dealing with 'all' legitimate right holders or if there is no encumbrance attached to the right (unpaid taxes, disputes, etc.) [16,17]. Thus, the buyer would be safe when paying or, later on, investing in the acquired land as there is no fear for probable eviction by other right holders who were not informed about the transaction.

With updated records on land value, banks may trust the system and rely on its information to issue loans by accepting land as collateral [18–20]. The loan can even be based on the recorded value, in the database, of the land/property being mortgaged. However, if the banks do not trust the system, using land to get a loan may be difficult, even impossible, albeit it is stipulated in the law [21].

The taxation process may be done smoothly when the records on location and value are up to date. It may be possible to identify taxable land/property and their holders. Thus, the government would rely on the revenues from land taxes/leases when planning the yearly national budget.

The updated records about how the land is used are crucial for planning purpose and for efficient land management [3,18]. It allows monitoring changes. The national land use plan may be developed based on the actual situation and avoids unnecessary expropriation and/or relocation of people. For all those places for which the use has to be changed, a proper implementation plan may be done based on updated evidence. Encroachment into sensitive areas (wetlands, forests, water bodies) or national utilities (airport, roads) may not occur as any change from the existing situation is controlled. Well-controlled land use may lead to land development. The development projects or private investments are carried out on suitable and/or vacant land. It may be known beforehand that the suitable land for a particular investment/project is (not) vacant, thus proper measures must be taken for a smooth implementation. Besides, in the case of negotiation, expropriation, or relocation, information on the legitimate right holders, their rights, what they use the land for, and the value of their properties is obtained from the system and the planning done accordingly. It may save money and time supposed to be spent on new surveys.

The information on *who is the person* (natural/non-natural) that has *what type of right* on *which piece of land* has to be known as it is. Unfortunately, the dynamicity of people-to-land relationship is sometimes forgotten as it is even the most challenging part in the establishment of the system [22]. In the initial phase, mobilization is done to incite landholders to register their rights; the government, very often supported by donors, avails money and other resources to get the people's rights registered [23]. However, once the records are collected and legal documents issued, there is less pressure to keep the system up to date [5]. Despite the fact that many establishment efforts are an initial success, some attempts fail in the end because no one thought about what comes next after the project donor and expert team have left [15,22,23]. No adequate planning for the updating phase is made as the team and the budget were concentrated on the establishment phase. Yet, to update the system requires ongoing resources, energy, and skills [15].

Three perspectives are identified in the literature regarding how to design an updating system. The first one suggests what to not ignore (such as security, affordability, simplicity, speed) when establishing a LAS [4–7,24–27], whereas the second perspective focuses on providing frameworks or ways to evaluate or measure how the established system is successful or performing [16,28–30], and the last view explicitly explains what to consider in the updating of land records [1,5,17,31–34]. Although these three perspectives are specifically related to the establishment, evaluation, and maintenance of the LAS, it is argued that some of them are cross cutting and very useful in the updating of land records. This is because the initial design of the system has an impact on its future performance and how easy the updating will be. For instance, the registration procedure can be simple, short, and affordable during the initial registration. The same will help when reporting changes in the recorded information, thus helping the system to perform better. Hence, parameters from the above perspectives that are believed to be useful when updating an LIS are summarized in Table 2. Corresponding indicators are added to indicate what is assessed in the case countries.

Table 2. Parameters and indicators to be considered when updating an LIS.

#	Parameters	Assessed Indicators	References
1	*Simplicity of the system* made of simple procedures, simple language understood by citizens and flexible ICT which does not require high expertise.	Procedure followed by citizens; Spoken language used in the database; Format of the database; Design and maintain the database;	[1,4–7,16,17,26,27,29,31,33]
2	*Affordability of the cost* to be paid by the applicant when reporting changes or of the service to be provided to the applicant. However, the received fees should allow cost recovery.	Source of funds for the initial registration; Fee for the initial registration versus fee for registering a change in land records;	[4,7,16,17,26,27,29–34]
3	*Speed* in processing the application supported by shorter procedures, completed in less days, and also faster processing capacity of the infrastructure used.	Time required to get new land documents;	[4,7,26,27,32,33]
4	*Decentralization of land services* to make them closer to the people in order to reduce the time and the cost.	Administrative location of offices and their empowerment in the registration process;	[1,7,16,26,27,29,31]
5	*Accessibility to the database* for both the citizens and local offices.	Possibility of accessing land database by local offices and citizens	[5–7,17,26,27,30]
6	*Security of the system* and for those who rely on the database to transact, to issue loan, etc.	Legal value of the issued land document; Protection of the system from deliberate abuses; Back-up the database;	[4,16,17,29]
7	*Suitability to the circumstance* by addressing existing needs and responding to the societal goals.	System addresses societal needs;	[4,28,32]
8	*System interconnectivity* to facilitate institutions cooperation and information sharing.	Land database is linked to other databases;	[5,6,28,31]
9	*Mobilization* to raise awareness about the importance/benefits of registration.	Communicate registration procedure to the citizens;	[1,5,7,26,27]
10	*Sustainability of the system* characterized by its upgradability, technological improvement as well as financially.	Source of funds for updating the system; Upgrading system structure;	[1,5,6,16,28–30,32,33]
11	*Compulsory registration* so that people have to comply with the law.	Obligation to register change in land records	[1,5]
12	*Incentives* to motivate people to register changes without delays.	Existence of incentives when updating land records	[5]
13	*Well organized administration* supported by well-defined goals and tasks of everyone involved	Institutions participating in land registration	[28,31]
14	*Continuous capacity building* of involved staff	Existence of training offered to land professionals	[24,28]
15	*Continuous assessment of the system* to know user needs, societal needs and for customer satisfaction	Availability of tools to assess the effectiveness of the land registration	[24,28,32]
16	*Political support* as decision makers	Type of support offered by the government in land registration (initial and updating phase)	[24]

4. Land Registration Programs in the Selected Countries

Land registration aims at specifying who holds what type of right on which piece of land. At the end of the exercise, collected land records include, among others, the person holding right, the form of the right, and the asset on which the right is held [18]. Developing and updating land records is a public need, which will prevent potential challenges that could lead to unsecure tenure thus hinder land investment [3]. Unfortunately, with 70–75% of land rights mostly reported as undocumented in the developing world [19,33], recording land updates is hardly done in the majority of LISs making them outdated and unreliable [35].

The land registration programs considered in this research (see Table 3) mainly aimed to ensure tenure security to rightful landholders with a particularity to vulnerable groups, boost land market and investment in land, issue legal land documents, improve the effectiveness of LAS and of the land registration specifically, prevent and reduce land disputes, and produce land records. The land records were systematically collected either for some areas (Cambodia, Ethiopia, Kyrgyzstan, Nepal, and Thailand) or for a specific category of land (Kenya, Mexico, Namibia) or for the entire territory (Rwanda). Whilst there are few particularities, these programs adopted an unconventional approach: Local people participated in the registration process, neighbors were present during boundary demarcation, images were used to collect cadastral data, and provisional land records were displayed for public inspection before the issuance of legal land documents.

In Mexico, the *ejido* (communal lands) members had to agree through a vote that the registration program can start in their area and also to approve the sketch map delimitating the community's outer boundary [36–38]. Though the land records were collected by land professionals in Kyrgyzstan, community representatives together with government officials formed a specialized land commission that received objections during public display [39]. Under the ongoing second-level rural land registration, Ethiopia is upgrading its systems by including cadastral data that were not collected during the first phase of the registration [9,40]. Kenya and Namibia are also still doing land registration, though, in the former country, the term 'native reserves' is no longer used, as the naming was intended to differentiate land held by Kenyans with the ones occupied by British settlers (Interview done on 16 January 2021).

An overview of the characteristics per country program is given in Table 3.

Table 3. Case study countries with their respective land registration programs.

Country	Program	Purpose	Approach for Collecting Land Data
Ethiopia	1st level rural land registration (1998 up to the 2nd level registration) [9,40–43]	Provide tenure security to rural farmers, reduce land disputes and address land issues had by vulnerable groups	Land records collected by the *kebele* land administration committee (elected from local people) supported by land professionals at *woreda* level. Landholders had to apply for registration and the provided information were verified in the field (in the presence of neighbors). The registered parcels were spatially located referring to neighbors in the north, south, east and west. No cadastral map was produced.
	2nd level rural land registration (2013 up to now) [44]	Upgrade the 1st level land records with cadastral records	The *kebele* land administration committee, para-surveyors and the field data recorder, all selected from the local community, formed a registration team. The parcel boundaries are drawn on aerial images in the presence of the registration team, landholders and neighbors.
Kenya	Registration of Native Reserves (1965–1966 up to now) [45–47]	Ensure tenure security of these communal land, promote intensive agriculture and put together land records	The adjudication team (appointed by the minister) and the adjudication committee (local people familiar with customary holdings) collected land records using printed aerial images. The images were taken after hedges were planted by landholders over their parcel boundaries. Within the area under registration, land claims were received before proceeding with parcel demarcation.
Rwanda	Land Tenure Regularization (2009 to 2013) [12,48–51]	Register and issue legal land documents; minimize and prevent land disputes; promote land investment and ensure tenure security to vulnerable groups.	The land committee and para-surveyors (both selected from the local people) collected land records by using printed aerial images on which were drawn parcel boundaries in the presence of landholders and their neighbors. The parcels were systematically registered without exception and claims were posted for public inspection before issuing the final land certificate.
Thailand	Land Titling Program (1984 to 2004) [52,53]	Issue title deed and tenure security to rightful owners, improve the effectiveness and sustainability of land administration institutions and produce cadastral maps	By using concrete markers (offered by the project), landholders had to mark the corner of their parcel in the presence of the village leader and the registration officer. Afterward, a technician would pass and draw the marked boundaries on a printed aerial image. All land was registered with exception to forest and land under reform.
Kyrgyzstan	Registration of Land and Real Estates (2000 to 2007) [39,54,55]	Put in place a consistent land registration, ease the existing procedures used in land and real estate transactions.	The registration team (legal and mapping specialist) was going house to house demarcating boundaries and collecting proof of ownership. In the absence of proof, the registration team tried to find the real owner in the archive. If nothing was found, a provisional registration was done until the landholder was able to prove their right. The provisional lists and maps were displayed for thirty days where land commission received objections.
Namibia	Communal Land Registration (2003 up to now) [56–60]	Register and verify customary land rights, introduce long term leases over communal land, encourage land investment and provide infrastructure support	The landholder has to apply and their claims are assessed by traditional authority (TA) and approved by the communal land board (CLB). Together with the ministry regional staff, the TA and CLB verify if the parcel is unassigned and conduct a systematic parcel mapping using GPS and aerial images. The claims are posted for seven days before the registration certificate can be issued to the claimant.

Table 3. *Cont.*

Country	Program	Purpose	Approach for Collecting Land Data
Nepal	Systematic Cadastral Registration (1965 to 1995) [61,62]	Record landowners and tenants, and what they respectively held so as to limit the ownership to land holding ceiling and protect the right of tenant farmers	The cabinet is the one to decide what area/district to be registered. After officially notifying the concerned area, land records were collected and handed to the office in charge for any claim. After settling disputes and claims, ownership certificates were issued.
Mexico	Certification of *Ejido* (communal land) (1992–2007) [36–38]	Provide tenure security to land holders in order to increase investment	Prior to the registration, a general assembly made of *ejido* members had to approve it through voting. Once approved, an *ejidal* commission drafted a map showing the inner and outer boundaries of the *ejidos* and the map was again approved by the general assembly through a vote. The INEGI (*Instituto Nacional de Estadística y Geográfica*—National Institute of Statistics and Geography) and the *ejido* members conducted a detailed registration by using a GPS. Produced maps were displayed for two weeks for objections and approved by the *ejido* assembly through voting. The RAN (*Registro Agrario Nacional*—National Agrarian Registry) would use the map as basis to issue certificates to individuals whose holdings were approved.
Cambodia	Land Management and Administration Project (2002 to 2013) [63,64]	Improve land tenure security and promote the development of efficient land markets	Cadastral staff, supported by the local authorities, filled the registration form with ownership information and demarcated parcel boundaries using ortho-photo. The lists of owners and village maps were displayed for thirty days for public inspection. Cadastral staff and local authorities tried to resolve disputing claims. The updated ownership list and maps are printed and approved by provincial governor and land titles issued based on the approved documents.

5. Good Practices in Updating Land Records from the Case Study Countries

Though ground experiences and most literature reveal that the updating of the land records seems to be a forgotten activity [15,22,23], first and foremost, resources supporting updating phase (procedures, legal, and institutional frameworks) have to be part of the land registration project plan and ready to be used [5]. This argument backs our discussion and specifically the identification of good practices in updating land records from the case-study countries. Good practices are those suggestions from land experts (see Table 2) that were made workable. Parameters considered when updating the land records are used to analyze updating procedures from countries for which land registration programs were studied (see Table 3).

Though obstacles hindering updating systems in the case study countries exist, the good news is that resources supporting the updating of land records are in place and this shows that updating is considered as important and needed. The available resources include the formal updating procedures and ways to communicate them to citizens; institutions handling land registration; government involvements in registration activities; digital database of land records; legal framework supporting registration activities; capacity building of staff involved in land registration; and means to assess the effectiveness of the registration of land. These are the first good practices to highlight. Further good practice analyses from the case study countries from the perspective of the sixteen assessment criteria identified in Table 2 are discussed here.

5.1. *Simplicity of the System*

To encourage the use of the updating system, procedures should be made simple and easier, developed in a language understood by citizens [4,16,29,31], and the technology used in the system design should be adapted to the local knowledge [5,6] as people tend to fear what they do not understand [4]. Once worked on, they are among the strongest sides to realize the simplicity of the updating system.

Based on the interviews, all the case-study countries possess formal procedures to be followed by citizens when updating land records. However, most of these procedures are not that simple as citizens need support to go through them. Fortunately, the support can be found at lower level where the services are sought. For Kenya, Nepal, and Mexico, procedures are somewhat complex as citizens have to consult land professionals (e.g., lawyers, skilled *ejidatario*, licensed people) to help them to go through the process. Rwanda is the only country possessing an online application system, which could make the process easier for those who are familiar with technology but somehow complex for unskilled citizens.

The findings show that for all countries considered here, both the design and the maintenance of the databases is done by local and foreign experts except in Nepal and Thailand where only local technicians do the whole work. This is positive as involving local people can help to create a sustainable system that can operate even if there is no external support or in cases of delay. According to the results, apart from Kenya where it is mainly a paper-based dataset, the designed land record databases are digital. With regard to the languages used, the findings are that most of the systems are designed considering the native language widely spoken in the country. This again facilitates the accessibility of the system by many citizens.

With the existence of updating procedures, the inclusion of native language to design the land records dataset, and the involvement of local experts in the design and the maintenance of the system, all of these combined could enable the updating of the system to be simplified. This is a strong point towards encouraging the reporting of change in land records.

5.2. *Affordability of the Cost*

The cost of land registration determines its future acceptance [4]. They can be kept low, but not too low to be a burden to the government [1,7,26,27], as landholders have to bear a part or all the cost for registering subsequent changes to support the self-financing of the system [1]. The difference between the updating fees and the initial registration fees should not be too big, at least in the beginning, in order to increase public awareness on the advantages of registration [1].

The results show that for the initial land registration, no payment was required for registering land in Ethiopia, Kenya, Nepal, and Mexico, while it was equal or less than US$6 in Thailand, Namibia, and Rwanda. In Cambodia, the fees were charged based on the size, being US$2.50 for one hectare of rural agriculture land, whereas in Kyrgyzstan, it is US$ 65 (unfortunately, no additional information was obtained that would explain the criteria for this payment). The cost is reasonable for citizens as confirmed by the respondents, though it turns out to be a burden to the government, which funded the initial registration activities and is often supported by donors in the majority of countries.

Though they can vary depending on the type of transaction, the findings are that the fees for registering change in land records are less than US$10 in Ethiopia, Mexico, Namibia, Kyrgyzstan, and Thailand but beyond US$30 in Kenya and Rwanda. They depend on land value in Nepal, whereas in Cambodia, they are calculated based on location and use of land (e.g., in rural area, they are almost US$25 for agriculture land and US$50 for residential).

Here one can see that the majority of country considered in this study managed to keep the fees low (below US$10). As says one interviewee from Cambodia, "There [are] usually informal fees which dissuade people from engaging in this process". These informal fees may include transport, per diem (in case witnesses are needed in the process), or fees to speed up the process—the calculating of which may not be easy as some are a secret between the giver and the receiver.

5.3. Speed

Speed refers to the time it takes to get a new land document when applying for registering a change in the land records [32]. The updating process should be completed in less days [7,26,27,34] to motivate the applicant to come again, otherwise they may opt for shortcuts leading to informality [3].

In most case-study countries, the results show that the updating period is equal or above one month. In Namibia, it takes three months minimum as that is when the communal land board meets to assess the applications. In Nepal, the process requires a maximum of five days, while in Kyrgyzstan, one day is enough to get a new land document. The good thing is that when all the documents are complete in Ethiopia and Kenya or during the land registration campaign in Rwanda (known as 'land week'), the process is completed in a single day.

The possibility of processing the application in one day (when all the requirements are in order) proves that the process can be shortened if there is proper and regular communication about requirements, available resources (human and technical) supporting updating, and continuous training for the staff. Fortunately, means to sensitize citizens about updating procedures and the attached benefits as well as trainings of involved staff are done in all the case countries. With more effort in availing resources, where the speed is not currently achieved, it may be achieved as well.

5.4. Decentralization

In order to improve the correctness of the records [7,26,27] and allow equitable access to all [17,29], land registration services have to be closer to the people: The distance and the time taken to register change in land records are reduced [1] and motivation to register changes in land records is increased.

Based on the results, the land registration services are available at a local level in all the case countries except in Mexico where the RAN (*Registro Agrario Nacional*—National Agrarian Register) that administer the *ejido* (communal land) operates only at the state (province) level. For these countries, there is a closer communication of local administration with the upper levels during the registration process either when processing the application (as everyone has a role to play depending on their position) or when seeking information to process it. For instance, in Ethiopia, the court decision is needed prior to processing inheritance, whereas in Kenya, Namibia, and Rwanda, the application is received at lower level and continues to be processed at upper levels by different actors.

Although the move from one desk to another and sometimes from one building to another may delay the registration process, the availability of information and services at a lower level is a positive asset that may encourage the updating of land records.

5.5. Accessibility to the Database

A database open to its users is one of the things facilitating the updating as anyone has the possibility to check if the information registered in their names are correct or not [5,6,17].

It is very positive that all countries under consideration in this study thought about opening the database to other people apart from those managing it. In all these countries, the findings are that the database is accessed by citizens, with the exception of Cambodia, and by local officers in all countries excluding Mexico (no local offices). Citizens from these countries can access the data in different ways: By physically going to the office in charge (all the case countries excluding Cambodia); through a specific website (Kyrgyzstan, Mexico, Rwanda, and Thailand); or using the mobile phone (Rwanda and Thailand). Based on the results, the fees charged vary from free to US$2. The same as citizens, providers of land services at the local level access the land database via website or phone or by going or sending an email to the institution in charge.

5.6. Security of the System

Anyone participating in land transactions needs to be sure that they are dealing with the right person and that the law will protect their rights when needed [4,17]. The legal value of the certificate/title issued after registering land rights and the existence of a backup of records used to produce a land document is key [16].

In all of the case countries, the findings are that the land document issued to landholders has a legal value, i.e., it can be used as ownership evidence in case of disputing claims. Except for Kenya where the database is paper based, backup that will help to restore the records in case of problems exists in all the other countries. As specified by the respondents from Ethiopia, Nepal, Rwanda, and Thailand, there are different access levels to the digital database for local officers, and what they can do on the database depends on their role in the process and the assigned area. The issuance of legal land documents after registration and the existence of backup for the stored land records are among good practices leading to a secured system.

5.7. Suitability to the Circumstance

To be accepted and used, the registration programs will have to respond to societal needs [28,32]. In case there is no link between the designed system and what people need, updating changes in land records may be difficult as no one is seeing the relevance of the system.

As discussed under Section 4, ensuring tenure security, boosting land market and land investment, issuing land certificate/title, improving the effectiveness of land registration, handling land disputes, and producing land records are the main purposes for which the explored land registration programs were conducted. Normally, the degree to which these purposes were achieved or how the developed system is addressing the current societal needs is assessed through specific research. As some of the land experts interviewed undertook or participated in such assessments, their thoughts are relevant. The good thing is that most respondents think that these goals were achieved in the registration programs and also that the established system is responding to the current societal needs. It is not surprising that there are also some who said that the purposes were partially achieved and that the system is not really addressing the actual societal needs. Designing an LAS and assessing its achievement are continuous activities because the system has to be adapted to the dynamicity of people-to-land relations. As such, hitting the target precisely and accurately seems to be hard as this target is continuously moving based on the societal needs.

5.8. System Interconnectivity

When there is connection between databases hosted by different institutions (and not as individual silos), some processes are shortened or removed, and the service provision is sped up, which facilitates the system updating [31]. Besides, this connectivity is crucial to create a data sharing platform (Spatial Data Infrastructure—SDI) leading to faster communication among actors about decisions affecting land registration [1].

It is unfortunate to find that, with these benefits, only Kyrgyzstan and Rwanda have the land records database linked with other databases (no information for Thailand). As an example, in Rwanda, the database of land records is connected to tax collection, building permits, population registers, court, mortgage registrations (used by banks), and subsidized agriculture inputs.

Before data sharing, the priority may be to have a well-functioning land registration and secured database and to have these may take time. Thus, system interconnection can be considered as one step further when developing an LAS. This could explain why the majority of the studied case countries do not have this interconnection, although willingness to do so is another parameter that should not be ignored.

5.9. Mobilization

Information is power. Thus, informing citizens about what to do and where to go when seeking land registration services is essential to encourage the updating of the database but also raise awareness about its benefits [5,7,26,27].

In addition to having formal procedures for updating land records, it is worthy to mention that, in all the explored countries, there are mechanisms in place to communicate these procedures to citizens. The results show that the communication is done through different means: Media (TV or radio), public meeting, written materials (e.g., brochures, manuals), physically going to the institution in charge of registration activities or their website, social media, or informing local leaders (village headman, traditional leader) who will dispatch the information where they live. This mobilization may be an indication that the willingness to have updated databases exists.

5.10. Sustainability of the System

Financial and technical sustainability is critical for land information system [28,30]. Sustainability should be ensured through mechanisms ensuring a continuous maintenance [17,29]. If the system is not self-financed, the government should at least be able to allocate budget for its operation.

Respondents from the case-study countries indicated that the government (supported by donors for many of them) allocates budget for updating land records. In Rwanda, the collected land revenues go in the national treasury and come back as allocated budget, whereas Cambodia and Kyrgyzstan affirmed having already self-financed land registration. Ethiopia and Rwanda have a plan to introduce a self-financed system.

The technical sustainability depends on the available budget to cover the cost of upgrading the system so as to adapt it to new technologies and societal needs [5,6]. The results show that, except in Kyrgyzstan and Cambodia, other countries upgraded, in one way or the other, the technical structure of the land database designed during the considered initial land registration programs (no information for Namibia). Ethiopia and Rwanda changed the version of their LIS (currently at version 5.1.3 and version 4 respectively); Kenya adapted laws related to technological advancement and did minimal computerization; Mexico updated the software used to store the data; Nepal moved from LAN (Local Area Network) to WAN (Wide Area Network) and is digitizing the database; and Thailand changed their mapping technology.

Financial and technical sustainability of the updating system seems to be uncertain as the main funds are from donors, and considering that, a few countries have a self-financed system or are planning to introduce it.

5.11. Compulsory Registration

Generally, there is a legal framework backing land registration activities that specify whether it is compulsory or not to register land rights (including those rights obtained through transfers) [5]. As the registered information is the foundation to make decisions on the legally accepted landholder, registration has to be compulsory [1]. Even though the law is not enough to increase the number of applications for change registration, it is believed that more people would register change when the registration is compulsory than when it is not. There is a kind of motivation and/or pressure as there is a legal protection for those whose rights are registered and whose information are up to date.

Fortunately the findings show that, except in Cambodia, Kenya, and Mexico, in all other countries, it is compulsory to register land transfers or other changes in land records. This is a strong stimulus in the updating of land records.

5.12. Incentives

In business, there are promotional periods to attract buyers or to empty stores. If this technique works well in business, why not in land services? Incentives can be given to those who register certain changes (e.g., succession) to motivate people to register them and on time [5].

It is worthy to report that, in the countries under study (no information for Cambodia), half of them offer incentives for some registration activities while others do not offer anything. It is free to register co-ownership between spouses or parcel merging in Ethiopia and boundary or area correction in Rwanda. Additionally, in Rwanda, during land registration campaigns (or land week), there is a mobile team of land professionals going from place to place, offering land services and issuing land certificates on the same day for those transferring land. In Thailand, they lower taxes or fees during promotional periods, while in Mexico, an interviewee explained that "they are regular actualization campaigns organized by the *Procuraduria Agraria* (Agrarian Procuracy) aiming at motivating and facilitating the registration of successor lists (testament) by *ejidatarios*, without any cost and in their village". Though there are no additional data collected from other countries under study, experience from Rwanda is that the received applications to update land records increase during land week campaigns (personal experience of the first author). This is a recommendable good practice.

5.13. Well Organized Administration

Institutions involved in land registration should each have clearly defined tasks in the process, and cooperate and communicate well with each other [28]. Overlapping responsibilities hinder service provision to the public and it may negatively affect the number of received applications [1]. Besides, there can be shortage of competent staff when there are too many offices [31]. The registration process can be long due to duplication or contradiction in the activities performed during application processing; citizens may be confused about what service to find where; conflicts between institutions may make citizens victims, and so forth.

Unfortunately, the results show that the majority of countries under study (Kenya, Mexico, Namibia, Nepal, and Rwanda) have more than one institutions on the same level (ministerial, regional or district) dealing with land registration. Only few countries have one institution handling land matters represented from national to local levels. Cambodia has the Ministry of Land Management, Urban Planning, and Construction at the national level and the Cadaster & Land Department at the province and district level. In Thailand, the Department of Land under the Ministry of Interior has branches at province and at district levels in Bangkok metropolitan, while in Kyrgyzstan, the State Agency on Land

Resources has branches at a regional level; and in Ethiopia, the land registration services are offered at *woreda* level.

Though it was not possible to determine all the respective responsibilities of institutions carrying out land registration in the countries considered, it is believed that the higher the number of involved institutions, the higher the difficulty in specifying the role of each of them. There is a high probability of having overlaps when there are many institutions. In addition, as said one interviewee from Namibia: "It could actually be a smooth process, but due to fact that there are so many stakeholders involved . . . , it lengthens the process". The number of involved institutions in the registration may also affect the registration speed.

5.14. Continuous Capacity Building

Updating land records is a daily exercise requiring adequate level of trained professionals [15]. The provider of land services at all levels needs to be updated about new technologies in their field as well as be informed about how the system is coping with the changing environment [28]. This can be done through workshops or short/long term trainings.

It is amazing to find that, in all the countries explored, trainings are offered to staff working in land registration. The subjects addressed include land administration and management, land valuation, land laws, cadastral surveying, IT-based courses, customer care, geo-information sciences, land use planning, and so on. This is a strength to keep. Referring to Mexico, which mentioned that the trainings have currently been stopped due to lack of resources, it is argued that due to the importance of having capacitated staff involved in the land registration, continuous capacity building has to be budgeted for when allocating funds to registration activities.

5.15. Continuous Assessment of the System

One way to know if the system is operating effectively is to get information from its users. User satisfaction should be aimed for in service provision including land registration services, otherwise they may be discouraged to report the changes in land records [32]. Regular review of the registration process should take place to assess the system efficiency and effectiveness [28].

Fortunately, the results show that methods to assess the effectiveness of the land registration exist in all the case countries except in Ethiopia (no information for Namibia). They include suggestion boxes, monitoring and evaluation, change management systems, audits, customer satisfaction surveys, conferences, seminars, research, governance score cards, business reports, claims received on social media or in land registration campaigns, field visits by senior management, and so forth. It is believed that these tools are very useful and should be used properly in order to know what to improve in the registration activities.

5.16. Political Support

It is hard, even impossible, to undertake a program if there is no buy-in from policy makers. Political support is among the requirements to implement the land registration programs [24].

Based on the fact that these programs were implemented, there is no doubt that the government gave its consent. This was ascertained by respondents from all of the programs studied who said that the government supported the initial land registration, and the updating of the system is fully under its responsibility. The government offered and looked for financial and technical support; availed infrastructures; obtained human resources and trained them; elaborated laws and regulation to support the programs; and so on. This is a strong incentive in the updating of land records.

6. Conclusions

Keeping up-to-date land records used to be challenging, and the existing literature focuses mainly on weaknesses in the updating process. This paper proves that there are good, recommendable practices in the nine countries considered. They include the existence of infrastructure supporting updating available in all of the case countries; for the majority, the systems are simplified and registration fees are reasonable; services are decentralized; the digital database is secured and accessible by citizens and local officers; mobilization is done to raise awareness about registration; the legal framework is there to emphasize the importance of registration; incentives are offered for some transactions to motivate people to report them on time; staff are trained; the system effectiveness is assessed; and the political support is offered.

Efforts are still needed in order to shorten updating procedures, introduce data-sharing platforms, ensure financial and technical sustainability, and reduce the number of institutions involved in the registration process. Though it is worthy to adapt the registration program to the societal needs, it was found to be difficult to assess whether the target was hit as the needs are continuously changing.

Whilst several countries' situations in terms of updating land records are depicted in this paper based on the literature and informants' information, further research may be conducted based on field observations per case country.

Author Contributions: The manuscript is part of an ongoing research done by M.B. under the guidance of her supervisors J.Z. and B.K.A. Conceptualization; methodology, formal analysis; writing—original draft preparation, M.B.; writing—review and editing, J.Z. and B.K.A. All authors have read and agreed to the published version of the manuscript.

Funding: Funding of the publication costs for this article has kindly been provided by the School of Land Administration Studies, University of Twente, in combination with Kadaster International, the Netherlands. The ongoing PhD research is funded by DAAD (*Deutscher Akademischer Austauschdienst*—German Academic Exchange Service) under its In-Country/In-Region Scholarship Program SLGA (Strengthening Capacities for Land Governance in Africa), 2017 (57377251).

Institutional Review Board Statement: Not applicable.

Informed Consent Statement: Not applicable.

Data Availability Statement: The data presented in this study are available on request from the corresponding author. The data are not publicly available in order to respect the privacy of the respondents.

Acknowledgments: The authors thank the anonymous respondents from the case study countries who provided information that lead to the production of this manuscript.

Conflicts of Interest: The authors declare no conflict of interest.

Appendix A List of Countries That Used Systematic Land Registration Program

Africa	America	Asia	Europe
Benin Burkina Faso Côte d'Ivoire Ethiopia Ghana Kenya Lesotho Madagascar Mozambique Namibia Rwanda Tanzania Uganda	Argentina Mexico	Brunei India Israel Malaysia Nepal Pakistan Philippines Sri Lanka Turkey Kyrgyzstan Thailand Vietnam Cambodia	Belgium Poland Spain Switzerland Armenia
13	2	13	5

References

1. Larsson, G. *Land Registration and Cadastral Systems: Tools for Land Information and Management*; Longman Scientific & Technical: New York, NY, USA, 1991.
2. Dale, P.F.; McLaughlin, J.D. *Land Administration*; Oxford University Press: Oxford, UK, 1999.
3. Williamson, I.; Enemark, S.; Wallace, J.; Rajabifard, A. *Land Administration for Sustainable Development*; ESRI Press Academic: Redlands, CA, USA, 2010.
4. Simpson, S.R. *Land Law and Registration*; Cambridge University Press: New York, NY, USA, 1976.
5. Enemark, S.; McLaren, R.; Lemmen, C. *Fit.For-Purpose Land Administration Guiding Principles for Country Implementation Securing Land*; United Nations Human Settlements Programme (UN-Habitat): Nairobi, Kenya, 2016; p. 132.
6. Enemark, S.; Bell, C.K.; McLaren, R. *Fit-For-Purpose Land Administration*; Joined Publication FIG & World Bank: Copenhagen, Denmark, 2014.
7. Zevenbergen, J.; Augustinus, C.; Antonio, D.R.; Bennett, R.M. Pro-Poor Land Administration: Principles for Recording the Land Rights of the Underrepresented. *Land Use Policy* **2013**, *31*, 595–604. [CrossRef]
8. Williamson, I.P. Cadastral Survey Techniques in Developing Countries—with Particular Reference to Thailand. *Aust. Surv.* **1983**, *31*, 496–512. [CrossRef]
9. Deininger, K.; Ali, D.A.; Holden, S.; Zevenbergen, J. Rural Land Certification in Ethiopia: Process, Initial Impact, and Implications for Other African Countries. *World Dev.* **2008**, *36*, 1786–1812. [CrossRef]
10. Deininger, K.; Augustinus, C.; Enemark, S.; Munro-Faure, P. *Innovations in Land Rights Recognition, Administration, and Governance*; The World Bank: Washington, DC, USA, 2010; p. 251.
11. Byamugisha, F.F.K. *Securing Africa's Land for Shared Prosperity: A Program to Scale Up Reforms and Investments*; Africa Development Forum; Agence Française de Développement and the World Bank: Washington, DC, USA, 2013.
12. Ngoga, T.H. *Rwanda's Land Tenure Reform: Non-Existent to Best Practice*; CABI Publishing: Boston, MA, USA, 2018.
13. Wohlin, C. Guidelines for Snowballing in Systematic Literature Studies and a Replication in Software Engineering. In Proceedings of the 18th International Conference on Evaluation and Assessment in Software Engineering, London, UK, 13–14 May 2014. [CrossRef]
14. Danglot, B.; Vera-Perez, O.; Yu, Z.; Zaidman, A.; Monperrus, M.; Baudry, B. A Snowballing Literature Study on Test Amplification. *J. Syst. Softw.* **2019**, *157*, 110398. [CrossRef]
15. Jing, Y.; Bennett, R.; Zevenbergen, J. Up-to-Date' in Land Administration: Setting the Record Straight. In *Innovative Cadastre and Land Rights Management*; FIG: Abuja, Nigeria, 2013.
16. FIG. FIG Statement on the Cadastre. Available online: https://www.fig.net/resources/publications/figpub/pub11/figpub11.asp (accessed on 26 June 2018).
17. FAO. *Voluntary Guidelines on the Governance of Tenure: At a Glance*; FAO: Rome, Italy, 2012.
18. Henssen, J. *Land Registration and Cadastre Systems: Principles and Related Issues*; Technische Universität München: Munich, Germany, 2010.
19. Unger, E.-M.; Bennett, R.M. *Fit-For-Purpose Land Administration for All: A Guide for Surveyors on Adoption and Adaptation of Fit-For-Purpose Land Administration*; Netherlands' Ministry of Foreign Affairs and Kadaster: The Hague, The Netherlands, 2019.
20. Ali, D.A.; Deininger, K.; Mahofa, G.; Nyakulama, R. Sustaining Land Registration Benefits by Addressing the Challenges of Reversion to Informality in Rwanda. *Land Use Policy* **2019**, 104317. [CrossRef]
21. Simbizi, M.C.D.; Bennett, R.M.; Zevenbergen, J. Land Tenure Security: Revisiting and Refining the Concept for Sub-Saharan Africa's Rural Poor. *Land Use Policy* **2014**, *36*, 231–238. [CrossRef]
22. Van der Molen, P. The Dynamic Aspect of Land Administration: An Often-Forgotten Component in System Design. *Comput. Environ. Urban Syst.* **2002**, *26*, 361–381. [CrossRef]
23. Magis, M.; Zevenbergen, J. Towards Sustainable Land Administration Systems: Designing for Long-Term Value Creation. In *Engaging the Challenges—Enhancing the Relevance*; FIG: Kuala Lumpur, Malasia, 2014.
24. Dale, P.F.; McLaughlin, J.D. *Land Information Management: An Introduction with Special Reference to Cadastral Problems in Third World Countries*; Oxford University Press: New York, NY, USA, 1988.
25. Williamson, I.P. Land Administration "Best Practice" Providing the Infrastructure for Land Policy Implementation. *Land Use Policy* **2001**, *18*, 297–307. [CrossRef]
26. Zevenbergen, J.; Augustinus, C.; Antonio, D. *Designing a Land Records System for the Poor*; United Nations Human Settlements Programme (UN-Habitat): Nairobi, Kenya, 2012.
27. Hendriks, B.; Zevenbergen, J.; Bennett, R.; Antonio, D. Pro-Poor Land Administration: Towards Practical, Coordinated, and Scalable Recording Systems for All. *Land Use Policy* **2019**, *81*, 21–38. [CrossRef]
28. Steudler, D. A Framework for the Evaluation of Land Administration Systems. Ph.D. Thesis, The University of Melbourne, Melbourne, Australia, 2004.
29. Burns, T. Land Administration Reform: Indicators of Success and Future Challenges. *Agric. Rural Dev.* **2007**, *37*, 244.
30. Deininger, K.; Selod, H.; Burns, A. *The Land Governance Assessment Framework: Identifying and Monitoring Good Practice in the Land Sector*; The World Bank: Washington, DC, USA, 2012. [CrossRef]
31. Hanstad, T. Designing Land Registration Systems for Developing Countries. *Am. Univ. Int. Law Rev.* **1998**, *13*, 647–703.

32. Chimhamhiwa, D.; van der Molen, P.; Mutanga, O.; Rugege, D. Towards a Framework for Measuring End to End Performance of Land Administration Business Processes—A Case Study. *Comput. Environ. Urban Syst.* **2009**, *33*, 293–301. [CrossRef]
33. Enemark, S. *Fit for Purpose: Building Spatial Frameworks for Sustainable and Transparent Land Governance*; The World Bank: Washington, DC, USA, 2013; p. 18.
34. World Bank. Methodology for Registering Property. Available online: https://www.doingbusiness.org/en/methodology/registering-property (accessed on 26 June 2019).
35. Moreri, K.; Fairbairn, D.; James, P. Issues in Developing a Fit for Purpose System for Incorporating VGI in Land Administration in Botswana. *Land Use Policy* **2018**, *77*, 402–411. [CrossRef]
36. Dower, P.C.; Pfutze, T. Land Titles and Violent Conflict in Rural Mexico. *J. Dev. Econ.* **2020**, *144*, 102431. [CrossRef]
37. Kutzman, D. *Land Certification and Deforestation: In The Impact of Procede on Land Use in Mexico*; The World Bank: Washington, DC, USA, 2012; p. 30.
38. Barnes, G.; DiGiano, M.; Augustinus, C. *Ejido Land Tenure and Registration System: Mexico Case Study—Synthesis Report*; Synthesis Report Report; United Nations Human Settlements Programme (UN-HABITAT): Nairobi, Kenya, 2015; p. 30.
39. Gainer, M. *Cementing the Right of Ownership: Land Registration in Kyrgyzstan, 1999–2009. Innovations for Successful Societies*; Princeton University: Princeton, NJ, USA, 2017.
40. Bezu, S.; Holden, S. Demand for Second-Stage Land Certification in Ethiopia: Evidence from Household Panel Data. *Land Use Policy* **2014**, *41*, 193–205. [CrossRef]
41. Deininger, K.; Zevenbergen, J.; Ali, D.A. Assessing the Certification Process of Ethiopia's Rural Lands. In *Colloque International "Les Frontières de la Question Foncière—At the Frontier of Land Issues"*; Montpellier, France, 2006; p. 14. Available online: https://land.igad.int/index.php/documents-1/countries/ethiopia/rural-development-1/282-assessing-the-certification-process-of-ethiopia-s-rural-lands/file (accessed on 6 December 2018).
42. Bureau of Environment Protection, Land Administration and Use (BoEPLAU). *Land Registration and Certification: Experiences from the Amhara National Regional State in Ethiopia. Bureau of Environment Protection, Land Administration and Use*; BoEPLAU: Amhara, Ethiopia, 2010.
43. Bezabih, M.; Kohlin, G.; Mannberg, A. Trust, Tenure Insecurity, and Land Certification in Rural Ethiopia. *J. Socio Econ.* **2011**, *40*, 833–843. [CrossRef]
44. Land Investment For Transformation (LIFT). *Second-Level Land Certification Manual*; Version 3.1; DAI Europe: London, UK, 2018.
45. Caukwell, R.A. *Cadastral Survey in Kenya and Its Role in the Development of the Country. Master of Sciences*; University of Nairobi: Nairobi, Kenya, 1977.
46. Wayumba, R.N. Impacts of Different Land Registration Systems on Communal Tenure in Kenya. Ph.D. Thesis, University of Otago, Dunedin, New Zealand, 2014.
47. Wayumba, G. A Historical Review of Land Tenure Reforms in Kenya. *IJSRES* **2015**, *2*, 45–51.
48. Sagashya, D.; English, C. Designing and Establishing a Land Administration System for Rwanda: Technical and Economic Analysis. In *Innovations in Land Rights Recognition, Administration, and Governance*; Joint Organizational Discussion Paper; The World Bank: Washington, DC, USA, 2010; pp. 43–59.
49. Republic of Rwanda. National Land Policy. Ministry of Lands, Environment, Forests, Water and Mines. 2004. Available online: https://rema.gov.rw/rema_doc/Policies/National_land_policy_english_version_.pdf (accessed on 24 August 2018).
50. Simbizi, M.C.D.; Bennett, R.M.; Zevenbergen, J. Pro-Poor Land Administration. In *Advances in Responsible Land Administration*; Taylor & Francis: Boca Raton, FL, USA, 2015; pp. 17–36.
51. Biraro, M.; Bennett, R.M.; Lemmen, C. Accelerated Land Administration Updates. In *Advances in Responsible Land Administration*; Taylor & Francis: Boca Raton, FL, USA, 2015; pp. 145–161.
52. Williamson, I.P. Cadastral and Land Information Systems in Developing Countries. *Aust. Surv.* **1986**, *33*, 27–43. [CrossRef]
53. Rattanabirabongse, V.; Eddington, R.A.; Burns, A.F.; Nettle, K.G. The Thailand Land Titling Project—Thirteen Years of Experience. *Land Use Policy* **1998**, *15*, 3–23. [CrossRef]
54. The World Bank. *Kyrgyz Republic-Land and Real Estate Registration Project (English): Project Appraisal*; The World Bank: Washington, DC, USA, 2000.
55. The World Bank. *Kyrgyz Republic-Land and Real Estate Registration Project (English): Project Performance Assessment Report*; The World Bank: Washington, DC, USA, 2010.
56. Werner, W. Tenure Reform in Namibia's Communal Areas. *J. Namib. Stud. Hist. Politics Cult.* **2015**, *18*, 67–87.
57. Werner, W. Land, Livelihoods and Housing Programme 2015–2018 Working Paper. *Work. Pap.* **2015**, *1*, 17.
58. Meijs, M.; Kapitango, D. *Namibia Land Management Series Number 2: Communal Land Registration*; Namibia Institute for Democracy, Ministry of Lands and Resettlement: Windhoek, Namibia, 2009.
59. Kapitango, D.; Meijs, M. Land Registration Using Aerial Photography in Namibia: Costs and Lessons. In *Innovations in Land Rights Recognition, Administration, and Governance*; Joint Organizational Discussion Paper; The World Bank: Washington, DC, USA, 2010; pp. 60–72.
60. Matthaei, E. What Does Registration of Communal Land Mean to Namibians? Available online: http://www.focusonland.com/countries/case-study-what-does-registration-of-communal-land-mean-to-nambians-/ (accessed on 26 November 2020).

61. Acharya, B.R. Spatial information management in Nepal. In *Spatial Information Management for Sustainable Real Estate Market—Best Practice Guidelines on Nation-Wide Land Administration*; UN/ECE-Working Party on Land Administration: Athens, Greece, 2003; Available online: https://www.oicrf.org/-/spatial-information-management-in-nepal (accessed on 18 December 2018).
62. Acharya, B.R.; Bhattarai, M.P. Cadastre in Nepal: Towards the Beginning of a Land Information System. *FIG Newsl.* **1996**, *14*, 8–10. Available online: https://www.oicrf.org/-/cadastre-in-nepal-towards-the-beginning-of-a-land-information-system (accessed on 18 December 2020).
63. Biddulph, R. Cambodia's Land Management and Administration Project. *UNU WIDER* **2015**, 1–14. [CrossRef]
64. Boramy, S. Systematic Land Titling in Cambodia Is the Component 3 of the Land Management and Adminsitration Project (LMAP). In *FIG Commission 7 Annual Meeting 2008 and Open Symposium on Environment and Land Administration "Big Works for the Defence of the Territory"*; OICRF: Gelderland, The Netherlands, 2008; Available online: https://www.oicrf.org/-/systematic-land-titling-in-cambodia-is-the-component-3-of-the-land-management-and-adminsitration-project-lmap (accessed on 20 January 2021).

MDPI
St. Alban-Anlage 66
4052 Basel
Switzerland
Tel. +41 61 683 77 34
Fax +41 61 302 89 18
www.mdpi.com

Land Editorial Office
E-mail: land@mdpi.com
www.mdpi.com/journal/land